Springer Series in Information Sciences 2

Editor: T. S. Huang

Springer Series in Information Sciences

Editors: Thomas S. Huang   Teuvo Kohonen   Manfred R. Schroeder
Managing Editor: H.K.V. Lotsch

1 **Content-Addressable Memories**
   By T. Kohonen   2nd Edition

2 **Fast Fourier Transform and Convolution Algorithms**
   By H. J. Nussbaumer   2nd Edition

3 **Pitch Determination of Speech Signals**
   Algorithms and Devices   By W. Hess

4 **Pattern Analysis and Understanding**
   By H. Niemann   2nd Edition

5 **Image Sequence Analysis**
   Editor: T.S. Huang

6 **Picture Engineering**
   Editors: King-sun Fu and T.L. Kunii

7 **Number Theory in Science and Communication**   With Applications in Cryptography, Physics, Digital Information, Computing, and Self-Similarity   By M.R. Schroeder
   2nd Edition

8 **Self-Organization and Associative Memory**   By T. Kohonen   3rd Edition

9 **Digital Picture Processing**
   An Introduction   By L.P. Yaroslavsky

10 **Probability, Statistical Optics and Data Testing**   A Problem Solving Approach   By B.R. Frieden

11 **Physical and Biological Processing of Images**   Editors: O.J. Braddick and A.C. Sleigh

12 **Multiresolution Image Processing and Analysis**   Editor: A. Rosenfeld

13 **VLSI for Pattern Recognition and Image Processing**   Editor: King-sun Fu

14 **Mathematics of Kalman-Bucy Filtering**
   By P.A. Ruymgaart and T.T. Soong
   2nd Edition

15 **Fundamentals of Electronic Imaging Systems**   Some Aspects of Image Processing   By W.F. Schreiber

16 **Radon and Projection Transform-Based Computer Vision**
   Algorithms, A Pipeline Architecture, and Industrial Applications   By J.L.C. Sanz, E.B. Hinkle, and A.K. Jain

17 **Kalman Filtering** with Real-Time Applications   By C.K. Chui and G. Chen

18 **Linear Systems and Optimal Control**
   By C.K. Chui and G. Chen

19 **Harmony: A Psychoacoustical Approach**   By R. Parncutt

20 **Group Theoretical Methods in Image Understanding**   By Ken-ichi Kanatani

21 **Linear Prediction Theory**
   A Mathematical Basis for Adaptive Systems   By P. Strobach

Henri J. Nussbaumer

Fast Fourier Transform
and
Convolution Algorithms

Second Corrected and Updated Edition

With 38 Figures

Springer-Verlag Berlin Heidelberg New York
London Paris Tokyo Hong Kong

Professor Henri J. Nussbaumer

Département d'Informatique, Laboratoire d'Informatique Technique,
Ecole Polytechnique Fédérale de Lausanne, CH-1015 Lausanne, Switzerland

Series Editors:

Professor Thomas S. Huang

Department of Electrical Engineering and Coordinated Science Laboratory,
University of Illinois, Urbana IL 61801, USA

Professor Teuvo Kohonen

Laboratory of Computer and Information Sciences, Helsinki University of Technology,
SF-02150 Espoo 15, Finland

Professor Dr. Manfred R. Schroeder

Drittes Physikalisches Institut, Universität Göttingen, Bürgerstrasse 42–44,
D-3400 Göttingen, Fed. Rep. of Germany

Managing Editor: Helmut K. V. Lotsch

Springer-Verlag, Tiergartenstrasse 17
D-6900 Heidelberg, Fed. Rep. of Germany

Second Printing 1990

ISBN 3-540-11825-X 2. Auflage   Springer-Verlag Berlin Heidelberg New York
ISBN 0-387-11825-X 2nd edition Springer-Verlag New York Berlin Heidelberg

ISBN 3-540-10159-4 1. Auflage Springer-Verlag Berlin Heidelberg New York
ISBN 0-387-10159-4 1st edition Springer-Verlag New York Berlin Heidelberg

Library of Congress Cataloging in Publication Data. Nussbaumer, Henri J., 1931–. Fast Fourier transform and convolution algorithms. (Springer series in information sciences; 2) Bibliography: P. Includes index. 1. Fourier transformations – Data processing. 2. Convolutions (Mathematics) – Data processing. 3. Digital filters (Mathematics) I. Title. II. Series. QA403.5.N87 1982   515.7′23   82-10650

This work is subject to copyright. All rights are reserved, whether the whole or part of the material is concerned, specifically the rights of translation, reprinting, reuse of illustrations, recitation, broadcasting, reproduction on microfilms or in other ways, and storage in data banks. Duplication of this publication or parts thereof is only permitted under the provisions of the German Copyright Law of September 9, 1965, in its current version and a copyright fee must always be paid. Violations fall under the prosecution act of the German Copyright Law.

© Springer-Verlag Berlin Heidelberg 1981 and 1982
Printed in Germany

The use of registered names, trademarks, etc. in this publication does not imply, even in the absence of a specific statement, that such names are exempt from the relevant protective laws and regulations and therefore free for general use.

Offset printing and bookbinding: Brühlsche Universitätsdruckerei, 6300 Giessen
2154/3150-543210 – Printed on acid-free paper

Preface to the Second Edition

In the first edition of this book, we covered in Chapter 6 and 7 the applications to multidimensional convolutions and DFT's of the transforms which we have introduced, back in 1977, and called polynomial transforms. Since the publication of the first edition of this book, several important new developments concerning the polynomial transforms have taken place, and we have included, in this edition, a discussion of the relationship between DFT and convolution polynomial transform algorithms. This material is covered in Appendix A, along with a presentation of new convolution polynomial transform algorithms and with the application of polynomial transforms to the computation of multidimensional cosine transforms.

We have found that the short convolution and polynomial product algorithms of Chap. 3 have been used extensively. This prompted us to include, in this edition, several new one-dimensional and two-dimensional polynomial product algorithms which are listed in Appendix B.

Since our book is being used as part of several graduate-level courses taught at various universities, we have added, to this edition, a set of problems which cover Chaps. 2 to 8. Some of these problems serve also to illustrate some research work on DFT and convolution algorithms.

I am indebted to Mrs A. Schlageter who prepared the manuscript of this second edition.

Lausanne
April 1982

HENRI J. NUSSBAUMER

Preface to the First Edition

This book presents in a unified way the various fast algorithms that are used for the implementation of digital filters and the evaluation of discrete Fourier transforms.

The book consists of eight chapters. The first two chapters are devoted to background information and to introductory material on number theory and polynomial algebra. This section is limited to the basic concepts as they apply to other parts of the book. Thus, we have restricted our discussion of number theory to congruences, primitive roots, quadratic residues, and to the properties of Mersenne and Fermat numbers. The section on polynomial algebra deals primarily with the divisibility and congruence properties of polynomials and with algebraic computational complexity.

The rest of the book is focused directly on fast digital filtering and discrete Fourier transform algorithms. We have attempted to present these techniques in a unified way by using polynomial algebra as extensively as possible. This objective has led us to reformulate many of the algorithms which are discussed in the book. It has been our experience that such a presentation serves to clarify the relationship between the algorithms and often provides clues to improved computation techniques.

Chapter 3 reviews the fast digital filtering algorithms, with emphasis on algebraic methods and on the evaluation of one-dimensional circular convolutions.

Chapters 4 and 5 present the fast Fourier transform and the Winograd Fourier transform algorithm.

We introduce in Chaps. 6 and 7 the concept polynomial transforms and we show that these transforms are an important tool for the understanding of the structure of multidimensional convolutions and discrete Fourier transforms and for the design of improved algorithms. In Chap. 8, we extend these concepts to the computation of one-dimensional convolutions by replacing finite fields of polynomials by finite fields of numbers. This facilitates introduction of number theoretic transforms which are useful for the fast computation of convolutions via modular arithmetic.

Convolutions and discrete Fourier transforms have many uses in physics and it is our hope that this book will prompt some additional research in these areas and will help potential users to evaluate and apply these techniques. We also feel that some of the methods presented here are quite general and might someday find new unexpected applications.

Part of the material presented here has evolved from a graduate-level course taught at the University of Nice, France. I would like to express my thanks to Dr. T. A. Kriz from IBM FSD for kindly reviewing the manuscript and for making many useful suggestions. I am grateful to Mr. P. Bellot, IBM, C.E.R., La Gaude, France, for his advice concerning the introductory chapter on number theory and polynomial algebra, and to Dr. J. W. Cooley, from IBM Research, Yorktown Heights, for his comments on some of the work which led to this book. Thanks are also due to Dr. P. Quandalle who worked with me on polynomial transforms while preparing his doctorate degree and with whom I had many fruitful discussions. I am indebted to Mrs. C. De Backer for her aid in improving the English and to Mrs. C. Chevalier who prepared the manuscript.

La Gaude
November 1980

HENRI J. NUSSBAUMER

Contents

Chapter 1 Introduction
1.1 Introductory Remarks. 1
1.2 Notations 2
1.3 The Structure of the Book 3

Chapter 2 Elements of Number Theory and Polynomial Algebra
2.1 Elementary Number Theory 4
    2.1.1 Divisibility of Integers. 4
    2.1.2 Congruences and Residues 7
    2.1.3 Primitive Roots. 11
    2.1.4 Quadratic Residues 17
    2.1.5 Mersenne and Fermat Numbers 19
2.2 Polynomial Algebra. 22
    2.2.1 Groups 23
    2.2.2 Rings and Fields 24
    2.2.3 Residue Polynomials 25
    2.2.4 Convolution and Polynomial Product Algorithms
          in Polynomial Algebra. 27

Chapter 3 Fast Convolution Algorithms
3.1 Digital Filtering Using Cyclic Convolutions 32
    3.1.1 Overlap-Add Algorithm 33
    3.1.2 Overlap-Save Algorithm 34
3.2 Computation of Short Convolutions and Polynomial Products . . 34
    3.2.1 Computation of Short Convolutions by the
          Chinese Remainder Theorem. 35
    3.2.2 Multiplications Modulo Cyclotomic Polynomials 37
    3.2.3 Matrix Exchange Algorithm 40
3.3 Computation of Large Convolutions by Nesting of Small
    Convolutions. 43
    3.3.1 The Agarwal-Cooley Algorithm. 43
    3.3.2 The Split Nesting Algorithm 47
    3.3.3 Complex Convolutions 52
    3.3.4 Optimum Block Length for Digital Filters 55
3.4 Digital Filtering by Multidimensional Techniques 56
3.5 Computation of Convolutions by Recursive Nesting of Polynomials 60
3.6 Distributed Arithmetic 64

3.7 Short Convolution and Polynomial Product Algorithms 66
    3.7.1 Short Circular Convolution Algorithms 66
    3.7.2 Short Polynomial Product Algorithms. 73
    3.7.3 Short Aperiodic Convolution Algorithms 78

Chapter 4   The Fast Fourier Transform
4.1 The Discrete Fourier Transform 80
    4.1.1 Properties of the DFT. 81
    4.1.2 DFTs of Real Sequences. 83
    4.1.3 DFTs of Odd and Even Sequences 84
4.2 The Fast Fourier Transform Algorithm 85
    4.2.1 The Radix-2 FFT Algorithm. 87
    4.2.2 The Radix-4 FFT Algorithm 91
    4.2.3 Implementation of FFT Algorithms 94
    4.2.4 Quantization Effects in the FFT 96
4.3 The Rader-Brenner FFT. 99
4.4 Multidimensional FFTs 102
4.5 The Bruun Algorithm 104
4.6 FFT Computation of Convolutions 107

Chapter 5   Linear Filtering Computation of Discrete Fourier Transforms
5.1 The Chirp z-Transform Algorithm. 112
    5.1.1 Real Time Computation of Convolutions and DFTs
           Using the Chirp z-Transform. 113
    5.1.2 Recursive Computation of the Chirp z-Transform. 114
    5.1.3 Factorizations in the Chirp Filter 115
5.2 Rader's Algorithm 116
    5.2.1 Composite Algorithms. 118
    5.2.2 Polynomial Formulation of Rader's Algorithm 120
    5.2.3 Short DFT Algorithms 123
5.3 The Prime Factor FFT 125
    5.3.1 Multidimensional Mapping of One-Dimensional DFTs. . . 125
    5.3.2 The Prime Factor Algorithm 127
    5.3.3 The Split Prime Factor Algorithm. 129
5.4 The Winograd Fourier Transform Algorithm (WFTA). 133
    5.4.1 Derivation of the Algorithm 133
    5.4.2 Hybrid Algorithms 138
    5.4.3 Split Nesting Algorithms. 139
    5.4.4 Multidimensional DFTs 141
    5.4.5 Programming and Quantization Noise Issues 142
5.5 Short DFT Algorithms 144
    5.5.1 2-Point DFT. 145
    5.5.2 3-Point DFT. 145

Contents XI

    5.5.3   4-Point DFT 145
    5.5.4   5-Point DFT 146
    5.5.5   7-Point DFT 146
    5.5.6   8-Point DFT 147
    5.5.7   9-Point DFT 148
    5.5.8   16-Point DFT 149

Chapter 6   Polynomial Transforms

6.1   Introduction to Polynomial Transforms 151
6.2   General Definition of Polynomial Transforms 155
    6.2.1   Polynomial Transforms with Roots in a Field of
            Polynomials 157
    6.2.2   Polynomial Transforms with Composite Roots 161
6.3   Computation of Polynomial Transforms and Reductions 163
6.4   Two-Dimensional Filtering Using Polynomial Transforms 165
    6.4.1   Two-Dimensional Convolutions Evaluated by Polynomial
            Transforms and Polynomial Product Algorithms 166
    6.4.2   Example of a Two-Dimensional Convolution Computed
            by Polynomial Transforms 168
    6.4.3   Nesting Algorithms 170
    6.4.4   Comparison with Conventional Convolution Algorithms . . 172
6.5   Polynomial Transforms Defined in Modified Rings 173
6.6   Complex Convolutions 177
6.7   Multidimensional Polynomial Transforms 178

Chapter 7   Computation of Discrete Fourier Transforms by Polynomial Transforms

7.1   Computation of Multidimensional DFTs by Polynomial Transforms 181
    7.1.1   The Reduced DFT Algorithm 182
    7.1.2   General Definition of the Algorithm 186
    7.1.3   Multidimensional DFTs 193
    7.1.4   Nesting and Prime Factor Algorithms 194
    7.1.5   DFT Computation Using Polynomial Transforms Defined
            in Modified Rings of Polynomials 196
7.2   DFTs Evaluated by Multidimensional Correlations and Polynomial
    Transforms 201
    7.2.1   Derivation of the Algorithm 201
    7.2.2   Combination of the Two Polynomial Transform Methods . 205
7.3   Comparison with the Conventional FFT 206
7.4   Odd DFT Algorithms 207
    7.4.1   Reduced DFT Algorithm. $N=4$ 209
    7.4.2   Reduced DFT Algorithm. $N=8$ 209
    7.4.3   Reduced DFT Algorithm. $N=9$ 209
    7.4.4   Reduced DFT Algorithm. $N=16$ 210

Chapter 8 Number Theoretic Transforms
8.1 Definition of the Number Theoretic Transforms 211
    8.1.1 General Properties of NTTs 213
8.2 Mersenne Transforms 216
    8.2.1 Definition of Mersenne Transforms 216
    8.2.2 Arithmetic Modulo Mersenne Numbers 219
    8.2.3 Illustrative Example 221
8.3 Fermat Number Transforms 222
    8.3.1 Definition of Fermat Number Transforms 223
    8.3.2 Arithmetic Modulo Fermat Numbers 224
    8.3.3 Computation of Complex Convolutions by FNTs 227
8.4 Word Length and Transform Length Limitations 228
8.5 Pseudo Transforms 230
    8.5.1 Pseudo Mersenne Transforms 231
    8.5.2 Pseudo Fermat Number Transforms 234
8.6 Complex NTTs 236
8.7 Comparison with the FFT 239

Appendix A Relationship Between DFT and Convolution Polynomial Transform Algorithms
A.1 Computation of Multidimensional DFT's by the Inverse Polynomial Transform Algorithm 241
    A.1.1 The Inverse Polynomial Transform Algorithm 241
    A.1.2 Complex Polynomial Transform Algorithms 244
    A.1.3 Round-off Error Analysis 246
A.2 Computation of Multidimensional Convolutions by a Combination of the Direct and Inverse Polynomial Transform Methods 247
    A.2.1 Computation of Convolutions by DFT Polynomial Transform Algorithms 248
    A.2.2 Convolution Algorithms Based on Polynomial Transforms and Permutations 249
A.3 Computation of Multidimensional Discrete Cosine Transforms by Polynomial Transforms 251
    A.3.1 Computation of Direct Multidimensional DCT's 251
    A.3.2 Computation of Inverse Multidimensional DCT's 253

Appendix B Short Polynomial Product Algorithms 255

Problems 263
References 269
Subject Index 275

1. Introduction

1.1 Introductory Remarks

The practical applications of the digital convolution and of the discrete Fourier transform (DFT) have gained growing importance over the last few years. This is a direct consequence of the major role played by digital filtering and DFTs in digital signal processing and by the increasing use of digital signal processing techniques made possible by the rapidly declining cost of digital hardware. The motivation for developing fast convolution and DFT algorithms is strongly rooted in the fact that the direct computation of length-N convolutions and DFTs requires a number of operations proportional to N^2 which becomes rapidly excessive for long dimensions. This, in turn, implies an excessively large requirement for computer implementation of the methods.

Historically, the most important event in fast algorithm development has been the fast Fourier transform (FFT), introduced by Cooley and Tukey in 1965, which computes DFTs with a number of operations proportional to $N \log N$ and therefore reduces drastically the computational complexity for large transforms. Since convolutions can be computed by DFTs, the FFT algorithm can also be used to compute convolutions with a number of operations proportional to $N \log N$ and has therefore played a key role in digital signal processing ever since its introduction. More recently, many new fast convolution and DFT techniques fave been proposed to further decrease the computational complexity corresponding to these operations. The fast DFT algorithm introduced in 1976 by Winograd is perhaps the most important of these methods because it achieves a theoretical reduction of computational complexity over the FFT by a method which can be viewed as the converse of the FFT, since it computes a DFT as a convolution. Indeed, as we shall see in this book, the relationship between convolution and DFT has many facets and its implications go far beyond a mere algorithmic procedure.

Another important factor in the development of new algorithms was the recognition that convolutions and DFTs can be viewed as operations defined in finite rings and fields of integers and of polynomials. This new point of view has allowed both derivation of some lower computational complexity bounds and design of new and improved computation techniques such as those based on polynomial transforms and number theoretic transforms.

In addition to their practical implications, many convolution and DFT algorithms are also of theoretical significance because they lead to a better understanding of mathematical structures which may have many applications in areas other than convolution and DFT. It is likely, for instance, that polynomial transforms will appear as a very general tool for mapping multidimensional problems into one-dimensional problems.

The matter of comparing different algorithms which perform the same functions is pervasive throughout this book. In many cases, we have used the number of arithmetic operations required to execute an algorithm as a measure of the computational complexity. While there is some rough relationship between the overall complexity of an algorithm and its algebraic complexity, the practical value of a computation method depends upon a number of factors. Apart from the number of arithmetic operations, the efficiency of an algorithm is related to many parameters such as the number of data moves, the cost of ancillary operation, the overall structural complexity, the performance capabilities of the computer on which the algorithm is executed, and the skill of the programmer. Therefore, ranking different algorithms as a function of actual efficiency expressed in terms of computer execution times is a difficult art so that the comparisons based on the number of arithmetic operations must be weighted as a function of the particular implementation.

1.2 Notations

It is always difficult to avoid the proliferation of different symbols and subscripts when presenting the various DFT and convolution algorithms. We have adopted here some conventions in order to simplify the presentation. Discrete data sequences are usually represented by lower case letters such as x_n. We have not used the representation $\{x_n\}$ for data sequences, because this simplifies the notation and because the context information prevents confusion between the sequence and the nth element of the sequence. Thus, in our representation, a discrete-time signal x_n is a sequence of the values of a continuous signal $x(t)$, sampled at times $t = nT$ and represented by a number. Polynomials are represented by capital letters such as

$$X(z) = \sum_{n=0}^{N-1} x_n z^n. \tag{1.1}$$

For transforms, we use the notation $\bar{X}_k$, which, for a DFT, has the form

$$\bar{X}_k = \sum_{n=0}^{N-1} x_n W^{nk}. \tag{1.2}$$

We have also sometimes adopted Rader's notation $\langle x \rangle_p$ for the residue of x modulo p.

1.3 The Structure of the Book

Chapter 2 presents introductory material on number theory and polynomial algebra. This covers in an intuitive way various topics such as the divisibility of integers and polynomials, congruences, roots defined in finite fields and rings. This background in mathematics is required to understand the rest of the book and may be skipped by the readers who are already familiar with number theory and modern algebra.

Fast convolution algorithms are discussed in Chap. 3. It is shown that most of these algorithms can be represented in polynomial algebra and can be considered as various forms of nesting.

The fourth chapter gives a simple development of the conventional fast Fourier transform algorithm and presents new versions of this method such as the Rader-Brenner algorithm.

Chapter 5 is devoted to the computation of discrete Fourier transforms by convolutions and deals primarily with Winograd Fourier transform which is an extremely efficient algorithm for the computation of the discrete Fourier transform.

In Chaps. 6 and 7, we introduce the polynomial transforms which are DFTs defined in finite rings and fields of polynomials. We show that these transforms are computed without multiplications and provide an efficient tool for computing multidimensional convolutions and DFTs.

In Chap. 8, we turn our attention to algorithms implemented in modular arithmetic and we present the number theoretic transforms which are DFTs defined in finite rings and fields of numbers. We show that these transforms may have important applications when implemented in special purpose hardware.

2. Elements of Number Theory and Polynomial Algebra

Many new digital signal processing algorithms are derived from elementary number theory or polynomial algebra, and some knowledge of these topics is necessary to understand these algorithms and to use them in practical applications.

This chapter introduces the necessary background required to understand these algorithms in a simple, intuitive way, with the intent of familiarizing engineers with the mathematical principles that are most frequently used in this book. We have made here no attempt to give a complete rigorous mathematical treatment but rather to provide, as concisely as possible, some mathematical tools with the hope that this will prompt some readers to study further, with some of the many excellent books that have been published on the subject [2.1–4].

The material covered in this chapter is divided into two main parts: elementary number theory and polynomial algebra. In elementary number theory, the most important topics for digital signal processing applications are the Chinese remainder theorem and primitive roots. The Chinese remainder theorem, which yields an unusual number representation, is used for number theoretic transforms (NTT) and for index manipulations which serve to map one-dimensional problems into multidimensional problems. The primitive roots play a key role in the definition of NTTs and are also used to convert discrete Fourier transforms (DFT) into convolutions, which is an important step in the development of the Winograd Fourier transform algorithm.

In the polynomial algebra section, we introduce briefly the concepts of rings and fields that are pervasive throughout this book. We show how polynomial algebra relates to familiar signal processing operations such as convolution and correlation. We introduce the Chinese remainder theorem for polynomials and we present some complexity theory results which apply to convolutions and correlations.

2.1 Elementary Number Theory

In this section, we shall be essentially concerned with the properties of integers. We begin with the simple concept of integer division.

2.1.1 Divisibility of Integers

Let a and b be two integers, with b positive. The division of a by b is defined by

$$a = bq + r, \quad 0 \leqslant r < b, \tag{2.1}$$

where q is called the *quotient* and r is called the *remainder*. When $r = 0$, b and q are *factors* or divisors of a, and b is said to *divide a*, this operation being denoted by $b|a$. When a has no other divisors than 1 and a, a is a *prime*. In all other cases, a is *composite*.

When a is composite, it can always be factorized into a product of powers of prime numbers $p_i^{c_i}$, where c_i is a positive integer, with

$$a = \prod_i p_i^{c_i}. \tag{2.2}$$

The *fundamental theorem of arithmetic* states that this factorization is unique.

The largest positive integer d which divides two integers a and b is called the *greatest common divisor* (GCD) and denoted

$$d = (a, b). \tag{2.3}$$

When $d = (a, b) = 1$, a and b have no common factors other than 1 and they are said to be *mutually prime* or *relatively prime*.

The GCD can be found easily by a division algorithm known as the *Euclidean algorithm*. In discussing this algorithm, we shall assume that a and b are positive integers. This is done without loss of generality, since $(a, b) = (-a, b) = (a, -b) = (-a, -b)$. Dividing a by b yields

$$a = bq_1 + r_1, \quad r_1 < b \tag{2.4}$$

by definition, $d = (a, b) \leqslant a$ or b. Therefore, if $r_1 = 0$, $b|a$ and $(a, b) = b$. If $r_1 \neq 0$, we obtain, by continuation of this procedure, the following system of equations:

$$\begin{aligned} b &= r_1 q_2 + r_2, & r_2 &< r_1 \\ r_1 &= r_2 q_3 + r_3, & r_3 &< r_2 \\ &\cdots \cdots \cdots \cdots \\ r_{k-2} &= r_{k-1} q_k + r_k, & r_k &< r_{k-1} \\ r_{k-1} &= r_k q_{k+1}. & & \end{aligned} \tag{2.5}$$

Since $r_1 > r_2 > r_3 \ldots$, the last remainder is zero. Thus, by the last equation, $r_k | r_{k-1}$. The preceding equation implies that $r_k | r_{k-2}$, since $r_k | r_{k-1}$. Finally, we obtain $r_k | b$ and $r_k | a$. Hence, r_k is a divisor of a and b. Suppose now that c is any divisor of a and b. By (2.4), c also divides r_1. Then, (2.5) implies that c divides r_2, $r_3 \ldots r_k$. Thus, any divisor c of a and b divides r_k and therefore $c \leqslant r_k$. Hence, r_k is the GCD of a and b.

An important consequence of Euclid's algorithm is that the GCD of two integers a and b is a linear combination of a and b. This can be seen by rewriting (2.4) and (2.5) as

$$r_1 = a - bq_1$$
$$r_2 = b - r_1 q_2$$
$$\cdots\cdots\cdots\cdots$$
$$r_k = r_{k-2} - r_{k-1} q_k. \tag{2.6}$$

The first equation shows that r_1 is a linear combination of a and b. The second equation shows that r_2 is a linear combination of b and r_1 and therefore of both a and b. Finally, the last equation implies that r_k is a linear combination of a and b. Since $r_k = (a, b)$, we have

$$(a, b) = ma + nb, \tag{2.7}$$

where m and n are integers. When a and b are mutually prime, (2.7) reduces to Bezout's relation

$$1 = ma + nb. \tag{2.8}$$

We now change our point of view by considering a linear equation with integer coefficients a, b, and c

$$ax + by = c \tag{2.9}$$

where x and y are a pair of integers which are the solution of this *Diophantine equation*. Such an equation has a solution if and only if $(a, b) | c$. To demonstrate this point, we note the following. It is obvious from (2.9) that for $a = 0$ or $b = 0$, we must have $b | c$ or $a | c$.

For $a \neq 0$, $b \neq 0$, it is apparent that if (2.9) holds for integers x and y, then $d = (a, b)$ is such that $d | c$. Conversely, if $d | c$, $c = c_1 d$ and (2.7) implies the existence of two integers m and n such that $d = ma + nb$. Hence $c = c_1 d = c_1 ma + c_1 nb$, and the solutions of the Diophantine equation are given by $x = c_1 m$, $y = c_1 n$. Thus, for $(a, b) | c$, the solution of the Diophantine equation is given by the Euclidean algorithm. The solution of the Diophantine equation is not unique, however. This can be seen by considering a particular solution $c = ax_0 + by_0$. Assuming x, y is another solution, we have

$$a(x - x_0) = b(y_0 - y) \tag{2.10}$$

and, by dividing this expression by d, we obtain

$$(a/d)(x - x_0) = (b/d)(y_0 - y) \tag{2.11}$$

Since $[(a/d), (b/d)] = 1$, this implies that $(b/d) | (x - x_0)$ and $x = x_0 + (b/d)s$ where s is an integer. Substituting into (2.11), we obtain

$$\begin{aligned} y &= y_0 - (a/d)s \\ x &= x_0 + (b/d)s. \end{aligned} \tag{2.12}$$

This defines a class of linearly related solutions for (2.9) which depend upon the integer s.

As a numerical example, consider the equation

$$15x + 9y = 21.$$

We first use Euclid's algorithm to determine the GCD d with $a = 15$ and $b = 9$,

$$15 = 9 \cdot 1 + 6$$
$$9 = 6 \cdot 1 + 3$$
$$6 = 3 \cdot 2$$

Hence $d = 3$. Since $3|21$, the Diophantine equation has a solution. We now define 3 as a linear combination of 15 and 9 by recasting the preceding set of equations as

$$6 = 15 - 9 \cdot 1$$
$$3 = 9 - 6 \cdot 1 = -15 + 2 \cdot 9$$

Thus, $m = -1$ and $n = 2$. Dividing $c = 21$ by $d = 3$ yields $c_1 = 7$. This gives a particular solution $x_0 = -7$, $y_0 = 14$. If we divide $a = 15$ and $b = 9$ by $d = 3$, we obtain $(a/d) = 5$ and $(b/d) = 3$. Hence, the general solution to the Diophantine equation becomes

$$y = 14 - 5s$$
$$x = -7 + 3s,$$

where s is any integer.

2.1.2 Congruences and Residues

In (2.1), the division of an integer a by an integer b produces a remainder r. All integers a which give the same remainder when divided by b can be thought as pertaining to the same *equivalence class* relative to the equivalence relation $a = bq + r$.

Two integers a_1 and a_2 pertaining to the same class are said to be *congruent modulo b* and the equivalence is denoted

$$a_1 \equiv a_2 \quad \text{modulo } b. \tag{2.13}$$

Thus, two numbers a_1 and a_2 are congruent modulo b if

$$b|(a_1 - a_2). \tag{2.14}$$

8   2. Elements of Number Theory and Polynomial Algebra

Underlying the concept of congruence is the fact that, in many physical problems, one is primarily interested in measuring relative values within a given range. This is apparent, for instance, when measuring angles. In this case, the angles are defined from 0 to 359° and two angles that differ by a multiple of 360° are considered to be equal. Hence angles are defined modulo 360.

Thus, in congruences, we are interested only in the remainder r of the division of a by b. This remainder is usually called the *residue* and is denoted by

$$r \equiv a \quad \text{modulo } b. \tag{2.15}$$

This representation is sometimes simplified to a form with the symbol $\langle \ \rangle$ [2.4],

$$r = \langle a \rangle_b, \tag{2.16}$$

where the subscript is omitted when there is no ambiguity on the nature of the modulus b.

It follows directly from the definition of residues given by (2.14) that additions and multiplications can be performed directly on residues

$$\begin{aligned}\langle a_1 + a_2 \rangle &= \langle\!\langle a_1 \rangle + \langle a_2 \rangle\!\rangle \\ \langle a_1 a_2 \rangle &= \langle\!\langle a_1 \rangle \langle a_2 \rangle\!\rangle.\end{aligned} \tag{2.17}$$

With congruences, division is not defined. We can however define something close to it by considering the linear congruence

$$a x \equiv c \quad \text{modulo } b \tag{2.18}$$

This linear congruence is the Diophantine equation $ax + by = c$ in which all terms are defined modulo b. Thus, we know by the results of the preceding section that we can find values of x satisfying (2.18) if and only if $d \,|\, c$, with $d = (a, b)$. In this case, the solutions can be derived from (2.12) and are given by

$$x \equiv x_0 + (b/d)s \quad \text{modulo } b, \tag{2.19}$$

where x_0 is a particular solution and s can be any integer smaller than b. However, there are only d distinct solutions since $(b/d)s$ has only d distinct values modulo b. An important consequence of this point is that the linear congruence $ax \equiv c$ modulo b always has a unique solution when $(a, b) = 1$. Thus, when $(a, b) \,|\, c$, the linear congruence $ax \equiv c$ modulo b can be solved and Euclid's algorithm provides a method for computing the values of x which satisfy this relation. We shall see later that Euler's theorem gives a more elegant solution to the linear congruence (2.18) when $(a, b) = 1$.

We consider now the problem of solving a set of simultaneous linear congruences with different moduli. Changing our notation, we want to find the integer x which satisfies simultaneously the k linear congruences

$$x \equiv r_i \text{ modulo } m_i, \quad i = 1,\ldots, k. \tag{2.20}$$

The solution to this problem plays a major role in many signal processing algorithms and is given by the *Chinese remainder theorem*

Theorem 2.1: Let m_i be k positive integers greater than 1 and relatively prime in pairs. The set of linear congruences $x \equiv r_i$ modulo m_i has a unique solution modulo M, with $M = \prod_{i=1}^{k} m_i$.

The proof of this theorem is established by using the relations

$$x \equiv \sum_{i=1}^{k} (M/m_i) r_i T_i \quad \text{modulo } M \tag{2.21}$$

$$(M/m_i) T_i \equiv 1 \quad \text{modulo } m_i. \tag{2.22}$$

Equation (2.22) defines k linear congruences. Since the m_i are mutually prime, $[m_i, (M/m_i)] = 1$ and each of these congruences has a unique solution T_i which can be computed by Euclid's algorithm or Euler's theorem (theorem 2.3). Let us now reduce x in (2.21) modulo m_u, one of the moduli m_i. Except for M/m_u, all the expressions M/m_i contain m_u as a factor and are therefore equal to zero modulo m_u. Hence, (2.21) reduces to

$$x \equiv (M/m_u) r_u T_u \quad \text{modulo } m_u \tag{2.23}$$

and, since (2.22) implies that $(M/m_u) T_u \equiv 1$ modulo m_u, (2.23) becomes

$$x \equiv r_u \quad \text{modulo } m_u. \tag{2.24}$$

It is seen easily that this operation can be repeated for all moduli m_i and therefore that (2.21) is the solution of the k linear congruences $x \equiv r_i$ modulo m_i.

As a simple application of the Chinese remainder theorem, let us find the solution to the simultaneous congruences

$$\langle x \rangle_3 = 2, \quad \langle x \rangle_4 = 1, \quad \langle x \rangle_5 = 3.$$

Here, we have $m_1 = 3$, $m_2 = 4$, $m_3 = 5$, $M = 60$, $M/m_1 = 20$, $M/m_2 = 15$, and $M/m_3 = 12$. The congruences $\langle 20\, T_1 \rangle_3 = 1$, $\langle 15\, T_2 \rangle_4 = 1$, and $\langle 12\, T_3 \rangle_5 = 1$ are solved, respectively, by $T_1 = 2$, $T_2 = 3$, and $T_3 = 3$. Hence

$$x \equiv 20 \cdot 2 \cdot 2 + 15 \cdot 1 \cdot 3 + 12 \cdot 3 \cdot 3 \text{ modulo } 60$$

$$x = 53.$$

The Chinese remainder theorem can be used to define residue number systems (RNS) which allow one to perform high-speed arithmetic operations without carry propagation from digit to digit. In such a system, an integer a is represented by its residues a_i modulo a set of relatively prime integers m_i

$$a = \{a_1, a_2, \ldots a_k\} \tag{2.25}$$

In this system, the addition or the multiplication of two integers a and b is done by adding or multiplying separately their various residues a_i and b_i, without any carry from one residue to another one. Thus, if $M = \prod_i m_i$ is chosen to be the product of many small relatively prime moduli m_i, the computation can accommodate large numbers although actual calculations are performed on a large set of small residues, without carry propagation. Hence, residue number systems are quite effective for high-speed multiplications and additions. Unfortunately, this advantage is usually offset by many practical difficulties related to the cost of translating from conventional number systems to RNS, the lack of a division operation, and the increased word length required for unambiguous operation in a modular system. Because of these limitations, the RNS is rarely used. We shall see, however, in Chap. 8 that modular arithmetic and the Chinese remainder theorem play an important role in the definition of number theoretic transforms and may have significant applications in these areas.

The Chinese remainder theorem is also often used to map an M-point one-dimensional data sequence x_n into a k-dimensional data array. This is done by noting that if n is defined modulo M, with $n = 0, \ldots, M - 1$, we can redefine n by the Chinese remainder theorem as

$$n \equiv \sum_{i=1}^{k} (M/m_i) n_i T_i \quad \text{modulo } M, \tag{2.26}$$

where the index n_i along dimension i takes the values $0, \ldots, m_i - 1$. This mapping, which is possible only when M is the product of relatively prime factors m_i, is very important for the computation of discrete Fourier transforms and convolutions, as will be seen in the following chapters.

We now introduce the concept of *permutation*. Let us consider again the set of M integers n, with $n = 0, \ldots, M - 1$. If we multiply modulo M, each element n_i of n by an integer a, we obtain a set of M numbers b_i defined by

$$b_i \equiv a \cdot n_i \quad \text{modulo } M. \tag{2.27}$$

The n_i are all distinct. We would like the b_i to be also all distinct in such a way that when the n_i span the M values $0, \ldots, M - 1$, the b_i span the same values, although in a different order.

Each equation (2.27) is a linear congruence and we already know from (2.19) that the solution of this congruence is unique if $(a, M) = 1$. Let us assume that $(a, M) = 1$ and consider two distinct values n_i and n_u pertaining to the set of the M integers n. Since $(a, M) = 1$, the linear congruence (2.27) defines two integers b_i and b_u corresponding, respectively, to n_i and n_u. Subtracting b_u from b_i yields

$$b_i - b_u \equiv a(n_i - n_u) \quad \text{modulo } M. \tag{2.28}$$

If $b_t = b_u$, this implies that $a(n_t - n_u) \equiv 0$ modulo M. This is impossible because a is relatively prime with M and $n_t - n_u < M$. Thus, for $(a, M) = 1$, the permutation defined by (2.27) maps all possible values of n.

As an example, consider the permutation defined by $b \equiv 5n$ modulo 6. 5 and 6 are mutually prime. When n takes successively the values 0, 1, 2, 3, 4, 5, the integers b take the corresponding values 0, 5, 4, 3, 2, 1.

We shall see in the following chapters that permutations are often used in signal processing to reorder a set of data samples. At this point, we return to the one-dimensional to multidimensional mapping using the Chinese remainder theorem to show that this method can be simplified by permutation. When M is the product of two mutually prime factors m_1 and m_2, (2.26) becomes

$$n \equiv m_2 n_1 T_1 + m_1 n_2 T_2 \quad \text{modulo } M. \tag{2.29}$$

Since T_1 and T_2 are mutually prime with m_1 and m_2, respectively, $m_2 n_1 T_1$ and $m_1 n_2 T_2$ can be viewed as the two permutations $n_1 T_1$ modulo m_1 and $n_2 T_2$ modulo m_2 of two sets of m_1 points and m_2 points, respectively. Hence the mapping defined by (2.29) can be replaced by the simpler mapping

$$n \equiv m_2 n_1 + m_1 n_2 \quad \text{modulo } M \tag{2.30}$$

and, in the case of more than two factors,

$$n \equiv \sum_{i=1}^{k} (M/m_i) n_i \quad \text{modulo } M. \tag{2.31}$$

The advantage of (2.30) over (2.26) is that the computation of the inverses T_i is no longer required.

As an example, consider $M = 6$, with $m_1 = 2$ and $m_2 = 3$. The sequence n is given by: $\{0, 1, 2, 3, 4, 5\}$. Since $T_1 = 1$ and $T_2 = 2$, (2.26) yields $n \equiv 3n_1 + 4n_2$ while (2.30) gives $n \equiv 3n_1 + 2n_2$. When the pair n_1, n_2 takes successively the values $\{(0, 0), (1, 0), (0, 1), (1, 1), (0, 2), (1, 2)\}$, the sequence n becomes $\{0, 3, 4, 1, 2, 5\}$ for the first equation and $\{0, 3, 2, 5, 4, 1\}$ for the second equation. Thus, both approaches span the complete set of values of n, although in a different order.

2.1.3 Primitive Roots

We have seen that defining integers modulo an integer m partitions these integers into m equivalence classes. Among these classes, those corresponding to integers which are relatively prime to m play a particularly important role and we shall often need to know how many integers are smaller than m and relatively prime to m. This quantity is usually denoted by $\phi(m)$ and called *Euler's totient function*.

We may observe that $\phi(1) = 1$ since $(1, 1) = 1$. When m is a prime, with $m = p$, all integers smaller than p are relatively prime to p. Thus,

$$\phi(p) = p - 1. \tag{2.32}$$

If $m = p^e$, the only numbers less than m and not prime with p are the multiples of p. Therefore,

$$\phi(p^e) = p^{e-1}(p - 1) = p^e(1 - 1/p). \tag{2.33}$$

In order to find $\phi(m)$ for any integer m, we first establish that Euler's totient function is multiplicative.

Theorem 2.2: If a and b are two mutually prime integers, $\phi(a \cdot b) = \phi(a) \phi(b)$.

The theorem is proved by considering all integers u smaller than $a \cdot b$ and defined by $u = aq + r, r = 0, 1, ..., a - 1$ and $q = 0, 1, ..., b - 1$. It is seen that u is relatively prime to a if r is one of the $\phi(a)$ integers r_i relatively prime to a. Thus, the $b\phi(a)$ integers u_i given by $(r_i), (a + r_i), (2a + r_i), ..., [(b - 1)a + r_i]$ are prime to a. If q is chosen among the $\phi(b)$ integers smaller than b and mutually prime with b, the corresponding integers u_i are relatively prime to b, since no factor of b can divide a or q. Thus, there are $\phi(a)\phi(b)$ integers relatively prime to a and b and therefore relatively prime to $a \cdot b$.

An immediate corollary of theorem 2.2 is that, if an integer N is given by its prime factorization $N = p_1^{c_1} p_2^{c_2} ... p_k^{c_k}$, then $\phi(N)$ becomes

$$\phi(N) = N \prod_{i=1}^{k} (1 - 1/p_i). \tag{2.34}$$

This follows from theorem 2.2 and from (2.33).

An important property of Euler's totient function is that the sum of $\phi(d)$ taken over all divisors d of N is equal to N,

$$\sum_{d|N} \phi(d) = N. \tag{2.35}$$

This property follows from the fact that N/d is a divisor of N when $d|N$. Thus,

$$\sum_{d|N} \phi(d) = \sum_{d|N} \phi(N/d). \tag{2.36}$$

We now consider the sets S of integers a such that $(a, N) = d, 1 \leq a \leq N$. Every integer from 1 to N belongs to one and only one set S. In each set, we have $(a, N) = d$ or $(a/d, N/d) = 1$; thus each set contains $\phi(N/d)$ integers. Since we have N integers,

$$N = \sum_{d|N} \phi(N/d) \tag{2.37}$$

and (2.35) follows from (2.36) and (2.37).

With Euler's totient function, we have specified the equivalence classes of integers that are defined modulo m and relatively prime with m. We shall now go further in the specification of equivalence classes by defining the order of an integer modulo m. This is done by considering an integer x_n defined by

$$x_n \equiv a^n \quad \text{modulo } m, \tag{2.38}$$

where a is an integer. If n takes successively the values 0, 1, 2, ..., then x_n will successively take the values $x_0, x_1, x_2, \ldots$. Since x_n can only take the m distinct values 0 to $m-1$, x_n will necessarily repeat a previously computed value x_i for some integer r, with $r > i$. Let r be the smallest value of n for which such a repetition occurs. We have

$$x_r \equiv x_i \quad \text{modulo } m, r > i. \tag{2.39}$$

Then, by definition, $x_{r+1} \equiv ax_r$ and $x_{i+1} \equiv ax_i$. Thus, $x_{r+1} \equiv x_{i+1}$ and, for any $n \geqslant r$,

$$x_n \equiv x_{n-r+i} \quad \text{modulo } m. \tag{2.40}$$

This means that, for $n \geqslant i$, the sequence x_n repeats itself cyclically with a period of $r - i$ elements. When $i = 0$, the sequence repeats itself from the beginning ($a^0 = 1$) and the cyclic group defined be (2.40) contains all the possible values of x_n corresponding to a and m. The conditions for this important case are given by Euler's theorem.

Theorem 2.3: If $(a, m) = 1$, then

$$a^{\phi(m)} \equiv 1 \quad \text{modulo } m. \tag{2.41}$$

This theorem is proved easily by considering the permutations $(a \cdot n)$ modulo m of the $\phi(m)$ integers n which are relatively prime to m. Since $(a, m) = 1$, all the permutation products are distinct. Moreover, since $(n, m) = 1$, $(a \cdot n, m) = 1$ and the permutation $\langle a \cdot n \rangle$ spans the complete set of integers mutually prime with m. Thus,

$$\prod_{(n,m)=1} n \equiv \prod_{(n,m)=1} a \cdot n \equiv a^{\phi(m)} \prod_{(n,m)=1} n \quad \text{modulo } (m). \tag{2.42}$$

We can cancel the product of n on both sides of the congruence because the various integers n are relatively prime to m. Thus, (2.42) yields (2.41) and the proof is completed. When m is a prime, with $m = p$, then $(a, p) = 1$ if p does not divide a and we have $\phi(p) = p - 1$. In this case, Euler's theorem reduces to Fermat's theorem

Theorem 2.4: If p is a prime, then, for every integer a,

$$a^{p-1} \equiv 1 \quad \text{modulo } p \tag{2.43}$$

or

$$a^p \equiv a \quad \text{modulo } p. \tag{2.44}$$

Theorems 2.3 and 2.4 give a simple alternative to Euclid's algorithm for solving linear congruences when $(a, m) = 1$. Consider the congruence

$$a x \equiv c \quad \text{modulo } m. \tag{2.45}$$

We know that, if $(a, m) = 1$, this congruence has always a unique solution. This solution is given by

$$x \equiv c \, a^{\phi(m)-1} \quad \text{modulo } m \tag{2.46}$$

or, when m is a prime with $m = p$,

$$x \equiv c \, a^{p-2} \quad \text{modulo } p. \tag{2.47}$$

An interesting application of Euler's theorem can be found for the Chinese remainder reconstruction process discussed in proving theorem 2.1. We have seen, with theorem 2.1, that if M is an integer which factors into k relatively prime integers m_i, then an integer x, defined modulo M can be reconstructed by (2.21) from the various residues r_i, with $r_i = x$ modulo m_i. One of the difficulties in using the Chinese remainder theorem consists in evaluating the k inverses T_i modulo m_i. However we note that, by using Euler's theorem, the Chinese remainder reconstruction defined by (2.21) and (2.22) can be replaced by a much simpler formulation which does not require the computation of inverses

$$x \equiv \sum_{i=1}^{k} (M/m_i)^{\phi(m_i)} r_i \quad \text{modulo } M. \tag{2.48}$$

This equation is established by noting that $(M/m_u)^{\phi(m_u)} \equiv 0$ modulo m_i for $i \neq u$ and that $(M/m_u)^{\phi(m_u)} \equiv 1$ modulo m_u by Euler's theorem.

Thus, Euler's theorem is sometimes used to solve linear congruences. The main interest of this theorem, however, lies in the specification of the *order* of an *integer*. We have seen above that the sequence $x_n \equiv a^n$ modulo m repeats itself with a periodicity $r - i$, from a value i. If $a^n \equiv 1$ for some value r of n, the sequence will repeat itself from its beginning, since $x_0 = 1$. Hence, if r is the smallest positive integer such that $a^r \equiv 1$ modulo m, the complete sequence of integers a^n modulo m will be periodic with period r.

Let us now determine the maximum value of r for a given m. We know by Euler's theorem that if $(a, m) = 1$, $r = \phi(m)$. If $(a, m) = d \neq 1$, we have $(a/d, m/d) = 1$ and $r_1 = \phi(m/d)$. Since $\phi(m/d) < \phi(m)$, the period r is maximum for $(a, m) = 1$. We shall call the element g which generates a sequence of length $\phi(m)$ a *primitive root*. An element g generating a shorter cyclic sequence of length $r < \phi(m)$ will be simply called a *root of order r*. We shall now consider the issue of how many roots of a given order can exist and define ways of finding these roots. The following theorems are used to support these objectives.

Theorem 2.5: If g is a root of order r modulo m, the r integers $g^0, g^1, ..., g^{r-1}$ are incongruent modulo m.

This theorem is proved by assuming that $g^{r_1} \equiv g^{r_2}$ for two distinct values r_1 and r_2 such that $r_2 < r_1 < r$. If this were the case, we would have

$$g^{r_1-r_2} \equiv 1 \quad \text{modulo } m.$$

This is impossible since $r_1 - r_2 < r$ and r is, by definition, the smallest integer such that $g^r \equiv 1$.

Theorem 2.6: If $(g, m) = 1$ and $g^b \equiv 1$ modulo m, the order r of the integer g must divide b.

If $g^b \equiv 1$ modulo m, $m|(g^b - 1)$. Let us assume that g is of order r. This means that r is the smallest integer such that $g^r \equiv 1$ modulo m. Since $m|(g^r - 1)$, $m|(g^d - 1)$, where $d = (r, b)$. Since d is the GCD of r and b, $d \leqslant r$. However, d cannot be less than r which is by definition the smallest integer such that $g^r \equiv 1$. Thus, $d = r$. Since $d|b$, $r|b$.

Theorem 2.7: If $(g, m) = 1$, the order r of the integer g must divide $\phi(m)$.

This theorem follows directly from theorem 2.6 and Euler's theorem, since $g^{\phi(m)} \equiv 1$ modulo m if $(g, m) = 1$.

It can also be shown that primitive roots exist only for $m = p^e$ or $m = 2p^e$, with p an odd prime. When $p = 2$, primitive roots exist only for $m = 2$ and $m = 4$. When $m = p$, with p an odd prime, the following theorem, first introduced by Gauss, specifies the number of roots of a given order.

Theorem 2.8: If $r|p - 1$, with p an odd prime, there are $\phi(r)$ incongruent integers which have order r modulo p.

Suppose that g has order r modulo p. Then, by theorem 2.5, the r integers $g^0, g^1, ..., g^{r-1}$ are incongruent modulo p and satisfy the equation $x^r \equiv 1$ modulo p. Thus, the sequence g^n modulo p is periodic and n is defined modulo r, with $n = 0, 1, ... r - 1$. Then, for $(b, r) = 1$, g^{bn} modulo p is a simple permutation of the sequence g^n modulo p. When $(b, r) \neq 1$ the sequence $b \cdot n$ modulo r will contain repetitions and therefore the corresponding integers g^b will be of order less than r. Thus, we have either zero or $\phi(r)$ incongruent roots of order r. Since all integers in the set $1, ..., p - 1$ have some order, the total number of roots, for all divisors of $p - 1$, is equal to $p - 1$. We note, with (2.35), that

$$\sum_{r|p-1} \phi(r) = p - 1. \tag{2.49}$$

Thus, there are $\phi(r)$ roots of order r for each divisor r of $p - 1$ and this completes the proof of the theorem.

The theory of primitive roots is quite complex and a complete treatment can be found in [2.1–3]. In practice, primitive roots modulo primes less than 10000 are given in [2.5]. When $m = p^e$ or $m = 2p^e$, for p an odd prime, the primitive roots are of order r, with $r = \phi(m)$. Thus, an integer will be primitive root if $g^n \not\equiv 1$ modulo m for $n < r$. Moreover, if $r = q_1^{e_1} q_2^{e_2} \ldots q_k^{e_k}$ is the prime factorization of r, any integer which is not primitive root will be a root of order r_i smaller than r, where r_i is a factor of r. Thus, a primitive root g satisfies the condition

$$g^{r/q_i^{e_i}} \not\equiv 1 \quad \text{modulo } m. \tag{2.50}$$

The use of (2.50) greatly simplifies the search for primitive roots as can be seen with the following example corresponding to $m = 41$. Since 41 is a prime, $r = \phi(41) = 40 = 5 \cdot 2^3$. A straightforward approach to checking whether an integer x is a primitive root would be to compute x^n modulo 41 for $n = 1, 2, \ldots, 39$ and to check that $x^n \not\equiv 1$ for all these values of n. When m is large, this method becomes rapidly impracticable and it is much simpler to check that $x^n \not\equiv 1$ modulo m only for the n being a factor of $\phi(m)$. In our case, we note that $2^{20} \equiv 1$, $3^8 \equiv 1$, $4^{20} \equiv 1$, and $5^{20} \equiv 1$. Thus 2, 3, 4, and 5 are ruled out as primitive roots. We note however that $6^8 \equiv 10$ and $6^{20} \equiv 40$ and, therefore, 6 is a primitive root modulo 41.

Once a primitive root g has been found, any root of order r_i, where $r_i | r$, can easily be found by raising g to the power r/r_i. Moreover, when m is composite, with $m = m_1, m_2, \ldots, m_k$, we know by the Chinese remainder theorem that any root of order r_i modulo m must also be a root of order r_i modulo $m_1, m_2, \ldots, m_k$. Thus, these roots are easily found once the primitive roots modulo p^e and modulo $2p^e$ are known.

Primitive roots play a very important role in digital signal processing. We shall see in Chap. 8 that they may be used to define number theoretic transforms. Another key application of primitive roots concerns the mapping of DFTs into circular correlations, which is a crucial step in the development of the Winograd Fourier transform algorithm. We shall discuss this point in detail in Chap. 5, but we give here the essence of the technique in the simple case of a p-point DFT, with p a prime,

$$\bar{X}_k = \sum_{n=0}^{p-1} x_n W^{nk}, \quad W = e^{-j2\pi/p}, \, j = \sqrt{-1}$$
$$k = 1, \ldots, p - 1 \tag{2.51}$$

and for $k = 0$,

$$\bar{X}_0 = \sum_{n=0}^{p-1} x_n. \tag{2.52}$$

The exponents and indices in (2.51) are defined modulo p. Thus, for $k \neq 0$ and $n \neq 0$, we can change the variables with

$$\begin{aligned} n &\equiv g^u \quad \text{modulo } p, \\ k &\equiv g^v \quad \text{modulo } p, \quad u, v = 0, \ldots, p - 2, \end{aligned} \tag{2.53}$$

where g is a primitive root modulo p.
Then, for $k \neq 0$, (2.51) becomes

$$\bar{X}_{gv} \equiv x_0 + \sum_{u=0}^{p-2} x_{g^u} W^{g^u+v}. \tag{2.54}$$

This demonstrates that the main part of the computation of $\bar{X}_k$ is a circular correlation of the sequence x_{g^u} with the sequence W^{g^u}.

2.1.4 Quadratic Residues

We have seen in the preceding section that primitive roots were closely analogous to exponentials. We shall discuss here the concept of quadratic residues. It is notable that this class of residues can be viewed as the equivalent of square roots defined in the set of integers modulo an integer m.

If $(a, m) = 1$, a is said to be a *quadratic residue* of m if $x^2 \equiv a$ modulo m has a solution. If this congruence has no solution, a is a *quadratic nonresidue* of m.

It is obvious, from the Chinese remainder theorem, that if a is a quadratic residue of m and if m is composite, a must be a quadratic residue of each mutually prime factor of m. Furthermore, it can be shown that a is a quadratic residue of p^e, with p an odd prime, if and only if a is a quadratic residue of p. When $m = 2^e$ and a is an odd integer, a is a quadratic residue of 2. Moreover, a is a quadratic residue of 4 if and only if $a \equiv 1$ modulo 4 and a is a quadratic residue of 2^k, $k \geqslant 3$ if and only if $a \equiv 1$ modulo 8. Thus, we can restrict our discussion to odd prime moduli since all other cases are deduced easily from this particular case.

In the following, we shall determine the number of distinct quadratic residues of p and show how to check integers for quadratic residue properties. We first establish the two following theorems.

Theorem 2.9: If p is an odd prime, the number $Q(p)$ of distinct quadratic residues is given by

$$Q(p) = 1 + (p-1)/2. \tag{2.55}$$

By definition, a is a quadratic residue if $x^2 \equiv a$ modulo p has a solution. $a = 0$ is a trivial solution. For $a \neq 0$, we note that, if x is a solution, $p - x$ is also a solution, since $x^2 \equiv a$ modulo p implies that $(p - x)^2 \equiv a$ modulo p. Thus, there are at most $(p-1)/2$ nonzero solutions: $1^2, 2^2, \ldots, [(p-1)/2]^2$. If two solutions were identical, we would have

$$x_1^2 \equiv x_2^2 \quad \text{modulo } p. \tag{2.56}$$

However, $x_1 + x_2 \leqslant p - 1$ since $x_1, x_2 \leqslant (p-1)/2$. Thus, $x_1 + x_2 \not\equiv 0$ modulo p and, since p is prime, (2.56) would imply that $x_1 \equiv x_2$. Under these conditions, all solutions are distinct and we have $(p-1)/2$ distinct quadratic residues different from zero.

Theorem 2.10: If p is an odd prime, the product of two quadratic residues or of two quadratic nonresidues is a quadratic residue. The product of a quadratic residue by a quadratic nonresidue is a quadratic nonresidue.

To prove this theorem, consider the quadratic residues $a_1, a_2, \ldots$ and the quadratic nonresidues $b_1, b_2, \ldots$. If $x_1^2 \equiv a_1$, $x_2^2 \equiv a_2$, then $(x_1 x_2)^2 \equiv a_1 a_2$. Moreover, since p is a prime, if a_1 is a quadratic residue, with $a_1 \neq 0$, the permutation $\langle a_1 n \rangle$ maps all values of n for $0 \leqslant n < p$. We already know that the $(p-1)/2$ terms $\langle a_i a_k \rangle$ are quadratic residues. Since there can only be $(p-1)/2$ quadratic residues of p, all the terms corresponding to $\langle a_i b_k \rangle$ are nonresidues.

Similarly, if we consider the permutations $\langle b_1 n \rangle$, the $(p-1)/2$ terms corresponding to $\langle a_i b_k \rangle$ are nonresidues and therefore the $(p-1)/2$ terms corresponding to $\langle b_i b_k \rangle$ are quadratic residues.

We now define the convenient symbol (a/p) due to Legendre. For $(a, p) = 1$ and p an odd prime,

$$(a/p) = \begin{cases} 1 & \text{if } a \text{ is a quadratic residue of } p \\ -1 & \text{if } a \text{ is a quadratic nonresidue of } p. \end{cases} \tag{2.57}$$

We can note immediately that the definition implies that $(a/p) = 1$ if $p\,|\,(x^2 - a)$. Hence $(1/p) = 1$ and $(a^2/p) = 1$, since $p\,|\,(a^2 - a^2)$.

In order to use Legendre's symbol for the determination of quadratic residues, we shall use a criterion introduced by Euler. We define this criterion here without proof.

Theorem 2.11: If p is an odd prime and a is an integer, then

$$p\,|\,[a^{(p-1)/2} - (a/p)]. \tag{2.58}$$

We note that, by theorem 2.10, the Legendre symbol is multiplicative

$$(ab/p) = (a/p)\,(b/p). \tag{2.59}$$

This symbol is also periodic, with period p, since if $x_0^2 \equiv a$ modulo p, we have obviously $x_0^2 \equiv a + kp$ modulo p. Thus

$$[(a + kp)/p] = (a/p). \tag{2.60}$$

We also give, without proof, the following two theorems.

Theorem 2.12 (Gauss): If p and q are distinct odd primes, then

$$(p/q)\,(q/p) = (-1)^{[(p-1)/2][(q-1)/2]}. \tag{2.61}$$

Theorem 2.13: If p an odd prime, then

$$(2/p) = (-1)^{(p^2-1)/8}. \tag{2.62}$$

We are now armed with enough material to determine rapidly whether an integer a is a quadratic residue. Consider, for instance, the case corresponding to $p = 53$. We want to know if $a = 33$ is a quadratic residue. Hence we want to compute (33/53). As a first step, we use the multiplicative properties of Legendre's symbol. Thus, $(33/53) = (3/53)(11/53)$, and the computation of (33/53) is reduced to the simpler problem of evaluating (3/53) and (11/53). Since 3 and 53 are odd prime, we compute (3/53) by theorem 2.12, with $p = 3$, $q = 53$. This implies $(p-1)/2 = 1$ and $(q-1)/2 = 26$. Thus, $(-1)^{[(p-1)/2][(q-1)/2]} = 1$ and $(3/53) = (53/3)$ by (2.61). We now use the periodicity property of Legendre's symbol to reduce 53 modulo 3 in (53/3). This gives the relation

$$(53/3) = (2/3)$$

and, finally

$$(3/53) = (2/3) = -1, \text{ by (2.62)}.$$

The symbol (11/53) is evaluated similarly;

$$(11/53) = (53/11) \text{ by (2.61)}$$
$$(11/53) = (9/11) \text{ by (2.60)}$$
$$= (3/11)(3/11) \text{ by (2.59)}.$$

Thus, $(11/53) = 1$. Since we have already shown that $(3/53) = -1$, $(33/53) = (3/53)(11/53) = -1$ and 33 is a quadratic nonresidue of 53. This means that it is impossible to find any integer x such that $x^2 \equiv 33$ modulo 53.

2.1.5 Mersenne and Fermat Numbers

Mersenne numbers are defined by

$$M_p = 2^p - 1 \tag{2.63}$$

with p an odd prime. Fermat numbers are defined by

$$F_t = 2^{2^t} + 1, \tag{2.64}$$

where t is any positive integer. These numbers are important in digital signal processing because arithmetic operations modulo Mersenne and Fermat numbers can be implemented relatively simply in digital hardware. This stems from the fact that the machine representation of numbers, usually given in binary notation by a B-bit number

$$a = \sum_{i=0}^{B-1} a_i 2^i, \quad a_i \in \{0, 1\} \tag{2.65}$$

can have an arithmetic overflow at or near the value of M_p and F_t.

If a is an integer, operations modulo a Mersenne number M_p are greatly simplified by noting that $2^p \equiv 1$. Therefore reduction of the integer a modulo M_p can be done without division using the expression

$$a = \sum_{i=0}^{p-1} (\sum_k a_{i+kp}) 2^i, \tag{2.66}$$

where i in (2.65) is replaced by $i + kp$. When two integers a and b, defined modulo M_p, are added together, this generates a $(p + 1)$-bit result, which is reduced modulo M_p by simply adding the most significant carry to the less significant bit position. Thus, operations modulo M_p are equivalent to the familiar one's complement arithmetic. Operations modulo a Fermat number are slightly more difficult to implement, but remain relatively simple when compared with other numbers.

We shall see in Chap. 8 that Mersenne and Fermat numbers play an important role in number theoretic transforms and we shall now establish some properties of these numbers which will be used in Chap. 8. A particularly important property is the order of a root modulo M_p or F_t since it acts to constrain transform length.

Starting with Mersenne numbers, we note that some numbers, like $M_3 = 7$, are prime, while others, like M_{11}, are composite. For composite Mersenne numbers, we have the following theorem.

Theorem 2.14: Every prime divisor q of a Mersenne number is given by

$$q = 2kp + 1. \tag{2.67}$$

By definition $q | 2^p - 1$. Fermat's theorem (theorem 2.4) also implies that $q | 2^{q-1} - 1$. Since $2^p \equiv 1$ and $2^{q-1} \equiv 1$, we have $2^d \equiv 1$ modulo q, with $d = (p, q - 1)$. Moreover, the condition $2^d \equiv 1$ modulo q implies that $d \neq 1$, since $q \neq 1$. Therefore, since p is a prime, $p | q - 1$ and we have $q = sp + 1$. However, s cannot be odd, because q would then be even, which is impossible. Thus $q = 2kp + 1$.

An immediate consequence of theorem 2.14 is the following theorem.

Theorem 2.15: All Mersenne numbers are relatively prime.

If two Mersenne numbers M_{p_1} and M_{p_2} were not relatively prime, this would imply that $2k_1 p_1 + 1 | 2k_2 p_2 + 1$. Since p_1, p_2, $2k_1 p_1 + 1$, and $2k_2 p_2 + 1$ are prime, this is possible only for $p_1 = p_2$ or $M_{p_1} = M_{p_2}$.

We see from theorem 2.14 that, for composite Mersenne numbers, every divisor q of M_p is such that $2p | q - 1$. Thus, for M_p composite, it is always possible to find roots of order p and $2p$ modulo M_p. Since $2^p \equiv 1$ modulo M_p, 2 and -2 are obviously roots of order p and $2p$. When M_p is a prime, the primitive roots are of order $\phi(2^p - 2)$. Thus, any root must be of order d such that $d | 2^p - 2$. Obviously $2 | 2^p - 2$ and any other divisor must be odd. By Fermat's theorem,

$p|2^p - 2$. Moreover, since a product of 3 consecutive integers is always divisible by 3, $3|[2^{(p-1)/2} - 1)]2^{(p-1)/2}[2^{(p-1)/2} + 1]$ and therefore, $3|(2^p - 2)$. Thus, any prime Mersenne number has roots which are factors of $6p$. In particular, the roots 2 and -2 are of orders p and $2p$, respectively. We also note that -1 is a root of order 2 for any Mersenne number and that there are no roots of order 4. Hence -1 is a quadratic nonresidue for any Mersenne number.

We consider now the Fermat numbers F_t. The first five Fermat numbers for $t = 0$ to $t = 4$ are prime. All other known Fermat numbers are composite. We shall give here some interesting properties of Fermat numbers.

Theorem 2.16: All Fermat numbers are mutually prime.

To prove this theorem, we note first that $F_2 = F_0 F_1 + 2$ and $F_3 = F_0 F_1 F_2 + 2$. Assume now that

$$F_t = F_0 F_1 F_2 \ldots F_{t-1} + 2. \tag{2.68}$$

If we multiply both sides of (2.68) by F_t, we have

$$F_t^2 = F_0 F_1 F_2 \ldots F_t + 2 F_t. \tag{2.69}$$

By the definition of Fermat numbers, $F_{t+1} = 2^{2^{t+1}} + 1 = (F_t - 1)^2 + 1$. Thus,

$$F_{t+1} = F_t^2 - 2F_t + 2 \tag{2.70}$$

and, by substituting F_t^2 from (2.69) into (2.70),

$$F_{t+1} = F_0 F_1 F_2 \ldots F_t + 2. \tag{2.71}$$

Hence, we have established (2.68) by induction. Suppose now that two Fermat numbers are not relatively prime. Then, $(F_m, F_k) = d$, with $d \neq 1$ and we would have $d|F_m$ and $d|F_k$. In this case, (2.71) would imply that $d|2$. This is impossible because d would have to be even and thus could not divide any Fermat number. Hence $d = 1$ and all Fermat numbers are mutually prime.

Theorem 2.17: 3 is a primitive root of all prime Fermat numbers.

Any primitive root g must be a quadratic nonresidue because if it were a quadratic residue, some powers of g would not be distinct. By theorem 2.9, the number $\bar{Q}(F_t)$ of distinct quadratic nonresidues is equal to 2^{2^t-1}. We also know, by theorem 2.8, that there are $\phi(F_t - 1) = 2^{2^t-1}$ distinct primitive roots modulo F_t. Since $\bar{Q}(F_t) = \phi(F_t - 1)$, all quadratic nonresidues are primitive roots and we need only to show that 3 is a quadratic nonresidue to prove the theorem. In order to show that 3 is a quadratic nonresidue for all Fermat numbers, we first note, by direct verification, that 3 is primitive root modulo $F_1 = 5$, since 3 is a root of order 4 modulo 5. We then show, by induction, that for any Fermat number,

$F_t \equiv 5$ modulo 12. (2.72)

This can be seen by noting that, if $F_t = 12k + 5$, then $F_{t+1} = (F_t - 1)^2 + 1 = (12k + 4)^2 + 1 = 12k_1 + 5$. Thus, we can check whether 3 is a quadratic nonresidue by computing Legendre's symbol $[3/(12k + 5)]$. We have

$$[3/(12k + 5)] = [(12k + 5)/3] = (2/3) = -1.$$

Hence, 3 is a quadratic nonresidue modulo F_t and, therefore, 3 is a primitive root modulo F_t.

When a Fermat number F_t is composite ($t > 4$), the following theorem is used for specifying the order of the various roots modulo F_t.

Theorem 2.18: Every prime factor of a composite Fermat number F_t is of the form $k2^{t+2} + 1$.

The proof of this theorem can be found in [2.6]. An immediate consequence of theorem 2.18 is that every Fermat number has roots of order $d = 2^n$, with $n \leqslant t + 2$. Integers 2 and -2 are obviously roots of order 2^{t+1}. A simple root of order 2^{t+2} can be found by noting that

$$[2^{2^{t-2}}(1 + 2^{2^{t-1}})]^2 \equiv -2 \text{ modulo } F_t. \tag{2.73}$$

Thus, in the ring of integers modulo F_t, $2^{2^{t-2}}(1 + 2^{2^{t-1}})$ is congruent to $\sqrt{-2}$ and is therefore a root of order 2^{t+2}.

We also note that, since $(2^{2^{t-1}})^2 \equiv -1$ modulo F_t, -1 is a quadratic residue of F_t. This means that $j = \sqrt{-1}$ is real in the ring of integers modulo F_t, with $j \equiv 2^{2^{t-1}}$. We shall use this property in Chap. 8 to simplify the computation of complex convolutions.

Mersenne and Fermat numbers have many other interesting properties that cannot be discussed in detail here. Some of these properties can be found in [2.7].

2.2 Polynomial Algebra

Polynomial algebra plays an important role in digital signal processing because convolutions and, to some extent, DFTs can be expressed in terms of operations on polynomials. This can be seen by considering the simple convolution y_l of two sequences h_n and x_m of N terms

$$y_l = \sum_{n=0}^{N-1} h_n x_{l-n}, \quad l = 0, \ldots, 2N - 2. \tag{2.74}$$

Now suppose that the N elements of h_n and x_m are assigned to be the coefficients of polynomials $H(z)$ and $X(z)$ of degree $N - 1$ in z, z being the polynomial variable. Hence we have

$$H(z) = \sum_{n=0}^{N-1} h_n z^n \qquad (2.75)$$

$$X(z) = \sum_{m=0}^{N-1} x_m z^m. \qquad (2.76)$$

If we multiply $H(z)$ by $X(z)$, the resulting polynomial $Y(z)$ will be of degree $2N - 2$, since $H(z)$ and $X(z)$ are of degree $N - 1$. Thus,

$$Y(z) = H(z)X(z) = \sum_{l=0}^{2N-2} a_l z^l. \qquad (2.77)$$

In the polynomial multiplication, each coefficient a_l of z^l is obtained by summing all products $h_n x_m$ such that $n + m = l$. Hence, $m = l - n$ and

$$a_l = y_l = \sum_{n=0}^{N-1} h_n x_{l-n} \qquad (2.78)$$

$$Y(z) = \sum_{l=0}^{2N-2} y_l z^l \qquad (2.79)$$

This means that the convolution of two sequences can be treated as the multiplication of two polynomials. Moreover, if the convolution defined by (2.74) is cyclic, the indices l, m, and n are defined modulo N. Thus, in N-term cyclic convolutions, we have $N \equiv 0$. This implies that $z^N \equiv 1$ and therefore that a cyclic convolution can be viewed as the product of two polynomials modulo the polynomial $z^N - 1$

$$Y(z) \equiv H(z)X(z) \text{ modulo } (z^N - 1). \qquad (2.80)$$

Thus, in order to deal with convolutions analytically, one must define various operations where the usual number sets are replaced by sets of polynomials. These operations on polynomials bear a strong relationship to operations on integers and can be treated in a unified way by using the concepts of groups, rings, and fields. In the following, we shall give only the flavor of these concepts, since full details are available in any textbook on modern algebra [2.8].

2.2.1 Groups

Consider a set A of N elements $a, b, c, \ldots$. These elements could be, for instance, positive integers or polynomials. Now suppose that we can relate elements in the set by an operation which is denoted $\oplus$. Again, this operation is quite general and could be, for example, an addition or a logical *or* operation, the only constraint at this stage being that a, b, and c pertain to the set A, with

$$c = a \oplus b \qquad a, b, c \in A. \qquad (2.81)$$

Then, any set which satisfies the following conditions is called a *group*

—Associative law: $a \oplus (b \oplus c) = (a \oplus b) \oplus c$
—Identity element. There is an element e of the, group which, for any element of the group, is such that $e \oplus a = a$.
—Inverse. Every element a of A has an inverse $\bar{a}$ which is an element of the group: $a \oplus \bar{a} = \bar{a} \oplus a = e$

When the operation is commutative, with $a \oplus b = b \oplus a$, the group is called *Abelian*. The *order* of a *group* is the number of elements of this group.

Now consider a group having a finite number of elements, and the successive operations $a \oplus a$, $a \oplus a \oplus a$, $a \oplus a \oplus a \oplus a$, ... Each of these operations produces an element of the group. Since the group is finite, the sequence will necessarily repeat itself with a period r. r is called the *order* of the *element a*. If the order of an element g is the same as the order of the group, all elements of the group are generated by g with the operations g, $g \oplus g$, $g \oplus g \oplus g$, ... In this case, g is called a *generator* and the group is called a *cyclic group*.

In order to illustrate these concepts, let us consider the set A of N integers 0, 1, ..., $N-1$. For addition modulo N, $a + b \equiv c$, with $a, b, c \in A$. Moreover, $a + (b + c) \equiv (a + b) + c$, $0 + a \equiv a$, and $a + (N - a) \equiv 0$. Thus, A is a group with respect to addition modulo N. This group is Abelian, since $a + b \equiv b + a$. It is also cyclic with the integer 1 as generator, since all elements of the group are generated by adding 1 to the preceding element. We now consider the set B of $N-1$ integers, 1, 2, ..., $N-1$ with identity element 1 and with the addition modulo N replaced by the multiplication modulo N. B is generally not a group with respect to multiplication modulo N, because some elements of the set have no inverse. For instance, if $N = 6$, only 1 and 5 have inverses. Thus, the set of integers 1, 2, 3, 4, 5 is not a group with respect to multiplication modulo 6. Note however that, when N is a prime, then B becomes a cyclic group. For instance, if $N = 5$, the inverses of 1, 2, 3, 4 are, respectively, 1, 3, 2, 4 and therefore the set of integers 1, 2, 3, 4 is a group, which is cyclic with generators 2 and 3. It can be seen that the group of the N integers 0, 1, ..., $N-1$ with addition modulo N has the same structure as the group of the N integers 1, 2, ..., N with multiplication modulo $(N + 1)$, $N + 1$ being a prime. Such a relation between two groups is called *isomorphism*.

2.2.2 Rings and Fields

A set A is a *ring* with respect to the two operations $\oplus$ and $\otimes$ if the following conditions are fulfilled:

—$(A, \oplus)$ is an Abelian group
—If $c = a \otimes b$, for $a, b \in A$, then $c \in A$
—Associative law: $a \otimes (b \otimes c) = (a \otimes b) \otimes c$
—Distributive law: $a \otimes (b \oplus c) = a \otimes b \oplus a \otimes c$ and $(b \oplus c) \otimes a = b \otimes a \oplus c \otimes a$

The *ring* is *commutative* if the law $\otimes$ is commutative and it is a *unit ring* if there is one (and only one) identity element u for the law $\otimes$.

It can be verified easily that the set of integers is a ring with respect to addition and multiplication.

If we now require that the operation $\otimes$ satisfies the additional condition that every element a has one (and only one) inverse ($a \otimes \bar{a} = u$), then a unit ring becomes a *field*. It can be verified easily that, for any prime p, the set of integers $0, 1, ..., p - 1$ form a field with addition and multiplication modulo p. This field is called a Galois field and denoted GF(p).

We shall give here several important results concerning fields.

Theorem 2.19: If a, b, c are elements of a field, the condition $a \otimes c = b \otimes c$ implies that $a = b$.

We have $a \otimes c = b \otimes c$. Thus, if $\bar{b}$ is the inverse of b, we have $\bar{b} \otimes a \otimes c = \bar{b} \otimes b \otimes c$. This implies that $\bar{b} \otimes a \otimes c = c$ and therefore that $b = a$.

A consequence of this theorem is that, if we consider the set S of the n distinct elements $a_1, a_2, ..., a_n$ of a finite field, then the n elements $a_i \otimes a_1$, $a_i \otimes a_2$, ..., $a_i \otimes a_n$ are all distinct. Since the result of the operation $\otimes$ is, by definition, an element of the field, the n elements $a_i \otimes a_1$, $a_i \otimes a_2$, ..., $a_i \otimes a_n$ are the set S. This generalizes to any field the concept of permutation that has been introduced by (2.27) for fields of integers modulo a prime p. By using an approach quite similar to that used for fields of integers, it is also possible to show that, for all finite fields, there are primitive roots g which generate all field elements, except e, by successive operations $g \otimes g$, $g \otimes g \otimes g$, ...

Another important property is that all finite fields have a number of elements which is p^d, where p is a prime. These fields are denoted $GF(p^d)$.

In the rest of this chapter, we shall restrict our discussion to rings and fields of polynomials. In these cases, the operations $\oplus$ and $\otimes$ usually reduce to additions and multiplications modulo polynomials. In order to simplify the notation, we shall replace special symbols $\oplus$ and $\otimes$ with the notation that has been defined for residue arithmetic. Using this notation, we first introduce residue polynomials and the Chinese remainder theorem.

2.2.3 Residue Polynomials

The theory of residue polynomials is closely related to the theory of integer residue classes. Thus, our presentation begins with the concept of polynomial division. In this presentation, we shall assume that all polynomial coefficients are defined in a field, in order to ensure that the usual arithmetic operations can be performed without restrictions on these coefficients. We commence with several basic definitions. A polynomial $P(z)$ divides a polynomial $H(z)$ if a polynomial $D(z)$ can be found such that $H(z) = P(z)D(z)$. $H(z)$ is said to be *irreducible* if its only divisors are of degree equal to zero. If $P(z)$ is not a divisor of $H(z)$, the division of $H(z)$ by $P(z)$ will produce a *residue* $R(z)$

$$H(z) = P(z)D(z) + R(z), \qquad (2.82)$$

where the degree of $R(z)$ is less than the degree of $P(z)$. This representation is unique. All polynomials having the same residue when divided by $P(z)$ are said to be *congruent* modulo $P(z)$ and the relation is denoted by

$$R(z) \equiv H(z) \quad \text{modulo } P(z). \qquad (2.83)$$

At this point, it is worth noting that when we deal with polynomials, we are mainly interested by the coefficients of the polynomials. Thus, if we have a set of N elements $a_0, a_1, \ldots, a_{N-1}$, arranging these elements in the form of a polynomial $H(z) = a_0 + a_1 z + a_2 z^2 \ldots + a_{N-1} z^N$ of the dummy variable z is essentially a convenient way of tagging the position of an element a_i relative to the others. This feature is very important in digital signal processing because each polynomial coefficient represents a sample of an analog signal stream and therefore defines its location and intensity.

Returning to the congruence relation (2.83), we see that two polynomials which differ only by a multiplicative constant are congruent. Thus, residue polynomials deal with the relative values of coefficients rather than with their absolute values. Equation (2.83) defines equivalence classes of polynomials modulo a polynomial $P(z)$. It can be verified easily, by referring to the definitions in the preceding section, that the set of polynomials defined with addition and multiplication modulo $P(z)$ is a ring and reduces to a field when $P(z)$ is irreducible.

When $P(z)$ is not irreducible, it can always be factorized uniquely into powers of irreducible polynomials. Note however that the factorization depends on the field of coefficients: $z^2 + 1$ is irreducible for coefficients in the field of rational numbers. If the coefficients are defined in the field of complex numbers, then $z^2 + 1 = (z - j)(z + j), j = \sqrt{-1}$.

Now suppose that $P(z)$ is the product of d polynomials $P_i(z)$ having no common factors (these polynomials are usually called relatively prime polynomials by analogy with relatively prime numbers)

$$P(z) = \prod_{i=1}^{d} P_i(z). \qquad (2.84)$$

Since each of these polynomials $P_u(z)$ is relatively prime with all the other polynomials $P_i(z)$, it has an inverse modulo every other polynomial. This means that we can extend the Chinese remainder theorem to the ring of polynomials modulo $P(z)$ and therefore express uniquely $H(z)$ as a function of the polynomials $H_i(z)$ obtained by reducing $H(z)$ modulo the various polynomials $P_i(z)$. The Chinese remainder theorem is then expressed as

$$H(z) \equiv \sum_{i=1}^{d} S_i(z) H_i(z) \quad \text{modulo } P(z), \qquad (2.85)$$

where, for every value u of i,

$$S_u(z) \equiv 0 \quad \text{modulo } P_i(z), \quad i \neq u \tag{2.86}$$
$$ \equiv 1 \quad \text{modulo } P_u(z)$$

and

$$S_u(z) \equiv T_u(z) \prod_{\substack{i=1 \\ i \neq u}}^{d} P_i(z) \quad \text{modulo } P(z) \tag{2.87}$$

with $T_u(z)$ defined by

$$T_u(z) \prod_{\substack{i=1 \\ i \neq u}}^{d} P_i(z) \equiv 1 \quad \text{modulo } P_u(z). \tag{2.88}$$

Note that (2.88) implies that $T_u(z) \not\equiv 0$ modulo $P_u(z)$. Thus, when $S_u(z)$ is reduced modulo the various polynomials $P_i(z)$, we obtain (2.86). Therefore, when $H(z)$, defined by (2.85), is reduced modulo $P_i(z)$, we obtain $H_i(z) \equiv H(z)$ modulo $P_i(z)$, which completes the proof of the theorem.

When computing $H(z)$ from the various residues $H_i(z)$ by the Chinese remainder theorem, one must determine the various polynomials $S_i(z)$. For a given $P(z)$, these polynomials are computed once and for all by (2.87). The most difficult part of calculating $S_i(z)$ relates to the evaluation of the inverses $T_i(z)$ defined by (2.88). This is done by using Euclid's algorithm, as described in Sect. 2.1, but with integers replaced by polynomials. The polynomials $S_i(z)$ can also be computed very simply by using computer programs for symbolic mathematical manipulation [2.9–10].

2.2.4 Convolution and Polynomial Product Algorithms in Polynomial Algebra

The Chinese remainder theorem plays a central role in the computation of convolutions because it allows one to replace the evaluation of a single long convolution by that of several short convolutions. We shall now show that the Chinese remainder theorem can be used to specify the lower bounds on the number of multiplications required to compute convolutions and polynomial products. In the case of an aperiodic convolution, these lower bounds are given by the Cook-Toom algorithm [2.11]:

Theorem 2.20: The aperiodic convolution of two sequences of lengths L_1 and L_2 is computed with $L_1 + L_2 - 1$ general multiplications.

In polynomial notation, the aperiodic convolution y_l of two sequences h_n and x_m is defined by

$$H(z) = \sum_{n=0}^{L_1-1} h_n z^n \tag{2.89}$$

$$X(z) = \sum_{m=0}^{L_2-1} x_m z^m \tag{2.90}$$

$$Y(z) = H(z)X(z) = \sum_{l=0}^{L_1+L_2-2} y_l z^l. \tag{2.91}$$

Since $Y(z)$ is of degree $L_1 + L_2 - 2$, $Y(z)$ is unchanged if it is defined modulo any polynomial $P(z)$ of degree equal to $L_1 + L_2 - 1$

$$Y(z) \equiv H(z)X(z) \quad \text{modulo } P(z). \tag{2.92}$$

We now assume that $P(z)$ is chosen to be the product of $L_1 + L_2 - 1$ first degree relatively prime polynomials

$$P(z) = \prod_{i=1}^{L_1+L_2-1} (z - a_i), \tag{2.93}$$

where the a_i are $L_1 + L_2 - 1$ distinct numbers in the field F of coefficients. Since $P(z)$ is the product of $L_1 + L_2 - 1$ relatively prime polynomials, we can apply the Chinese remainder theorem to the computation of (2.92). This is done by reducing the polynomials $H(z)$ modulo $(z - a_i)$, performing $L_1 + L_2 - 1$ polynomial multiplications $H_i(z)X_i(z)$ on the reduced polynomials, and reconstructing $Y(z)$ by the Chinese remainder theorem. We note however that the reductions modulo $(z - a_i)$ are equivalent to substitutions of a_i for z in $H(z)$ and $X(z)$. Thus, the reduced polynomials $H_i(z)$ and $X_i(z)$ are the simple scalars $H(a_i)$ and $X(a_i)$ so that the polynomial multiplications reduce to $L_1 + L_2 - 1$ scalar multiplications $H(a_i)X(a_i)$. This completes the proof of the theorem.

Note that this theorem provides not only a lower bound on the number of general multiplications, but also a practical algorithm for achieving this lower bound. However, the bound concerns only the number of general multiplications, that is to say, the multiplications where the two factors depend on the data. The bound does not include multiplications by constant factors which occur in the reductions modulo $(z - a_i)$ and in the Chinese remainder reconstruction. For short convolutions, the $L_1 + L_2 - 1$ distinct a_i can be chosen to be simple integers such as $0, +1, -1$ so that these multiplications are either trivial or reduced to a few additions. For longer convolutions, the a_i must be chosen among a larger set of distinct values. In this case, some of the a_i are no longer simple so that multiplications in the reductions and Chinese remainder operation are unavoidable. This means that the Cook-Toom algorithm is practical only for short convolutions. For longer convolutions, better algorithms can be obtained by using a transform approach which we now show to be closely related to the Cook-Toom algorithm and Lagrange interpolation.

This can be seen by noting that $Y(z)$ is reconstructed by the Chinese remainder theorem from the $L_1 + L_2 - 1$ scalars $Y(a_i)$ obtained by substituting a_i for z in $Y(z)$. Thus, the Cook-Toom algorithm expresses a Lagrange interpolation process [2.9]. Since the field F of coefficients and the interpolation values can be chosen at will, we can select the a_i to be the $L_1 + L_2 - 1$ successive powers of a number W, provided that all these numbers are distinct in the field F. In this case, $a_i = W^i$ and the reductions modulo $(z - a_i)$ are expressed by

$$H(W^i) = \sum_{n=0}^{L_1-1} h_n W^{in}, \quad i = 0, 1, ..., L_1 + L_2 - 2 \tag{2.94}$$

with similar relations for $X(W^i)$. Thus, with this particular choice of a_i, the Cook-Toom algorithm reduces to computing aperiodic convolutions with transforms having the DFT structure. In particular, if $W = 2$ and if F is the field of integers modulo a Mersenne number ($2^p - 1$, p prime) or a Fermat number ($2^v + 1$, $v = 2^t$), the Cook-Toom algorithm defines a Mersenne or a Fermat transform (Chap. 8).

When $W = e^{-2j\pi/(L_1+L_2-1)}$, $j = \sqrt{-1}$, the Cook-Toom algorithm can be viewed as the computation of an aperiodic convolution by DFTs. In this case, $P(z)$ becomes

$$P(z) = \prod_{i=0}^{L_1+L_2-2} (z - W^i) = z^{L_1+L_2-1} - 1. \tag{2.95}$$

Hence, if the interpolation points are chosen to be complex exponentials, the Cook-Toom algorithm is equivalent to computing with DFTs the circular convolution of two input sequences obtained by appending $L_2 - 1$ zeros to the sequence h_n and $L_1 - 1$ zeros to the sequence x_m. This computation method will be described more exhaustively in Chap. 3 and is known as the *overap-add* method.

The computational complexity results concerning convolutions have been extended by Winograd [2.12] to polynomial products modulo a polynomial $P(z)$.

Theorem 2.21: A polynomial product $Y(z) \equiv H(z)X(z)$ modulo $P(z)$ is computed with $2D-d$ general multiplications, where D is the degree of $P(z)$ and d is the number of irreducible factors $P_i(z)$ of $P(z)$ over the field F.

This theorem is proved by again using the Chinese remainder theorem. $Y(z)$ is computed by calculating the reduced polynomials $H_i(z) \equiv H(z)$ modulo $P_i(z)$, $X_i(z) \equiv X(z)$ modulo $P_i(z)$, evaluating the d polynomial products $Y_i(z) \equiv H_i(z)X_i(z)$ modulo $P_i(z)$, and reconstructing $Y(z)$ from $Y_i(z)$ by the Chinese remainder theorem. As before, the multiplications by scalars corresponding to the reductions and Chinese remainder reconstruction are not counted and the only general multiplications are those corresponding to the d products $Y_i(z)$ evaluated modulo polynomials $P_i(z)$ of degree D_i. Since $P(z)$ is given by

$$P(z) = \prod_{i=1}^{d} P_i(z), \tag{2.96}$$

we have

$$D = \sum_{i=1}^{d} D_i. \tag{2.97}$$

Each polynomial product $Y_i(z)$ can be computed as an aperiodic convolution of two sequences of D_i terms, followed by a reduction modulo $P_i(z)$

$$Y_i(z) \equiv H_i(z)X_i(z) \quad \text{modulo } P_i(z). \tag{2.98}$$

Thus, by theorem 2.20, $Y_i(z)$ is calculated with $2D_i - 1$ general multiplications and the total number of multiplications becomes $\sum_{i=1}^{d}(2D_i - 1) = 2D - d$. This completes the proof of theorem 2.21.

We have already seen, with (2.80), that an N-point circular convolution can be considered as a polynomial product modulo $(z^N - 1)$. If F is the field of complex numbers, $(z^N - 1)$ factors into N polynomials $(z - W^i)$ of degree 1, with $W = e^{-j2\pi/N}$, $j = \sqrt{-1}$. In this case, the computation technique defined by theorem 2.21 is equivalent to the DFT approach and requires only N general multiplications. Unfortunately, the W^i are irrational and complex so that the multiplications by scalars corresponding to DFT computation must also be considered as general multiplications.

When F is the field of rational numbers, $z^N - 1$ factors into polynomials having coefficients that are rational numbers. These polynomials are called *cyclotomic polynomials* [2.1] and are irreducible for coefficients in the field of rational numbers. The number d of distinct cyclotomic polynomials which are factors of $z^N - 1$ can be shown to be equal to the number of divisors of N, including 1 and N. Thus, we have one cyclotomic polynomial of degree D_i for each divisor N_i of N and D_i can be shown to be [2.1]

$$D_i = \phi(N_i), \tag{2.99}$$

where $\phi(N_i)$ is Euler's totient function (Sect. 2.1.3). Thus, for circular convolutions with coefficients in the field of rationals, theorem 2.21 reduces to theorem 2.22.

Theorem 2.22: An N-point circular convolution is computed with $2N-d$ general multiplications, where d is the number of divisors of N, including 1 and N.

Theorems 2.21 and 2.22 provide an efficient way of computing circular convolutions because the coefficients of the cyclotomic polynomials are simple integers and can be simply 0, +1, −1, except for very large cyclotomic polynomials. When N is a prime, for instance, $z^N - 1 = (z - 1)(z^{N-1} + z^{N-2} + \cdots + 1)$. Thus, the reductions and Chinese remainder reconstruction are implemented with a small number of additions and, usually, without multiplications. In order to illustrate this computation procedure, consider, for instance, a circular convolution of 3 points. Since 3 is a prime, we have $z^3 - 1 = (z - 1)(z^2 + z + 1)$. Reducing $X(z)$ modulo $(z - 1)$ is done with 2 additions by simply substituting 1 for z in $X(z)$. For the reduction modulo $(z^2 + z + 1)$, we note that $z^2 \equiv -z - 1$. Thus, this reduction is also done with 2 additions by subtracting the coefficient of z^2 in $X(z)$ from the coefficients of z^0 and z^1. When the sequence $H(z)$ is fixed, the Chinese remainder reconstruction can be considered as the inverse of the reductions and is done with a total of 4 additions. Moreover, the polynomial multiplication modulo $(z - 1)$ is a simple scalar multiplication and the polynomial multiplication modulo $(z^2 + z + 1)$ is done with 3 multiplications and 3 additions as shown in Sect. 3.7.2. Thus, when $H(z)$ is fixed, a 3-point circular convolution is computed with 4 multiplications and 11 additions as opposed to 9 multiplications and 6 additions for direct computation. We shall see in Chap. 3 that a systematic application of the methods defined by theorems 2.20–2.22 allows one to design very efficient convolution algorithms.

3. Fast Convolution Algorithms

The main objective of this chapter is to focus attention on fast algorithms for the summation of lagged products. Such problems are very common in physics and are usually related to the computation of digital filtering processes, convolutions, and correlations. Correlations differ from convolutions only by virtue of a simple inversion of one of the input sequences. Thus, although the developments in this chapter refer to convolutions, they apply equally well to correlations.

The direct calculation of the convolution of two N-point sequences requires a number of arithmetic operations which is of the order of N^2. For large convolutions, the corresponding processing load becomes rapidly excessive and, therefore, considerable effort has been devoted to devising faster computation methods. The conventional approach for speeding up the calculation of convolutions is based on the fast Fourier transform (FFT) and will be discussed in Chap. 4. With this approach, the number of operations is of the order of $N \log_2 N$ when N is a power of two.

The speed advantage offered by the FFT algorithm can be very large for long convolutions and the method is, by far, the most commonly used for the fast computation of convolutions. However, there are several drawbacks to the FFT, which relate mainly to the use of sines and cosines and to the need for complex arithmetic, even if the convolutions are real.

In order to overcome the limitations of the FFT method, many other fast algorithms have been proposed. In fact, the number of such algorithms is so large that an exhaustive presentation would be almost impossible. Moreover, many seemingly different algorithms are essentially identical and differ only in the formalism used to develop a description. In this chapter, we shall attempt to unify our presentation of these methods by organizing them into algebraic and arithmetic methods. We shall show that most algebraic methods reduce to various forms of nesting and yield computational loads that are often equal to and sometimes less than the FFT method while eliminating some of its limitations. We shall then present arithmetic methods which can be used alone or in combination with algebraic methods and which allow significant processing efficiency gains when implemented in special purpose hardware.

3.1 Digital Filtering Using Cyclic Convolutions

Most fast convolution algorithms, such as those based on the FFT, apply only to periodic functions and therefore compute only cyclic convolutions. However, practical filtering applications concern essentially the aperiodic convolution y_l of a limited sequence h_n, of length N_1, with a quasi-infinite data sequence x_m

$$y_l = \sum_{n=0}^{N_1-1} h_n x_{l-n}. \tag{3.1}$$

Thus, in order to take advantage of the various fast cyclic convolution algorithms to speed-up the calculation of digital filtering processes, we need some means to convert the aperiodic convolution y_l into a series of cyclic convolutions. This can be done in two ways. The first method is called the *overlap-add* method [3.1]. The second method, which is called the *overlap-save* method [3.2], is very similar to the first one and yields comparable results in terms of computational complexity.

3.1.1 Overlap-Add Algorithm

The overlap-add algorithm, as an initial step, sections the input sequence x_m into v contiguous blocks x_{u+vN_2} of equal length N_2, with $m = u + vN_2$, $u = 0, \ldots, N_2 - 1$, and $v = 0, 1, 2, \ldots$ for the successive blocks. The aperiodic convolution of each of these blocks x_{u+vN_2} with the sequence h_n is then computed and yields output sequences $y_{v,l}$ of $N_1 + N_2 - 1$ samples. In polynomial notations, calculating these aperiodic convolutions is equivalent to determining the coefficients of a polynomial $Y_v(z)$ defined by

$$Y_v(z) = H(z) X_v(z) \tag{3.2}$$

with

$$H(z) = \sum_{n=0}^{N_1-1} h_n z^n \tag{3.3}$$

$$X_v(z) = \sum_{u=0}^{N_2-1} x_{u+vN_2} z^u \tag{3.4}$$

$$Y_v(z) = \sum_{l=0}^{N_1+N_2-2} y_{v,l} z^l. \tag{3.5}$$

Since $Y_v(z)$ is a polynomial of degree $N_1 + N_2 - 2$, it can be computed modulo any polynomial of degree $N \geq N_1 + N_2 - 1$ and in particular, modulo ($z^N - 1$). In this case, the successive aperiodic convolutions $y_{v,l}$ are computed as circular convolutions of length N, with $N \geq N_1 + N_2 - 1$, in which the input blocks are augmented by adding $N - N_1$ zero-valued samples at the end of the sequence h_n and $N - N_2$ zero-valued samples at the end of the sequence x_{u+vN_2}.

The overlap-add method derives its name from the fact that the output of each section overlaps its neighbor by $N - N_2$ samples. These overlapping output samples must be added to produce the desired y_l. Thus, for $N = N_1 + N_2 - 1$, a continuous digital filtering process is evaluated with one circular convolution of N points for every N_2 output samples plus $(N_1 - 1)/(N - N_1 + 1)$ additions per output sample.

3.1.2 Overlap-Save Algorithm

The overlap-save algorithm sections the input data sequence into v overlapping blocks x_{u+vN_2} of equal length N, with $m = u + vN_2$, $u = 0, ..., N - 1$, and v taking the values 0, 1, 2, ... for successive blocks. In this method, each data block has a length N, instead of N_2 for the overlap-add algorithm, and overlaps the preceding block by $N - N_2$ samples. The output of the digital filter is constructed by computing the successive length-N circular convolutions of the blocks x_{u+vN_2} with the block of length N obtained by appending $N - N_1$ zero-valued samples to h_n. Hence, the output $y_{l_1+l_2N_2}$ of each circular convolution is given by

$$y_{l_1+l_2N_2} = \sum_{n=0}^{N-1} h_n x_{\langle l_1-n \rangle + vN_2}, \quad l_1 = 0,...,N-1$$

$$l_2 = 0,1,... \tag{3.6}$$

$$l = l_1 + l_2 N_2, \tag{3.7}$$

where $\langle l_1 - n \rangle$ is taken modulo N. This means that, for $l_1 \geq n$, $\langle l_1 - n \rangle = l_1 - n$ and, for $l_1 < n$, $\langle l_1 - n \rangle = N + l_1 - n$. We assume now that $N = N_1 + N_2 - 1$. For $l_1 \geq N_1 - 1$, $\langle l_1 - n \rangle$ is always equal to $l_1 - n$ because all samples h_n are zero-valued for $n > N_1 - 1$. Hence, the last $N - N_1 + 1$ output samples of each cyclic convolution are valid output samples of the digital filter, while the first $N_1 - 1$ output samples of the cyclic convolutions must be discarded because they correspond to interfering intervals.

It can be seen that the overlap-save algorithms produce $N - N_1 + 1$ output samples of the digital filter, without any final addition. Thus, the overlap-save algorithm is often preferred to the overlap-add algorithm because its implementation as a computer program is slightly simpler when standard computer programs are used for the calculation of convolutions.

3.2 Computation of Short Convolutions and Polynomial Products

In many fast convolution algorithms, the calculation of a large convolution is replaced by that of a large number of small convolutions and polynomial products. This means that the processing load is strongly dependent upon the efficiency of the algorithms used for the calculation of small convolutions and

3.2 Computation of Short Convolutions and Polynomial Products

polynomial products. In this section, we shall describe several techniques which allow one to optimize the design of such short algorithms.

3.2.1 Computation of Short Convolutions by The Chinese Remainder Theorem

We have seen in Chap. 2 that the number of multiplications required to compute a circular convolution is minimized by breaking the computation into that of polynomial products via the Chinese remainder theorem. This is done by noting that a circular convolution y_l of N terms

$$y_l = \sum_{n=0}^{N-1} h_n x_{l-n}, \quad l = 0, \ldots, N-1, \tag{3.8}$$

where h_n and x_m are the two input sequences of length N, can be viewed as a polynomial product modulo $(z^N - 1)$. In polynomial notation, we have

$$H(z) = \sum_{n=0}^{N-1} h_n z^n \tag{3.9}$$

$$X(z) = \sum_{m=0}^{N-1} x_m z^m \tag{3.10}$$

$$Y(z) \equiv H(z)X(z) \quad \text{modulo } (z^N - 1) \tag{3.11}$$

$$Y(z) = \sum_{l=0}^{N-1} y_l z^l. \tag{3.12}$$

For coefficients in the field of rational numbers, $z^N - 1$ is the product of d cyclotomic polynomials $P_i(z)$, where d is the number of divisors of N, including 1 and N

$$z^N - 1 = \prod_{i=1}^{d} P_i(z). \tag{3.13}$$

$Y(z)$ is computed by first reducing the input polynomials $H(z)$ and $X(z)$ modulo $P_i(z)$

$$H_i(z) \equiv H(z) \quad \text{modulo } P_i(z) \tag{3.14}$$

$$X_i(z) \equiv X(z) \quad \text{modulo } P_i(z). \tag{3.15}$$

Then, $Y(z)$ is obtained by computing the d polynomial products $H_i(z)X_i(z)$ modulo $P_i(z)$ and using the Chinese remainder theorem, as shown in Fig. 3.1 for N prime, to reconstruct $Y(z)$ from the products modulo $P_i(z)$,

$$Y(z) \equiv \sum_{i=1}^{d} S_i(z) H_i(z) X_i(z) \quad \text{modulo } (z^N - 1), \tag{3.16}$$

3. Fast Convolution Algorithms

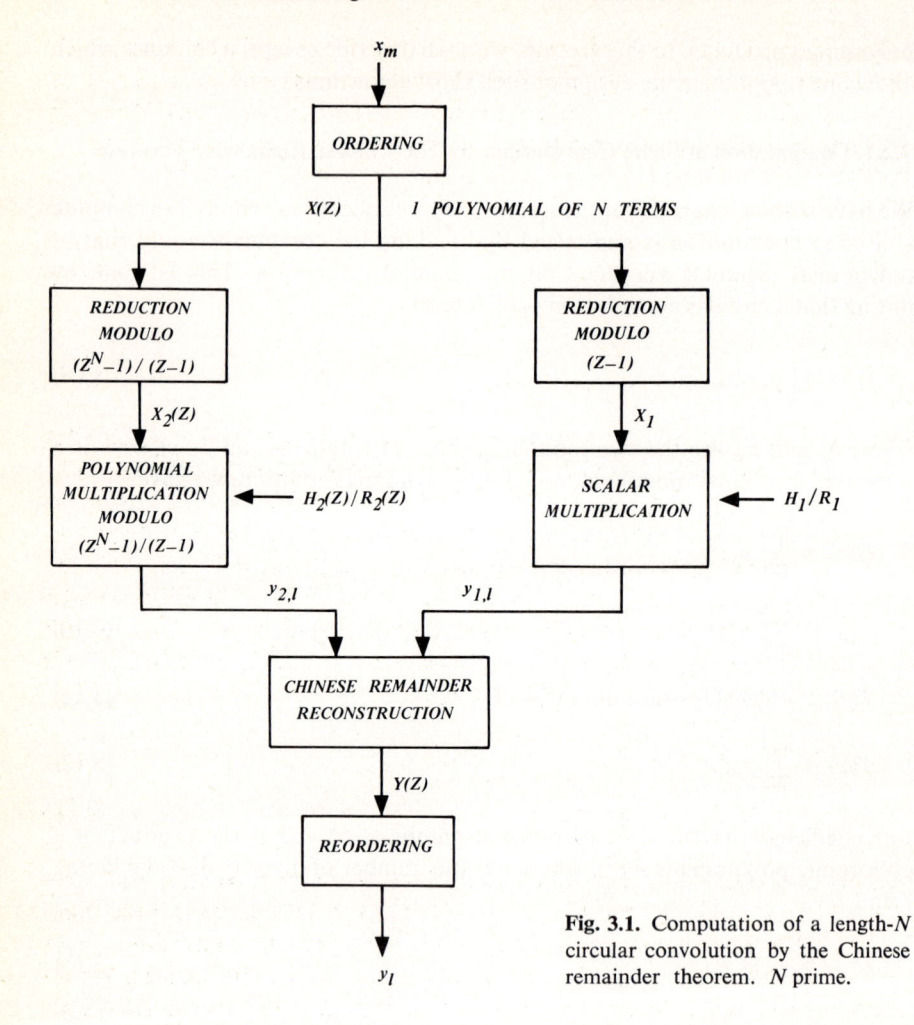

Fig. 3.1. Computation of a length-N circular convolution by the Chinese remainder theorem. N prime.

where, for each value u of i,

$$S_u(z) \equiv 0 \quad \text{modulo } P_i(z), \; i \neq u$$
$$ \equiv 1 \quad \text{modulo } P_u(z) \tag{3.17}$$

and

$$S_u(z) \equiv T_u(z)/R_u(z)$$
$$\equiv \left([\prod_{\substack{i=1 \\ i \neq u}}^{d} P_i(z)] / \{[\prod_{\substack{i=1 \\ i \neq u}}^{d} P_i(z)] \text{ modulo } P_u(z)\} \right) \text{ modulo } (z^N - 1). \tag{3.18}$$

Except for large values of N, the cyclotomic polynomials $P_i(z)$ are particularly simple, since the coefficients of z can only be 0 or ± 1. This means that the reductions modulo $P_i(z)$ and the Chinese remainder reconstruction are done without multiplications and with only a limited number of additions. When N is a prime, for instance, d is equal to 2, and $z^N - 1$ is the product of the two cyclotomic polynomials $P_1(z) = z - 1$ and $P_2(z) = z^{N-1} + z^{N-2} + \ldots + 1$. Thus, X_1 and $X_2(z)$ are computed with $N - 1$ additions by

$$X_1 = \sum_{m=0}^{N-1} x_m \qquad (3.19)$$

$$X_2(z) = \sum_{m=0}^{N-2} (x_m - x_{N-1}) z^m. \qquad (3.20)$$

For the Chinese remainder reconstruction, we note that $S_u(z) = T_u(z)/R_u(z)$, where $1/R_u(z)$ is defined modulo $P_u(z)$. Thus, multiplication by $1/R_u(z)$ can be combined with the multiplication $H_u(z) X_u(z)$ modulo $P_u(z)$. Moreover, in many practical cases, one of the input sequences h_n is fixed. In this case $H_u(z)/R_u(z)$ can be precomputed and the only operations relative to the Chinese remainder reconstructions are the multiplications by $T_u(z)$. For N prime, $H_1 X_1 / R_1$ is a scalar $y_{1,1}$, while $H_2(z) X_2(z) / R_2(z)$ modulo $P_2(z)$ is a polynomial of $N - 1$ terms with the coefficients of z^l given by $y_{2,l}$. Thus, the Chinese remainder reconstruction is done in this case with $2(N - 1)$ additions by

$$Y(z) = (y_{1,1} + y_{2,N-2}) z^{N-1} + y_{1,1} - y_{2,0}$$
$$+ \sum_{l=1}^{N-2} (y_{1,1} - y_{2,l} + y_{2,l-1}) z^l. \qquad (3.21)$$

It can be seen that the Chinese remainder operation requires the same number of additions as the reductions modulo the various cyclotomic polynomials. This result is quite general and applies to any circular convolution, with one of the sequences being fixed. Hence, the reductions and Chinese reconstructions are implemented very simply and the main problem associated with the computation of convolutions relates to the evaluation of polynomial products modulo the cyclotomic polynomials $P_i(z)$.

3.2.2 Multiplications Modulo Cyclotomic Polynomials

We note first that, since the polynomials $P_i(z)$ are irreducibles, they always can be computed by interpolation with $2D_i - 1$ general multiplications, D_i being the degree of $P_i(z)$ (theorem 2.21). Using this method for multiplications modulo $(z^2 + 1)$ and $(z^3 - 1)/(z - 1)$ yields algorithms with 3 multiplications and 3 additions as shown in Sect. 3.7.2.

For longer polynomial products, this method is not practical because it requires $2D_i - 1$ distinct polynomials $z - a_i$. The four simplest interpolation polynomials are $z, 1/z, (z - 1)$, and $(z + 1)$. Thus, when the degree D_i of $P_i(z)$ is larger than 2, one must use integers a_i different from 0 and ± 1, which implies that the corresponding reductions modulo $(z - u_i)$ and the Chinese remainder reconstructions use multiplications by powers of a_i which have to be implemented either with scalar multiplications or by a large number of successive additions. Thus one is led to depart somewhat from the interpolation method using real integers, in order to design algorithms with a reasonable balance between the number of additions and the number of multiplications.

One such technique consists in using complex interpolation polynomials, such as $(z^2 + 1)$, which are computed with more multiplications than polynomials with real roots but for which the reductions and Chinese remainder operations remain simple.

Another approach consists in converting one-dimensional polynomial products into multidimensional polynomial products. If we first assume that we want to compute the aperiodic convolution y_l of two length-N sequences h_n and x_m, this corresponds to the simple polynomial product $Y(z)$ defined by

$$Y(z) = H(z)X(z) \tag{3.22}$$

$$Y(z) = \sum_{l=0}^{2N-2} y_l z^l \tag{3.23}$$

$$H(z) = \sum_{n=0}^{N-1} h_n z^n \tag{3.24}$$

$$X(z) = \sum_{m=0}^{N-1} x_m z^m. \tag{3.25}$$

The polynomials $H(z)$ and $X(z)$ are of degree $N - 1$ and have N terms. If N is composite with $N = N_1 N_2$, $H(z)$ and $X(z)$ can be converted into two-dimensional polynomials by

$$n = N_1 n_2 + n_1, \quad m_2, n_2 = 0, \ldots, N_2 - 1 \tag{3.26}$$

$$m = N_1 m_2 + m_1, \quad m_1, n_1 = 0, \ldots, N_1 - 1 \tag{3.27}$$

$$z^{N_1} = z_1 \tag{3.28}$$

$$H(z) = H(z, z_1) = \sum_{n_1=0}^{N_1-1} H_{n_1}(z_1) z^{n_1} \tag{3.29}$$

$$X(z) = X(z, z_1) = \sum_{m_1=0}^{N_1-1} X_{m_1}(z_1) z^{m_1} \tag{3.30}$$

with

$$H_{n_1}(z_1) = \sum_{n_2=0}^{N_2-1} h_{N_1 n_2 + n_1} z_1^{n_2} \tag{3.31}$$

$$X_{m_1}(z_1) = \sum_{m_2=0}^{N_2-1} x_{N_1 m_2 + m_1} z_1^{m_2}. \tag{3.32}$$

Thus, $Y(z)$ is computed by evaluating a two-dimensional product $Y(z, z_1)$ and reconstructing $Y(z)$ from $Y(z, z_1)$ with $z_1 = z^{N_1}$

$$Y(z, z_1) = \sum_{m_1=0}^{N_1-1} \sum_{n_1=0}^{N_1-1} H_{n_1}(z_1) X_{m_1}(z_1) z^{m_1+n_1}. \tag{3.33}$$

This operation is equivalent to the computation of an aperiodic convolution of length-N_1 input sequences in which the scalars are replaced by polynomials of N_2 terms and each multiplication is replaced by the aperiodic convolution of two length-N_2 sequences. Thus, if M_1 and M_2 are the number of multiplications required to calculate the aperiodic convolutions of sequences of length N_1 and N_2, the aperiodic convolution of the two length-N sequences is evaluated with M multiplications where

$$M = M_1 M_2. \tag{3.34}$$

This approach can be used recursively to cover the case of more than two dimensions and it has the advantage of breaking down the computation of large polynomial products into that of smaller polynomial products. Hence, a polynomial product modulo a cyclotomic polynomial $P(z)$ of degree N can be computed with a multidimensional aperiodic convolution followed by a reduction modulo $P(z)$.

In many instances, $P(z)$ can be converted easily into a multidimensional polynomial by simple transformations such as $P(z) = P_2(z_1), z_1 = z^{N_1}$. A cyclotomic polynomial $P(z) = (z^9 - 1)/(z^3 - 1) = z^6 + z^3 + 1$ can be viewed, for example, as a polynomial $P_2(z_1) = z_1^2 + z_1 + 1$ in which z^3 is substituted to z_1. In these cases, the multidimensional approach can be refined by calculating the polynomial product modulo $P(z)$ as a two-dimensional polynomial product modulo $(z^{N_1} - z_1), P_2(z_1)$. With this method, a polynomial multiplication modulo $(z^4 + 1)$ is implemented with 9 multiplications and 15 additions (Sect. 3.7.2) by computing a polynomial product modulo $(z^2 - z_1)$ on polynomials defined modulo $(z_1^2 + 1)$. This is a significant improvement over the direct computation by interpolation of the same polynomial multiplication modulo $(z^4 + 1)$, which requires 7 multiplications and 41 additions.

The main advantage of this approach stems from the fact that the polynomial multiplications modulo $(z^{N_1} - z_1)$ can be computed by interpolation on powers of z_1. More precisely since $(z^{N_1} - z_1)$ is an irreducible polynomial of degree N_1, a polynomial multiplication modulo $(z^{N_1} - z_1)$ can be evaluated with $2N_1 - 1$ general multiplications by interpolation on $2N_1 - 1$ distinct points (theorem 2.21). The two simplest interpolation points are $z = 0$ and $1/z = 0$. For the $2N_1 - 3$ remaining interpolation points, we note that, if the degree N_2 of the polynomial $P_2(z_1)$ is larger than N_1, the $2N_1$ points $\pm 1, \ldots, \pm z_1^{N_2-1}$ are all distinct. Thus, for

$N_1 \leqslant N_2$ the $2N_1 - 3$ remaining interpolation points can be chosen to be powers of z_1. Under these conditions, the interpolation is given by

$$A_k(z_1) \equiv \sum_{m_1=0}^{N_1-1} X_{m_1}(z_1) z_1^{m_1 k} \text{ modulo } P_2(z_1) \quad k = 0, \ldots, 2N_1 - 2$$

$$k \neq N_1 - 1, N_1 - 2 \quad (3.35)$$

$$A_{N_1-2}(z_1) = X_0(z_1) \quad (3.36)$$

$$A_{N_1-1}(z_1) = X_{N_1-1}(z_1) \quad (3.37)$$

with similar relations for $B_k(z_1)$ corresponding to $H(z)$. Hence, $Y(z)$ is computed by evaluating $A_k(z_1)$, $B_k(z_1)$, calculating the $2N_1-1$ polynomial multiplications $A_k(z_1)B_k(z_1)$ modulo $P_2(z_1)$, and reconstructing $Y(z)$ by the Chinese remainder theorem. In (3.35), the multiplications by power of z_1 correspond to simple shifts of the input polynomials, followed by reductions modulo $P_2(z_1)$. When $P_2(z_1) = z_1^{2^{t_2}} + 1$, these operations reduce to a rotation by ($m_1 k$ modulo 2^{t_2}) words of the input polynomials, followed by a simple sign inversion of the overflow words. Thus, (3.35) is calculated without multiplications and, when N_1 is composite, the number of additions can be reduced by use of an FFT-type algorithm. In particular, if N_1 is a power of 2 and $P_2(z_1) = z_1^{N_2} + 1$, with $N_2 = 2^{t_2}$, the reductions corresponding to the computation of $z^{N_1} - z_1$ can be computed with only $N_2(2N_1 \log_2 N_1 - 5)$ additions. We shall see now that, when one of the input sequences is fixed, the Chinese remainder reconstruction can be done with approximately the same number of additions.

3.2.3 Matrix Exchange Algorithm

If we consider a polynomial multiplication modulo a polynomial $P(z)$ of degree N and with M scalar multiplications, the corresponding algorithm for such a process can usually be viewed as a computation by rectangular transforms,

$$\boldsymbol{H} = \boldsymbol{E}\boldsymbol{h} \quad (3.38)$$

$$\boldsymbol{X} = \boldsymbol{F}\boldsymbol{x} \quad (3.39)$$

$$\boldsymbol{Y} = \boldsymbol{H} \otimes \boldsymbol{X} \quad (3.40)$$

$$\boldsymbol{y} = \boldsymbol{G}\boldsymbol{Y}, \quad (3.41)$$

where $\boldsymbol{h}$ and $\boldsymbol{x}$ are column vectors of the input sequences h_n and x_m, $\boldsymbol{E}$ and $\boldsymbol{F}$ are the input matrices of size $M \times N$, $\otimes$ denotes the element by element product, $\boldsymbol{G}$ is the output matrix of size $N \times M$, and $\boldsymbol{y}$ is a column vector of the output sequence y_l. When the algorithm is designed by an interpolation method, the matrices $\boldsymbol{E}$ and $\boldsymbol{F}$ correspond to the various reductions modulo $(z - a_i)$ while the matrix G correspond to the Chinese remainder reconstruction. Thus, Eh, Fx, and GY are computed without multiplications, while $H \otimes X$ is evaluated with M multiplications. Since the Chinese remainder reconstruction can be viewed as the inverse operation of the reductions, it is roughly equivalent in complexity to the two input sequence reductions. Consequently, the matrix G is usually about twice as complex as matrices E and F. In most practical filtering applications, one of the input sequences, h_n, is constant so that H is precomputed and stored. Thus, the number of additions for the algorithm is dependent upon the complexity of F and G matrices but not upon that of the E matrix. In this circumstance, it is highly desirable to permute E and G matrices in order to reduce the number of additions. As a first step toward doing this, we express the process given by (3.38–41) in the form

$$y_l = \sum_{n=0}^{N-1} \sum_{m=0}^{N-1} h_n x_m \sum_{k=0}^{M-1} e_{n,k} f_{m,k} g_{k,l}, \qquad (3.42)$$

where $e_{n,k}$, $f_{m,k}$, and $g_{k,l}$ are, respectively, the elements of matrices E, F, and G. For a circular convolution algorithm modulo $P(z) = z^N - 1$, we must satisfy the condition

$$S = \sum_{k=0}^{M-1} e_{m,k} f_{n,k} g_{k,l} = 1 \text{ if } m + n - l \equiv 0 \quad \text{modulo } N$$

$$= 0 \text{ if } m + n - l \not\equiv 0 \quad \text{modulo } N. \qquad (3.43)$$

Similarly, polynomial product algorithms modulo $(z^N + 1)$ or $(z^N - z_1)$ correspond, respectively, to the conditions

$$S = 1 \qquad \text{if } m + n - l = 0$$
$$S = -1 \text{ or } z_1 \quad \text{if } m + n - l = N$$
$$S = 0 \qquad \text{if } m + n - l \not\equiv 0 \text{ modulo } N. \qquad (3.44)$$

We now replace the matrices E and G by matrices E^1 and G^1, with elements, respectively, $g_{k,N-m}$ and $e_{N-l,k}$. Subsequently, S becomes S^1 and (3.43) obviously implies

$$S^1 = \sum_{k=0}^{M-1} e_{N-l,k} f_{n,k} g_{k,N-m} \quad = 1 \text{ if } m + n - l \equiv 0 \text{ modulo } N$$

$$= 0 \text{ if } m + n - l \not\equiv 0 \text{ modulo } N. \qquad (3.45)$$

Thus, as pointed out in [3.3], the convolution property still holds when the matrices E and G are exchanged with simple transposition and rearrangement of lines and columns. The same general approach can also be used for polynomial products modulo $(z^N + 1)$ or $(z^N - z_1)$. However, in these cases, the conditions $S = 1$ and $S = -1$ or z_1 in (3.44) are exchanged for m or $l = 0$. Thus, the elements of E^1 and G^1 must be $g_{k,N-m}$ for $m \neq 0$, $-g_{k,N-m}$ or $(1/z_1)g_{k,N-m}$ for $m = 0$, and $e_{N-l,k}$ for $l \neq 0$, $-e_{N-l,k}$ or $z_1 e_{N-l,k}$ for $l = 0$. This approach has been used for the polynomial product algorithm modulo $(z^9-1)/(z^3-1)$ given in Sect. 3.7.2. Using this method for polynomial products modulo $(z^{2^t} + 1)$, with $(z^{N_1} - z_1)$, $(z_1^{N_2} + 1)$ and $N_1 = 2^{t_1} \leqslant N_2 = 2^{t_2}$ yields $(N_1 - 6 + 2N_1 \log_2 N_1) N_2$ additions for the Chinese remainder reconstruction and therefore $(N_1 - 11 + 4N_1 \log_2 N_1) N_2$ additions for the computation of the polynomial product modulo $(z^{N_1} - z_1)$.

Table 3.1. Number of multiplications and additions for short cyclic convolution algorithms.

Convolution size N	Number of multiplications M	Number of additions A
2	2	4
3	4	11
4	5	15
5	10	31
5	8	62
7	16	70
8	14	46
8	12	72
9	19	74
16	35	155
16	33	181

Table 3.2. Number of arithmetic operations for various polynomial products modulo $P(z)$.

Ring $P(z)$	Degree of $P(z)$	Number of multiplications	Number of additions
$z^2 + 1$	2	3	3
$(z^3 - 1)/(z - 1)$	2	3	3
$z^4 + 1$	4	9	15
$z^4 + 1$	4	7	41
$(z^5 - 1)/(z - 1)$	4	9	16
$(z^5 - 1)/(z - 1)$	4	7	46
$(z^7 - 1)/(z - 1)$	6	15	53
$(z^9 - 1)/(z^3 - 1)$	6	15	39
$z^8 + 1$	8	21	77
$(z_1^2 + z_1 + 1)(z_2^5 - 1)/(z_2 - 1)$	8	21	83
$(z_1^4 + 1)(z_2^2 + z_2 + 1)$	8	21	76
$(z_1^2 + 1)(z_2^5 - 1)/(z_2 - 1)$	8	21	76
$z^{16} + 1$	16	63	205
$(z^{27} - 1)/(z^9 - 1)$	18	75	267
$z^{32} + 1$	32	147	739
$z^{64} + 1$	64	315	1899

We list the details of a number of frequently used algorithms for the short convolutions and polynomial products in Sects. 3.7.1 and 3.7.2. Tables 3.1 and 3.2 summarize the corresponding number of operations for these algorithms and others. We have optimized these algorithms to favor a reduction of the number of multiplications. Thus, the algorithms lend themselves to efficient implementation on computers in which the multiplication execution time is much greater than that for addition and subtraction. When multiply execution time is about the same as addition, it is preferable to use other polynomial product algorithms in which the number of additions is reduced at the expense of an increased number of multiplications. Additional polynomial product algorithms are given in Appendix B.

3.3 Computation of Large Convolutions by Nesting of Small Convolutions

For large convolutions, the algorithms derived from interpolation methods become complicated and inefficient. We shall show here that the computation of long cyclic convolutions is greatly simplified by using a one-dimensional to multidimensional mapping suggested by Good [3.4], combined with a nesting approach proposed by Agarwal and Cooley [3.5].

3.3.1 The Agarwal-Cooley Algorithm

The Agarwal-Cooley method requires that the length N of the cyclic convolution y_l must be a composite number with mutually prime factors. In the following, we shall assume that $N = N_1 N_2$ with $(N_1, N_2) = 1$. The convolution y_l is given by the familiar expression

$$y_l = \sum_{n=0}^{N-1} h_n x_{l-n}, \quad l = 0, \ldots, N-1. \tag{3.46}$$

Since N_1 and N_2 are mutually prime and the indices k and n are defined modulo $N_1 N_2$, a direct consequence of the Chinese remainder theorem (theorem 2.1) is that l and n can be mapped into two sets of indices l_1, n_1 and l_2, n_2, with

$$\begin{cases} l \equiv N_1 l_2 + N_2 l_1 & \text{modulo } N_1 N_2, \; l_2, n_2 = 0, \ldots, N_2 - 1 \\ n \equiv N_1 n_2 + N_2 n_1 & \text{modulo } N_1 N_2, \; l_1, n_1 = 0, \ldots, N_1 - 1. \end{cases} \tag{3.47}$$

Thus, the one-dimensional convolution y_l becomes a two-dimensional convolution of size $N_1 \times N_2$

$$y_{N_1 l_2 + N_2 l_1} = \sum_{n_1=0}^{N_1-1} \sum_{n_2=0}^{N_2-1} h_{N_1 n_2 + N_2 n_1} x_{N_1(l_2-n_2)+N_2(l_1-n_1)}. \tag{3.48}$$

In polynomial notation, this convolution can be viewed as a one-dimensional polynomial convolution of length N_1

$$H_{n_1}(z) = \sum_{n_2=0}^{N_2-1} h_{N_1 n_2 + N_2 n_1} z^{n_2} \tag{3.49}$$

$$X_{m_1}(z) = \sum_{m_2=0}^{N_2-1} x_{N_1 m_2 + N_2 m_1} z^{m_2} \tag{3.50}$$

$$Y_{l_1}(z) = \sum_{l_2=0}^{N_2-1} y_{N_1 l_2 + N_2 l_1} z^{l_2} \tag{3.51}$$

$$Y_{l_1}(z) \equiv \sum_{n_1=0}^{N_1-1} H_{n_1}(z) X_{l_1-n_1}(z) \quad \text{modulo } (z^{N_2} - 1). \tag{3.52}$$

Each polynomial multiplication $H_{n_1}(z)X_{l_1-n_1}(z)$ modulo $(z^{N_2}-1)$ corresponds to a convolution of length N_2 which is computed with M_2 scalar multiplications and A_2 scalar additions. Thus, the convolution of length $N_1 N_2$ is computed by (3.52) as a convolution of length N_1 in which each scalar is replaced by a polynomial of N_2 terms and each multiplication is replaced by a convolution of length N_2. Under these conditions, if M_1 and A_1 are the number of multiplications and additions required to compute a convolution of length N_1, the number of multiplications M and additions A corresponding to the convolution of length $N_1 N_2$ reduces to

$$M = M_1 M_2 \tag{3.53}$$

$$A = A_1 N_2 + M_1 A_2. \tag{3.54}$$

By permuting the role of N_1 and N_2, the same convolution could have been computed as a convolution of N_2 terms in which the scalars would have been replaced by polynomials of N_1 terms. In this case, the number of multiplications would be the same, but the number of additions would be $A_2 N_1 + M_2 A_1$. Thus, the number of additions for the nesting algorithm is dependent upon the order of the operations. If the first arrangement gives fewer additions, we must have

$$A_1 N_2 + M_1 A_2 < A_2 N_1 + M_2 A_1 \tag{3.55}$$

or,

$$(M_1 - N_1)/A_1 < (M_2 - N_2)/A_2. \tag{3.56}$$

Therefore, the convolution to perform first is the one for which the quantity $(M_1 - N_1)/A_1$ is smaller.

When N is the product of more than two relatively prime factors N_1, N_2, N_3, ..., N_d, the same nesting approach can be used recursively by converting the convolution of length N into a convolution of size $N_1 \times N_2 \times N_3 ... \times N_d$ and

3.3 Computation of Large Convolutions by Nesting of Small Convolutions

by computing this multidimensional convolution as a convolution of length N_1 of arrays of size $N_2 \times N_3 \ldots \times N_d$, where all multiplications are replaced by convolutions of size $N_2 \times N_3 \ldots \times N_d$. The same process is repeated on the arrays of sizes $N_2 \times N_3 \ldots \times N_d$, $N_3 \times \ldots \times N_d$, until all convolutions are reduced to $M_1 M_2 \ldots M_{d-1}$ convolutions of length N_d, where the M_i are the number of multiplications corresponding to a convolution of N_i terms. Under these conditions, and assuming that A_i is the number of additions for a length-N_i one-dimensional convolution, the total number of operations for the convolution of dimension $N_1 N_2 \ldots N_d$ computed by the nesting algorithm becomes

$$M = M_1 M_2 \ldots M_d \qquad (3.57)$$

$$A = A_1 N_2 \ldots N_d + M_1 A_2 N_3 \ldots N_d + M_1 M_2 A_3 N_4 \ldots N_d + \ldots$$
$$+ M_1 M_2 \ldots M_{d-1} A_d. \qquad (3.58)$$

It should be noted that in these formulas the number of additions A_i corresponding to each convolution of length N_i contributes to only a fraction of the total number of additions A, while the number of multiplication M_i is a direct factor of M. This means that for large convolutions, it is generally advantageous to select short convolution algorithms minimizing the number of multiplications, even if this minimization is done at the expense of a relatively large increase in the number additions. This is illustrated by considering as an example a convolution of length 8. Such a convolution can be computed with two different algorithms, one with 14 multiplications and 46 additions, and the other with 12 multiplications and 72 additions. When these algorithms are used for calculating a simple convolution of 8 terms, the second algorithm is obviously less interesting than the first one, since it saves only 2 multiplications at the expense of 26 more additions. However, when a convolution of length 63 is nested with a convolution of length 8 to compute a convolution of 504 terms, the situation is completely reversed: using the first length-8 algorithm yields 4256 multiplications and 28240 additions, as opposed to 3648 multiplications and 26304 additions when the second length-8 algorithm is used.

Table 3.3 summarizes the number of arithmetic operations for a variety of one-dimensional cyclic convolutions computed via the Agarwal-Cooley algorithm by nesting the short convolutions corresponding to Table 3.1. It can be seen by comparison with Table 4.6 that the nesting approach yields a smaller number of operations than the conventional method using FFTs for short and medium-sized convolutions up to a length of about 200. One significant advantage of the nesting method over the FFT algorithms is that it does not use trigonometric functions and that real convolutions are computed with real arithmetic instead of complex arithmetic.

Multidimensional convolutions can also be calculated by the nesting algorithm. A convolution of size $N_1 N_2 \times N_1 N_2$ can, for instance, be calculated as a convolution of size $N_1 \times N_1$ in which each scalar is replaced by an array of size

3. Fast Convolution Algorithms

Table 3.3. Number of arithmetic operations for one-dimensional convolutions computed by the Agarwal-Cooley nesting algorithm.

Convolution size N	Number of multiplications M	Number of additions A	Multiplications per point M/N	Additions per point A/N
18	38	184	2.11	10.22
20	50	230	2.50	11.50
24	56	272	2.33	11.33
30	80	418	2.67	13.93
36	95	505	2.64	14.03
60	200	1120	3.33	18.67
72	266	1450	3.69	20.14
84	320	2100	3.81	25.00
120	560	3096	4.67	25.80
180	950	5470	5.28	30.39
210	1280	7958	6.10	37.90
360	2280	14748	6.33	40.97
420	3200	20420	7.62	48.62
504	3648	26304	7.24	52.19
840	7680	52788	9.14	62.84
1008	10032	71265	9.95	70.70
1260	12160	95744	9.65	75.99
2520	29184	241680	11.58	95.90

Table 3.4. Number of arithmetic operations for two-dimensional convolutions computed by the Agarwal-Cooley algorithm.

Convolution size $N \times N$	Number of multiplications M	Number of additions A	Multiplications per point M/N	Additions per point A/N
3×3	16	77	1.78	8.56
4×4	25	135	1.56	8.44
5×5	100	465	4.00	18.60
7×7	256	1610	5.22	32.86
8×8	196	1012	3.06	15.81
9×9	361	2072	4.46	25.58
12×12	400	3140	2.78	21.81
16×16	1225	7905	4.79	30.88
20×20	2500	15000	6.25	37.50
30×30	6400	41060	7.11	45.62
36×36	9025	62735	6.96	48.41
40×40	19600	116440	12.25	72.77
60×60	40000	264500	11.11	73.47
72×72	70756	488084	13.65	94.15
80×80	122500	767250	19.14	119.88
120×120	313600	1986240	21.78	137.93

$N_2 \times N_2$ and each multiplication is replaced by a convolution of size $N_2 \times N_2$. The number of arithmetic operations for various convolutions computed by this method and the short algorithms of Sect. 3.7.1 is given in Table 3.4. We shall see in Chap. 5 that the computation of multidimensional convolutions by a nesting method plays a key role in the calculation of DFTs by the Winograd algorithm. When the multidimensional convolutions have common factors in several dimensions, the multidimensional polynomial rings which correspond to these convolutions have special properties that can be used to simplify the calculations. In this case, the nesting method, which does not exploit these properties, becomes relatively inefficient and should be replaced by a polynomial transform approach, as discussed in Chap. 6.

3.3.2 The Split Nesting Algorithm

As indicated in the previous section, the nesting method has many desirable features for the evaluation of convolutions. The method is particularly attractive for short- and medium-length convolutions.

The main drawbacks of the nesting approach relate to the use of several relatively prime moduli for indexing the data, the excessive number of additions for large convolutions, and the amount of memory required for storing the short convolution algorithm. The first point is intrinsic to the nesting method, since one-dimensional convolutions are converted into multidimensional convolutions only if N is the product of several relatively prime factors. If N does not satisfy this condition, a one-dimensional to multidimensional mapping is feasible only at the expense of a length increase of the input sequences which translates into an increased number of arithmetic operations, as will be shown in Sect. 3.4. Thus, one cannot hope to eliminate relatively prime moduli in the computation of one-dimensional convolutions by the Agarwal-Cooley nesting approach.

However, the impact of the other limitations concerning storage requirements and the number of additions can be relieved by replacing the convolution nesting process with a nesting of polynomial products. In this method, which is called *split nesting* [3.6], the short convolutions of the Agarwal-Cooley method are computed as polynomial products. We shall restrict our discussion of this method to convolutions of length $N_1 N_2$, with N_1 and N_2 distinct odd primes, since all other cases can be deduced easily from this simple example.

Since N_1 is prime, $z^{N_1} - 1$ factors into the two cyclotomic polynomials $(z-1)$ and $P_2(z) = z^{N_1-1} + z^{N_1-2} + \ldots + 1$. The cyclic convolution y_l of two N_1-point sequences h_n and x_m can be computed as a polynomial product modulo $(z^{N_1} - 1)$ with

$$H(z) = \sum_{n=0}^{N_1-1} h_n z^n \qquad (3.59)$$

$$X(z) = \sum_{m=0}^{N_1-1} x_m z^m \qquad (3.60)$$

$$Y(z) = \sum_{l=0}^{N_1-1} y_l z^l \equiv H(z)X(z) \quad \text{modulo } (z^{N_1} - 1). \tag{3.61}$$

Using the Chinese remainder theorem, $Y(z)$ is calculated as shown in Fig. 3.1 by reducing $H(z)$ and $X(z)$ modulo $P_2(z)$ and $(z-1)$ to $H_2(z)$, H_1 and $X_2(z)$, X_1, respectively, computing the polynomial products $H_2(z)X_2(z)$ modulo $P_2(z)$ and $H_1 X_1$ modulo $(z-1)$, and reconstructing $Y(z)$ by

$$Y(z) \equiv S_1(z)H_1 X_1 + S_2(z)H_2(z)X_2(z) \quad \text{modulo } (z^{N_1} - 1) \tag{3.62}$$

$$S_2(z) \equiv 1, \quad S_1(z) \equiv 0 \quad \text{modulo } P_2(z)$$
$$S_2(z) \equiv 0, \quad S_1(z) \equiv 1 \quad \text{modulo } (z-1). \tag{3.63}$$

The reductions and the Chinese remainder operations always have the same structure, regardless of the particular numerical value of N_1. Therefore, these operations, when implemented in a computer, need not be stored as individual procedures for each value of N_1, N_2 ..., but can be defined by a single program structure. In particular, the reductions modulo $P_2(z)$ and $(z-1)$ are computed with $N_1 - 1$ additions by

$$H_2(z) = \sum_{n=0}^{N_1-2} (h_n - h_{N_1-1})z^n \tag{3.64}$$

$$H_1 = \sum_{n=0}^{N_1-1} h_n \tag{3.65}$$

Thus, the only part of each convolution algorithm which needs to be individually stored is that corresponding to the polynomial products. With this method, the savings in storage can be quite significant. If we consider for example a simple convolution of 15 terms, the calculation can be performed by nesting a convolution of 3 terms with a convolution of length 5. Since the 3-point and 5-point convolutions are computed, respectively, with 11 additions and 31 additions, a typical computer program would require about 42 instructions to implement the short convolution algorithms in the conventional nesting method. Alternatively, if the 3-point and 5-point convolutions are computed as polynomial products, the calculation breaks down into scalar multiplications and polynomial multiplications modulo $(z^3-1)/(z-1)$ and modulo $(z^5-1)/(z-1)$. Since these two polynomial products are calculated, respectively, with 3 additions and 16 additions, a program for the split nesting approach would require a general purpose program for reductions and Chinese remainder reconstruction modulo a prime number, plus about 19 instructions to implement the two polynomial products.

The implementation of a convolution of length $N_1 N_2$ by split nesting can be conveniently described using a polynomial representation. As a first step (similar to that used with conventional nesting), the convolution y_l of length $N_1 N_2$ is con-

3.3 Computation of Large Convolutions by Nesting of Small Convolutions

verted by index mapping into a convolution $y_{N_1 l_2 + N_2 l_1}$ of size $N_1 \times N_2$, as shown by (3.48). This two-dimensional convolution can be viewed as the product modulo $(z_1^{N_1} - 1)$, $(z_2^{N_2} - 1)$ of the two-dimensional polynomials $H(z_1, z_2)$ and $X(z_1, z_2)$ given by

$$H(z_1, z_2) = \sum_{n_1=0}^{N_1-1} \sum_{n_2=0}^{N_2-1} h_{N_1 n_2 + N_2 n_1} z_1^{n_1} z_2^{n_2} \qquad (3.66)$$

$$X(z_1, z_2) = \sum_{m_1=0}^{N_1-1} \sum_{m_2=0}^{N_2-1} x_{N_1 m_2 + N_2 m_1} z_1^{m_1} z_2^{m_2} \qquad (3.67)$$

$$Y(z_1, z_2) = \sum_{l_1=0}^{N_1-1} \sum_{l_2=0}^{N_2-1} y_{N_1 l_2 + N_2 l_1} z_1^{l_1} z_2^{l_2}$$

$$\equiv H(z_1, z_2) X(z_1, z_2) \text{ modulo } (z_1^{N_1} - 1), (z_2^{N_2} - 1). \qquad (3.68)$$

Since N_1 and N_2 are primes, $z_1^{N_1} - 1$ and $z_2^{N_2} - 1$ factor, respectively, into the cyclotomic polynomials $(z_1 - 1)$, $P_2(z_1) = z_1^{N_1-1} + z_1^{N_1-2} + \ldots + 1$ and $(z_2 - 1)$, $P_2(z_2) = z_2^{N_2-1} + z_2^{N_2-2} + \ldots + 1$. $Y(z_1, z_2)$ can therefore be computed by a Chinese remainder reconstruction from polynomial products defined modulo these cyclotomic polynomials with

$$Y(z_1, z_2) \equiv \sum_{i=1}^{2} \sum_{k=1}^{2} S_{i,k}(z_1, z_2) H_{i,k}(z_1, z_2) X_{i,k}(z_1, z_2)$$

$$\text{modulo } (z_1^{N_1} - 1), (z_2^{N_2} - 1) \qquad (3.69)$$

$$S_{2,2}(z_1, z_2) \equiv 1 \text{ , } S_{2,1}(z_1, z_2), S_{1,2}(z_1, z_2), S_{1,1}(z_1, z_2)$$
$$\equiv 0 \text{ modulo } P_2(z_1), P_2(z_2)$$

$$S_{1,2}(z_1, z_2) \equiv 1 \text{ , } S_{1,1}(z_1, z_2), S_{2,1}(z_1, z_2), S_{2,2}(z_1, z_2)$$
$$\equiv 0 \text{ modulo } (z_1 - 1), P_2(z_2)$$

$$S_{2,1}(z_1, z_2) \equiv 1 \text{ , } S_{1,1}(z_1, z_2), S_{1,2}(z_1, z_2), S_{2,2}(z_1, z_2)$$
$$\equiv 0 \text{ modulo } P_2(z_1), (z_2 - 1)$$

$$S_{1,1}(z_1, z_2) \equiv 1 \text{ , } S_{2,1}(z_1, z_2), S_{1,2}(z_1, z_2), S_{2,2}(z_1, z_2)$$
$$\equiv 0 \text{ modulo } (z_1 - 1), (z_2 - 1), \qquad (3.70)$$

where

$$X_{2,2}(z_1, z_2) \equiv X(z_1, z_2) \text{ modulo } P_2(z_1), P_2(z_2)$$
$$X_{1,2}(z_2) \equiv X(z_1, z_2) \text{ modulo } (z_1 - 1), P_2(z_2)$$
$$X_{2,1}(z_1) \equiv X(z_1, z_2) \text{ modulo } P_2(z_1), (z_2 - 1)$$
$$X_{1,1} \equiv X(z_1, z_2) \text{ modulo } (z_1 - 1), (z_2 - 1) \qquad (3.71)$$

and similar relations for $H_{i,k}(z_1, z_2)$. A detailed representation of the convolution of N_1N_2 points is given in Fig. 3.2. As shown, the procedure includes a succession of reductions modulo $(z_1 - 1)$, $(z_2 - 1)$, $P_2(z_1)$, $P_2(z_2)$ followed by one scalar multiplication, two polynomial multiplications modulo $P_2(z_1)$ and modulo $P_2(z_2)$, and one two-dimensional polynomial multiplication modulo $P_2(z_1)$, $P_2(z_2)$. The convolution product is then computed from these polynomial products by a series of Chinese remainder reconstructions. In this approach the polynomial product modulo $P_2(z_1)$, $P_2(z_2)$ is computed by a nesting technique identical to that used for ordinary two-dimensional convolutions, with a polynomial multiplication modulo $P_2(z_1)$ in which all scalars are replaced by a polynomial of $N_2 - 1$ terms and in which all scalar multiplications are replaced by polynomial multiplications modulo $P_2(z_2)$.

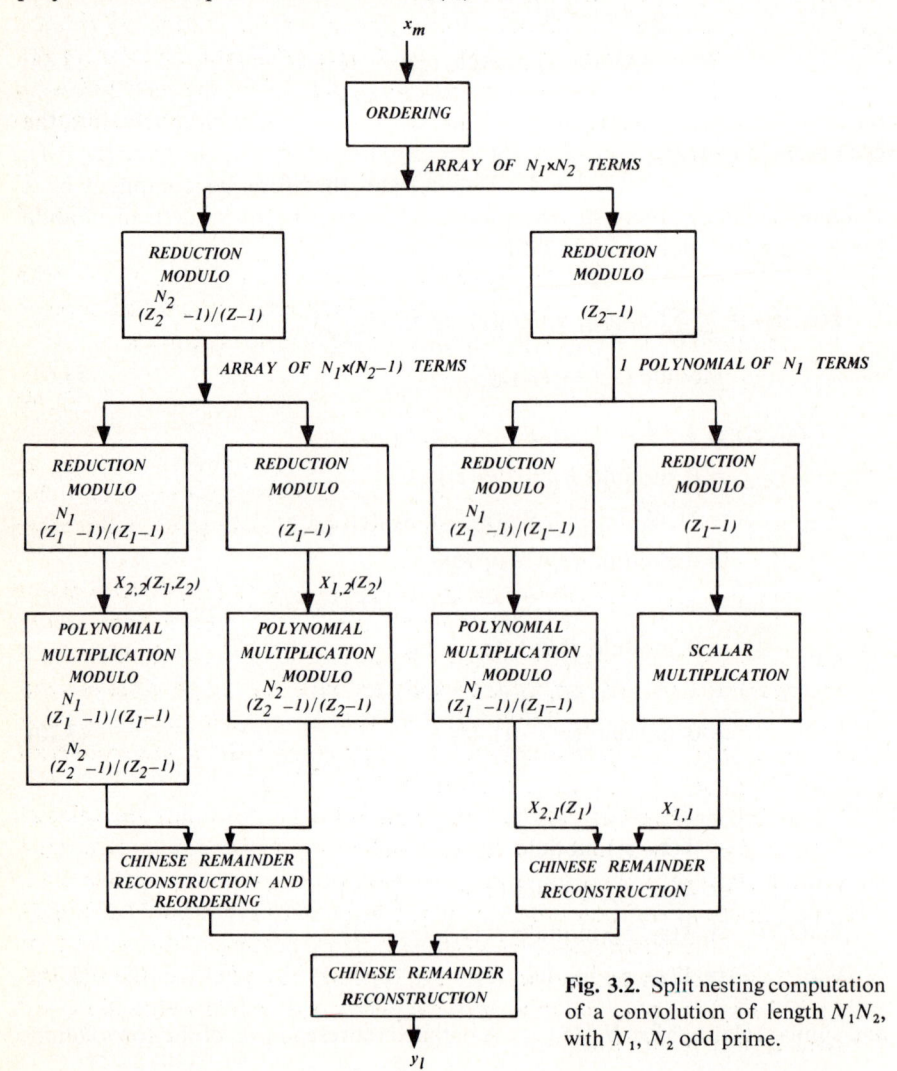

Fig. 3.2. Split nesting computation of a convolution of length N_1N_2, with N_1, N_2 odd prime.

3.3 Computation of Large Convolutions by Nesting of Small Convolutions

We shall see now that the split nesting method reduces the number of additions. Assuming that the short convolutions of lengths N_1 and N_2 are computed, respectively, with A_1 additions, M_1 multiplications and A_2 additions, M_2 multiplications, the polynomial products modulo $P_2(z_1)$ and $P_2(z_2)$ are calculated with $M_1 - 1$ and $M_2 - 1$ multiplications while the number of additions breaks down into

$$A_1 = A_{1,1} + A_{2,1}$$
$$A_2 = A_{1,2} + A_{2,2}, \qquad (3.72)$$

where $A_{2,1}$ and $A_{2,2}$ are the additions corresponding to the polynomial products and $A_{1,1}$, $A_{1,2}$ are the additions corresponding to the reductions and the Chinese remainder operations. Since the number of multiplications for the polynomial product modulo $P_2(z_1)$, $P_2(z_2)$ computed by nesting is $(M_1 - 1)(M_2 - 1)$, the total number of multiplications M for the convolution of dimension $N_1 N_2$ is given by

$$M = M_1 M_2. \qquad (3.73)$$

The computation of the polynomial product modulo $P_2(z_1)$, $P_2(z_2)$ is done with $(N_2 - 1)A_{2,1} + (M_1 - 1)A_{2,2}$ additions. Hence, the total number of additions A for the convolution of length $N_1 N_2$ computed by split nesting reduces to

$$A = N_2 A_1 + N_1 A_{1,2} + M_1 A_{2,2}. \qquad (3.74)$$

Since $M_1 > N_1$ and $N_1 A_{1,2} + M_1 A_{2,2} < M_1(A_{1,2} + A_{2,2}) = M_1 A_2$, it can be seen, by comparison with (3.54), that splitting the computations saves $(M_1 - N_1)A_{1,2}$ additions over the conventional nesting method, while using the same number of multiplications.

The split nesting technique can be applied recursively to larger convolutions of length $N_1 N_2 N_3 \ldots N_d$ and, with slight modifications, to factors that are powers of primes. In Table 3.5, we give the number of arithmetic operations for one-dimensional convolutions computed by split nesting of the polynomial product algorithms corresponding to Table 3.2. It is seen, by comparing with conventional nesting (Table 3.3), that the split nesting method reduces the number of additions by about 25% for large convolutions.

In the split nesting method, the computation breaks down into the calculation of one-dimensional and multidimensional polynomial products, where the latter products are evaluated by nesting. We have shown previously in Sect. 3.2.2 that such multidimensional polynomial products can be computed more efficiently by multidimensional interpolation than by nesting. Thus, additional computational savings are possible, at the expense of added storage requirements, if the split nesting multidimensional polynomial product algorithms are designed by interpolation.

Table 3.5. Number of arithmetic operations for cyclic convolutions computed by the split nesting algorithm.

Convolution size N	Number of multiplications M	Number of additions A	Multiplications per point M/N	Additions per point A/N
18	38	184	2.11	10.22
20	50	218	2.50	10.90
24	56	244	2.33	10.17
30	80	392	2.67	13.07
36	95	461	2.64	12.81
60	200	964	3.33	16.07
72	266	1186	3.69	16.47
84	320	1784	3.81	21.24
120	560	2468	4.67	20.57
180	950	4382	5.28	24.34
210	1280	6458	6.10	30.75
360	2280	11840	6.33	32.89
420	3200	15256	7.62	36.32
504	3648	21844	7.24	43.34
840	7680	39884	9.14	47.48
1008	10032	56360	9.95	55.91
1260	12160	72268	9.65	57.36
2520	29184	190148	11.58	75.46

For instance, the evaluation of a convolution of 120 points involves the calculation of a polynomial product modulo $(z_1^4 + 1)$, $(z_2^5 - 1)/(z_2 - 1)$, $(z_3^2 + z_3 + 1)$. Such a polynomial product can be computed with 189 multiplications and 640 additions instead of 243 multiplications and 641 additions, if it is computed by nesting a polynomial product modulo $(z_2^5 - 1)/(z_2 - 1)$ with the polynomial product modulo $(z_1^4 + 1)$, $(z_3^2 + z_3 + 1)$ designed by interpolation (Table 3.2). With this approach, a convolution of 120 terms is evaluated with only 506 multiplications and 2467 additions as opposed to 560 multiplications and 2468 additions for the conventional split nesting method. The price to be paid for these savings is the additional memory required for storing the polynomial product algorithm modulo $(z_1^4 + 1)$, $(z_3^2 + z_3 + 1)$.

3.3.3 Complex Convolutions

Complex convolutions can be computed via nesting methods by simply replacing real arithmetic with complex arithmetic. In this case, if M and A are the number of real multiplications and additions corresponding to a real convolution of N terms, the number of real multiplications and additions becomes $4M$ and $2A + 2M$ for a complex convolution of N terms. If complex multiplication is imple-

3.3 Computation of Large Convolutions by Nesting of Small Convolutions

mented with an algorithm using 3 multiplications and 3 additions, as shown in Sect. 3.7.2, the number of operations reduces to $3M$ multiplications and $2A + 3M$ additions.

However, a more efficient method can be devised by taking advantage of the fact that $j = \sqrt{-1}$ is real in certain fields [3.7]. In the case of fields of polynomials modulo $(z^q + 1)$, with q even, $z^q \equiv -1$ and j is congruent to $z^{q/2}$. Thus, for a complex convolution of length $N = N_1 N_2$, with N_1 odd and $N_2 = 2^t$, with the use of the index mapping given by (3.47–52), one can define a complex one-dimensional polynomial convolution by

$$Y_{l_1}(z) + j\hat{Y}_{l_1}(z) \equiv \sum_{n_1=0}^{N_1-1} [H_{n_1}(z) X_{l_1-n_1}(z) - \hat{H}_{n_1}(z) \hat{X}_{l_1-n_1}(z)]$$
$$+ j[\hat{H}_{n_1}(z) X_{l_1-n_1}(z) + H_{n_1}(z) \hat{X}_{l_1-n_1}(z)] \text{ modulo } (z^{N_2} - 1) \quad (3.75)$$

where $H_{n_1}(z)$, $X_{m_1}(z)$, and $Y_{l_1}(z)$ are the polynomials corresponding to the real parts of the input and output sequences, and $\hat{H}_{n_1}(z)$, $\hat{X}_{m_1}(z)$, and $\hat{Y}_{l_1}(z)$ are the polynomials corresponding to the imaginary parts of the input and output sequences with

$$\hat{H}_{n_1}(z) = \sum_{n_2=0}^{N_2-1} \hat{h}_{N_1 n_2 + N_2 n_1} z^{n_1} \quad (3.76)$$

$$\hat{X}_{m_1}(z) = \sum_{m_2=0}^{N_2-1} \hat{x}_{N_1 m_2 + N_2 m_1} z^{m_1} \quad (3.77)$$

$$\hat{Y}_{l_1}(z) = \sum_{l_2=0}^{N_2-1} \hat{y}_{N_1 l_2 + N_2 l_1} z^{l_2}. \quad (3.78)$$

Since $N_2 = 2^t$, $z^{N_2} - 1 = (z-1) \prod_{v=0}^{t-1} (z^{2^v} + 1)$, and $Y_{l_1}(z) + j\hat{Y}_{l_1}(z)$ can be computed by the Chinese remainder reconstruction from the various reduced polynomials $Y_{v,l_1}(z) + j\hat{Y}_{v,l_1}(z)$ defined by

$$Y_{v,l_1}(z) + j\hat{Y}_{v,l_1}(z) \equiv Y_{l_1}(z) + j\hat{Y}_{l_1}(z) \text{ modulo } (z^{2^v} + 1) \quad (3.79)$$

$$Y_{u,l_1} + j\hat{Y}_{u,l_1} \equiv Y_{l_1}(z) + j\hat{Y}_{l_1}(z) \text{ modulo } (z - 1), \quad (3.80)$$

the terms $Y_{u,l_1} + j\hat{Y}_{u,l_1}$ and $Y_{0,l_1} + j\hat{Y}_{0,l_1}$ correspond, respectively, to $z \equiv 1$ and $z \equiv -1$ and are therefore scalar convolutions of dimension N_1 which are computed with M_1 complex multiplications. Each of these convolutions is calculated with $3M_1$ real multiplications by using complex arithmetic with 3 real multiplications per complex multiplication. For $v \neq 0$, $j \equiv z^{2^{v-1}}$ and each complex polynomial convolution $Y_{v,l_1}(z) + j\hat{Y}_{v,l_1}(z)$ is computed with only two real convolutions

$$Q_{v,l_1}(z) \equiv \sum_{n_1=0}^{N_1-1} [H_{v,n_1}(z) + z^{2^{v-1}}\hat{H}_{v,n_1}(z)]$$
$$[X_{v,l_1-n_1}(z) + z^{2^{v-1}}\hat{X}_{v,l_1-n_1}(z)] \text{ modulo } (z^{2^v} + 1) \tag{3.81}$$

$$\hat{Q}_{v,l_1}(z) \equiv \sum_{n_1=0}^{N_1-1} [H_{v,n_1}(z) - z^{2^{v-1}}\hat{H}_{v,n_1}(z)]$$
$$[X_{v,l_1-n_1}(z) - z^{2^{v-1}}\hat{X}_{v,l_1-n_1}(z)] \text{ modulo } (z^{2^v} + 1), \tag{3.82}$$

where

$$Y_{v,l_1}(z) = [Q_{v,l_1}(z) + \hat{Q}_{v,l_1}(z)]/2 \tag{3.83}$$

$$\hat{Y}_{v,l_1}(z) \equiv -z^{2^{v-1}}[Q_{v,l_1}(z) - \hat{Q}_{v,l_1}(z)]/2 \text{ modulo } (z^{2^v} + 1). \tag{3.84}$$

The multiplications by $z^{2^{v-1}}$ modulo $(z^{2^v} + 1)$ in these expressions correspond to a simple rotation by 2^{v-1} words of the 2^v word polynomials, followed by sign inversion of the overflow words. Thus, for $v \neq 0$, complex multiplication is implemented with only two real multiplications and the total number of real multiplications for the convolution of size $N_1 N_2$ becomes $2M_1(M_2 + 1)$ instead of $3M_1 M_2$ or $4M_1 M_2$ with conventional approaches using 3 or 4 real multiplications per complex multiplication.

It should be noted that this saving in number of multiplications is achieved without an increase in number of additions. This can be seen by noting that the computation process is equivalent to that encountered in evaluating 2 real convolutions of length N, plus $6M_1$ additions corresponding to the complex multiplications required for the calculation of $Y_{u,l_1} + j\hat{Y}_{u,l_1}$ and $Y_{0,l_1} + j\hat{Y}_{0,l_1}$ and a total of $4N_1(N_2 - 2)$ additions for constructing the auxiliary polynomials. Therefore, if A is the number of additions for a real convolution of length N, the number of real additions for a complex convolution becomes $2A + 4N + 6M_1 - 8N_1$ if the sequence $h_n + j\hat{h}_n$ is fixed. The same complex convolution of N terms would have required $2A + 2M_1 M_2$ or $2A + 3M_1 M_2$ real additions if complex multiplication algorithms with 4 multiplications and 2 additions or 3 multiplications and 3 additions had been used. This demonstrates how, in most cases, computing complex convolutions by using the properties of j in a ring of polynomials can save multiplications and additions over conventional methods. With this approach, a complex convolution of 72 terms is computed with 570 real multiplications and 3230 real additions as opposed to 1064 multiplications and 3432 additions for the same convolution calculated with a complex multiplication algorithm using 4 real multiplications, and 798 multiplications, 3698 additions in the case of a complex multiplication algorithm using 3 multiplications.

We have, thus far, described the computation of complex convolutions in rings of polynomials modulo $(z^q + 1)$, q even. With some loss of efficiency, the same concept can also be applied to other rings. For example, in a field modulo $P(z) = (z^3 - 1)/(z - 1) = z^2 + z + 1$, we have $[(2z + 1)/\sqrt{3}]^2 \equiv -1$ modulo $P(z)$. Thus, in this case, $j \equiv (2z + 1)/\sqrt{3}$ modulo $P(z)$. Note however, that this approach is less attractive than when j is defined modulo $(z^q + 1)$, q even, since multiplications by $j = (2z + 1)/\sqrt{3}$ cannot be implemented with simple polynomial rotations and additions.

3.3.4 Optimum Block Length for Digital Filters

In many digital filtering applications, one of the sequence, h_n, is of limited length, N_1, and represents the impulse response of the filter. The other sequence, x_m, is usually the input data sequence and can be considered to be of infinite length. The noncyclic convolution of these sequences can be obtained by computing a series of circular convolutions of length N and reconstructing the digital filter output by the overlap-add or overlap-save techniques described in Sect. 3.1.

With the overlap-add technique, the data sequence x_m is sectioned into blocks of length N_2 and the aperiodic convolution of each block with the sequence h_n is computed by using a length-N cyclic convolution such that $N = N_1 + N_2 - 1$. In this cyclic convolution, the input blocks of length N are obtained by appending $N_2 - 1$ zeros to the sequence h_n and $N_1 - 1$ zeros to the data sequence blocks. If $M_1(N)$ and $M_2(N, N_1)$ are, respectively, the number of multiplications per output point for the cyclic convolutions of length N and for the N_1-tap digital filter, $M_2(N, N_1)$ is given by

$$M_2(N, N_1) = M_1(N)N/(N - N_1 + 1). \tag{3.85}$$

Similarly, $A_2(N, N_1)$, the number of additions for the digital filter, is given as a function of $A_1(N)$, the number of additions per output point for the cyclic convolutions of length N, by

$$A_2(N, N_1) = [A_1(N)N + N_1 - 1]/(N - N_1 + 1). \tag{3.86}$$

Since $M_1(N)$ is an increasing function of N and $N/(N - N_1 + 1)$ is a decreasing function of N, there is an optimum block size N which minimizes the number of multiplications. Table 3.6 lists optimum block sizes N and corresponding numbers of operations for digital filters of various tap length N_1 computed by circular convolutions and split nesting (Table 3.5). It can be seen that N is typically not much larger than N_1. This is due to the fact that $M_1(N)$ is a rapidly increasing function of N.

When compared to FFT filter methods, these results show that the split nesting method is preferable to the FFT approach for tap lengths up to about 256. FFT methods require larger block sizes for filter implementation because $M_1(N)$ increases much more slowly with N for FFTs than for split nesting.

56   3. Fast Convolution Algorithms

Table 3.6. Optimum block sizes and number of operations for digital filters computed by circular convolutions and split nesting.

Filter tap length N_1	Multiplications per point $M_2(N, N_1)$	Additions per point $A_2(N, N_1)$	Optimum block size N
2	2.23	10.88	18
4	2.53	12.46	18
8	3.29	14.77	24
16	4.44	21.76	60
32	6.04	34.25	84
64	8.12	37.98	180
128	9.78	51.36	360
256	12.10	72.17	1260
512	14.53	94.91	2520

3.4 Digital Filtering by Multidimensional Techniques

We have seen in the preceding sections that a filtering process can be computed as a sequence of cyclic convolutions which are evaluated by nesting. The nesting method maps one-dimensional cyclic convolutions of length-N into a multidimensional cyclic convolution of size $N_1 \times N_2 \ldots \times N_d$ provided $N_1, N_2, \ldots, N_d$, the factors of N, are relatively prime.

We shall now present a method introduced by Agarwal and Burrus [3.8] which maps directly the one-dimensional aperiodic convolution of two length-N sequences into a multidimensional aperiodic convolution of arrays of dimension $N_1 \times N_2 \ldots \times N_d$, where $N_1, N_2, \ldots, N_d$ are factors of N which need not be relatively prime. The aperiodic convolution of two length-N sequences h_n and x_m is given by

$$y_l = \sum_{n=0}^{N-1} h_n x_{l-n}, \quad l = 0, \ldots, 2N-2, \tag{3.87}$$

where the output sequence y_l is of length $2N - 1$ and the sequences h_n, x_m, and y_l are defined to be zero outside their definition length. In polynomial notation, this convolution can be considered as the product of two polynomials of degree $N - 1$ with

$$Y(z) = H(z)X(z) \tag{3.88}$$

$$H(z) = \sum_{n=0}^{N-1} h_n z^n \tag{3.89}$$

$$X(z) = \sum_{m=0}^{N-1} x_m z^m \tag{3.90}$$

$$Y(z) = \sum_{l=0}^{2N-2} y_l z^l. \tag{3.91}$$

We assume now that N is composite, with $N = N_1 N_2$. In this case, the one-dimensional polynomials $H(z)$ and $X(z)$ are mapped into the two-dimensional polynomials $H(z, z_1)$ and $X(z, z_1)$ by redefining indices n and m and introducing a new polynomial variable z_1, defined by $z^{N_1} = z_1$

$$n = N_1 n_2 + n_1 \quad n_2, m_2 = 0, \ldots, N_2 - 1 \tag{3.92}$$

$$m = N_1 m_2 + m_1 \quad n_1, m_1 = 0, \ldots, N_1 - 1 \tag{3.93}$$

$$H(z, z_1) = \sum_{n_2=0}^{N_2-1} z_1^{n_2} \left(\sum_{n_1=0}^{N_1-1} h_{N_1 n_2 + n_1} z^{n_1} \right) = \sum_{n_2=0}^{N_2-1} H_{n_2}(z) z_1^{n_2} \tag{3.94}$$

$$X(z, z_1) = \sum_{m_2=0}^{N_2-1} z_1^{m_2} \left(\sum_{m_1=0}^{N_1-1} x_{N_1 m_2 + m_1} z^{m_1} \right) = \sum_{m_2=0}^{N_2-1} X_{m_2}(z) z_1^{m_2}. \tag{3.95}$$

Multiplying $H(z, z_1)$ by $X(z, z_1)$ yields a new two-dimensional polynomial $\tilde{Y}(z, z_1)$ defined by

$$\tilde{Y}(z, z_1) = H(z, z_1) X(z, z_1) \tag{3.96}$$

$$\tilde{Y}(z, z_1) = \sum_{s_2=0}^{2N_2-2} z_1^{s_2} \left(\sum_{s_1=0}^{2N_1-2} \tilde{y}_{N_1 s_2 + s_1} z^{s_1} \right) = \sum_{s_2=0}^{2N_2-2} Y_{s_2}(z) z_1^{s_2}. \tag{3.97}$$

The various samples of y_l are then obtained as the coefficients of z in $\tilde{Y}(z, z_1)$, after setting $z_1 = z^{N_1}$. Hence, the aperiodic convolution y_l can be considered as a two-dimensional convolution of arrays of size $N_1 \times N_2$. More precisely, this two-dimensional convolution is a one-dimensional convolution of length $2N_2 - 1$ where the N_2 input samples are replaced by N_2 polynomials of N_1 terms and all multiplications are replaced by aperiodic convolutions of two length-N_1 sequences

$$Y_{s_2}(z) = \sum_{n_2=0}^{N_2-1} H_{n_2}(z) X_{s_2 - n_2}(z). \tag{3.98}$$

Under these conditions, the convolution y_l is computed by a method somewhat similar to the nesting approach used for cyclic convolutions, but with short cyclic convolution algorithms replaced by aperiodic convolution algorithms. Thus, if M_1 and M_2 are the number of multiplications required for the aperiodic convolutions of length $2N_1 - 1$ and $2N_2 - 1$, then M, the number of multiplications corresponding to the aperiodic convolution y_l, is given by

$$M = M_1 M_2. \tag{3.99}$$

The number of additions is slightly more difficult to evaluate than for conventional nesting because here the nested convolutions are noncyclic. Hence, the input additions corresponding to the length $2N_2 - 1$ convolution algorithm are performed on polynomials of N_1 terms, while the output additions are done on polynomials of $2N_1 - 1$ terms. Moreover, since each polynomial $Y_{s_2}(z)$ of degree $2N_1 - 1$ is multiplied by $z_1^{s_2} = z^{N_1 s_2}$, the various polynomials $Y_{s_2}(z)$ overlap by $N_1 - 1$ samples. Let $A_{1,2}$ and $A_{2,2}$ be the number of input and output additions required for a length-$(2N_2 - 1)$ aperiodic convolution and A_1 be the total number of additions corresponding to the aperiodic convolution of length $2N_1 - 1$. Then, A, the total number of additions for the aperiodic convolution y_l, is given by

$$A = M_2 A_1 + A_{1,2} N_1 + A_{2,2}(2N_1 - 1) + (N_1 - 1)(2N_2 - 2). \tag{3.100}$$

When N is the product of more than two factors, the same formulation can be used recursively. Since the factors of N need not be relatively prime, N can be chosen to be a power of a prime, usually 2 or 3. In this case, with $N = 2^t$ or $N = 3^t$, the aperiodic convolution is computed by t identical stages calculating aperiodic convolutions of length 3 or 5. Only one short convolution algorithm needs be stored and computer implementation is greatly simplified by use of the regular structure of the radix-2 or radix-3 nesting algorithm.

We give, in Sect. 3.7.3, short algorithms which compute aperiodic convolution of sequences of lengths 2 and 3. For the first algorithm, the noncyclic convolution of length 2 is calculated with 3 multiplications when one of the sequences is fixed. Using this algorithm, the aperiodic convolution of two sequences of length $N = 2^t$ is computed with M multiplications, M being given by

$$M = 3^t. \tag{3.101}$$

The number of multiplications per input point is therefore equal to $(1.5)^t$. If the same aperiodic convolution is calculated by FFT, with 2 real sequences per FFT of dimension 2^{t+1}, as shown in Sect. 4.6, the number of real multiplications per input point is $3(2 + t)$. Since $3(2 + t)$ increases much more slowly with t than $(1.5)^t$, the FFT approach is preferred for long sequences. However, for small values of t, up to $t = 8$, $(1.5)^t$ is smaller than $3(2 + t)$ and the multidimensional noncyclic nesting method yields a lower number of multiplications than the conventional radix-2 FFT approach. This fact restricts the region of preferred applicability for the algorithm to aperiodic convolution lengths of up 256 input samples or less.

The radix-2 nesting algorithm can also be redefined to be a radix-3 nesting process which uses the short algorithm in Sect. 3.7.3. In this case, the aperiodic convolution of two sequences of length 3 is calculated with 5 multiplications and an aperiodic convolution of $N = 3^{t_1}$ input samples is computed in t_1 stages with $M = 5^{t_1}$ multiplications. Thus, the number of multiplications per input point is equal to $(5/3)^{t_1}$. This result can be compared, for convolutions of equal length, to that of the radix-2 algorithm by setting $3^{t_1} = 2^t$ which yields $t_1 = t \log_3 2$. Consequently, the number of multiplications per input point reduces to $(5/3)^{t \log_3 2} = (1.38)^t$ for an aperiodic convolution computed with the radix-3 nesting algorithm, as opposed to $(1.5)^t$ for a convolution of equal length calculated by a radix-2 algorithm. Thus, the number of multiplications increases less rapidly with convolution size for a radix-3 algorithm than for a radix-2 algorithm, as can be seen in Table 3.7 which lists the number of operations for various convolutions computed by radix-2 and radix-3 algorithms.

Table 3.7. Number of arithmetic operations for aperiodic convolutions computed by radix-2 and radix-3 one-dimensional to multidimensional mapping.

Length of input sequences N	Number of multiplications M	Number of additions A	Multiplications per input point M/N	Additions per input point A/N
2	3	3	1.50	1.50
3	5	20	1.67	6.67
4	9	19	2.25	4.75
8	27	81	3.37	10.12
9	25	194	2.78	21.56
16	81	295	5.06	18.44
27	125	1286	4.63	47.63
32	243	993	7.59	31.03
64	729	3199	11.39	49.98
81	625	7412	7.72	91.51
128	2187	10041	17.09	78.45
243	3125	40040	12.86	164.77
256	6561	31015	25.63	121.15

It can be seen, for instance, that a convolution of about 256 input terms is computed by a radix-3 algorithm with approximately half the number of multiplications corresponding to a radix-2 algorithm. Unfortunately, this saving is achieved at the expense of an increased number of additions so that the advantage of the radix-3 algorithm over the radix-2 algorithm is debatable.

A comparison with cyclic nesting techniques can be made by noting that the output of an N-taps digital filter can be computed by sectioning the input data sequence into successive blocks of N samples, calculating a series of aperiodic convolutions of these blocks with the sequence of tap values, and adding the overlapping output samples of these convolutions. With this method, the number of multiplications per output sample of the digital filter is the same as the number of multiplications per input point of the aperiodic convolutions, while the numbers of additions per point differ only by $(N - 1)/N$. Therefore, it can be seen that the implementation of a digital filtering process by noncyclic nesting methods usually requires a larger number of arithmetic operations than when cyclic methods are used. This difference becomes very significant in favor of cyclic nesting for large convolution lengths. Thus, while an aperiodic nesting method using a set of identical computation stages is inherently simpler to implement than the mixed radix method used with cyclic nesting, this advantage is offset by an increase in number of arithmetic operations.

The aperiodic nesting method described above can be considered as a generalization of the overlap-add algorithm. An analogous multidimensional formulation can also be developed for the overlap-save technique [3.8]. This yields aperiodic nesting algorithms that are very similar to those derived from the overlap-add algorithm and give about the same number of operations.

3.5 Computation of Convolutions by Recursive Nesting of Polynomials

The calculation of a convolution by cyclic or noncyclic nesting of small length convolutions has many desirable attributes. It requires fewer arithmetic operations than the FFT approach for sequence lengths of up to 200 samples and does not require complex arithmetic with sines and cosines. For large convolution lengths, however, cyclic and noncyclic nesting techniques are of limited interest because the number of operations increases exponentially with the number of stages, instead of linearly as for the FFT. In this section, we shall describe a computation method, based on the recursive nesting of irreducible polynomials [3.9], which retains to some extent the regular structure of the aperiodic nesting approach but, nevertheless, yields a number of operations which increases only linearly with the number of stages. We first discuss the case of a length-N circular convolution, with $N = 2^t$,

$$y_l = \sum_{n=0}^{N-1} h_n x_{l-n}, \quad l = 0, \ldots, N-1. \tag{3.102}$$

In polynomial notation, each output sample y_l corresponds to the coefficient of z^l in the product modulo $(z^N - 1)$ of two polynomials $H(z)$ and $X(z)$

$$Y(z) \equiv H(z)X(z) \text{ modulo } (z^N - 1) \tag{3.103}$$

$$H(z) = \sum_{n=0}^{N-1} h_n z^n \tag{3.104}$$

$$X(z) = \sum_{m=0}^{N-1} x_m z^m \tag{3.105}$$

$$Y(z) = \sum_{l=0}^{N-1} y_l z^l \tag{3.106}$$

3.5 Computation of Convolutions by Recursive Nesting of Polynomials

In the field of rational numbers, $z^N - 1$ factors into $t + 1$ irreducible polynomials

$$z^N - 1 = (z - 1) \prod_{v=0}^{t-1} (z^{2^v} + 1). \tag{3.107}$$

Thus, the computation of $Y(z)$ can be accomplished via the Chinese remainder theorem. In this case, the main part of the calculation consists in evaluating the $t + 1$ polynomial products modulo $(z^{2^{t-1}} + 1), \ldots, (z + 1), (z - 1)$. For the polynomial products modulo $(z + 1), (z - 1)$, however, we have $z \equiv \pm 1$ and the polynomial multiplications reduce to simple scalar multiplications.

We shall show now that, for higher order polynomials, the process can be greatly simplified by using a recursive nesting technique. We have seen in Sect. 3.2.2 that if $N = N_1 N_2$, a polynomial product modulo $(z^N + 1)$ could be computed as a polynomial product modulo $(z^{N_1} - z_1)$ in which the scalars are replaced by polynomials of N_2 terms evaluated modulo $(z_1^{N_2} + 1)$. If $N = 2^t$ and $N_1 \leq N_2$, the polynomial product modulo $(z^{N_1} - z_1)$ can be computed by interpolation on $z = 0$, $1/z = 0$ and powers of z_1, with $2N_1 - 1$ multiplications and $N_1 - 11 + 4N_1 \log_2 N_1$ additions. This means that, if M_1 and A_1 are the number of multiplications and additions corresponding to the polynomial product modulo $(z_1^{N_2} + 1)$, and if $N_1 = 2^{t_1}$, $N_2 = 2^{t_2}$, the number of multiplications M and additions A required to evaluate the polynomial product modulo $(z^N + 1)$ is

$$M = (2^{t_1+1} - 1)M_1 \tag{3.108}$$

$$A = (2^{t_1} - 11 + t_1 2^{t_1+2})2^{t_2} + (2^{t_1+1} - 1)A_1. \tag{3.109}$$

The same method can be extended to compute a polynomial product modulo $(z^{N_1^2 N_2} + 1)$ by nesting a polynomial product modulo $(z^{N_1} - z_1)$ with a polynomial product modulo $(z_1^{N_2 N_1} + 1)$ which is computed as indicated above. Thus, a polynomial product modulo $(z^N + 1)$, with $N = N_1^d N_2$ and $N_2 < N_1$, is calculated recursively by using d identical stages implementing polynomial products modulo $(z^{N_1} - z_1)$, where the scalars are replaced in the first stage by polynomials of $N_1^{d-1} N_2$ terms, in the second stage by polynomials of $N_1^{d-2} N_2$ terms, and in the last stage by polynomials of N_2 terms.

In this case, the total number of multiplications M for the polynomial product of dimension $N = 2^t = N_1^d N_2$, computed by a radix-N_1 algorithm, is

$$M = (2^{t_1+1} - 1)^d M_1 \tag{3.110}$$

or, with $dt_1 + t_2 = t$,

$$M = M_1 (2^{t_1+1} - 1)^{(t-t_2)/t_1}. \tag{3.111}$$

Since $(2^{t_1+1} - 1)^{1/t_1}$ is larger than 2, the number of multiplications M increases exponentially with t. If we take, for instance, $t_1 = t_2 = 1$ and $M_1 = 3$, then $M = 3^t$ and $M/N = (1.5)^t$. This result is very similar to that obtained for aperiodic convolutions.

The foregoing recursive method, however, can be greatly improved by using a mixed radix approach based upon a set of exponentially increasing radices. Consider, for example, the computation of a polynomial product modulo $(z^N + 1)$, with $N = 2^t$ and $t = 2^d$. We begin the calculation of this polynomial product with a polynomial product modulo $(z^2 + 1)$ evaluated with 3 multiplications as shown in Sect. 3.7.2. Then, as shown above, a polynomial product modulo $(z^4 + 1)$ can be computed modulo $(z^2 - z_1)$, $(z_1^2 + 1)$ with 9 multiplications. Next, instead of calculating a polynomial product modulo $(z^8 + 1)$ as in the fixed radix scheme, we compute directly a polynomial product modulo $(z^{16} + 1)$ as a polynomial product modulo $(z^4 - z_1)$ of polynomials defined modulo $(z^4 + 1)$. This procedure is then continued recursively, each stage being implemented with a radix which is the square of the radix of the preceding stage. Finally, at the stage v, the polynomial product modulo $(z^{2^{2^v+1}} + 1)$ is computed modulo $(z^{2^{2^v}} - z_1)$, $(z_1^{2^{2^v}} + 1)$ with $(2^{2^v+1} - 1) M_1$ multiplications, where M_1 is the number of multiplications for the polynomial product modulo $(z^{2^{2^v}} + 1)$. Therefore, the total number of multiplications for the polynomial product modulo $(z^N + 1)$ is

$$M = 3 \prod_{v=0}^{d-1} (2^{2^v+1} - 1) \tag{3.112}$$

Thus,

$$M < 9 \cdot 2^{d-3} \cdot 2^{2^d} \tag{3.113}$$

and, since $N = 2^{2^d}$, the number of multiplications per output sample, M/N, satisfies the inequality

$$M/N < (9/8) \log_2 N. \tag{3.114}$$

Thus, with this approach, the number of multiplications increases much more slowly with N than for other methods which are based upon the use of constant radices, or conventional cyclic or noncyclic nesting. Note that, except for the constant multiplicative factor 9/8, the law defined by (3.114) is essentially the same as that for convolution via the FFT method.

The mixed radix nesting technique is not limited to dimensions N such that $N = 2^t$, $t = 2^d$. Vector lengths with $t \neq 2^d$ can be accommodated with a slight loss of efficiency by using an initial polynomial other than $(z^2 + 1)$ and a set of increasing radices that are an approximation of an exponential law. We list in Table 3.8 the arithmetic operation count for various polynomial products that can be computed by this technique.

3.5 Computation of Convolutions by Recursive Nesting of Polynomials

Table 3.8. Number of arithmetic operations for polynomial products modulo $(z^N + 1)$, $N = 2^t$, computed by mixed radix nesting.

Ring	Radices	Number of multiplications	Number of additions
$z^2 + 1$	2	3	3
$z^4 + 1$	2, 2	9	15
$z^8 + 1$	2, 2, 2	27	57
$z^{16} + 1$	2, 2, 4	63	205
$z^{32} + 1$	2, 2, 2, 4	189	599
$z^{64} + 1$	2, 2, 2, 8	405	1599
$z^{128} + 1$	2, 2, 4, 8	945	4563
$z^{256} + 1$	2, 2, 4, 16	1953	10531
$z^{512} + 1$	2, 2, 2, 4, 16	5859	26921
$z^{1024} + 1$	2, 2, 2, 4, 32	11907	58889
$z^{2048} + 1$	2, 2, 2, 8, 32	25515	143041
$z^{4096} + 1$	2, 2, 2, 8, 64	51435	304769

The number of operations for circular convolutions of length N, with $N = 2^t$, can easily be deduced recursively from the data given in Table 3.8 by noting that a circular convolution of length $2N$ is computed by reducing the input sequences modulo $(z^N - 1)$ and $(z^N + 1)$, evaluating a circular convolution of N terms and a polynomial product modulo $(z^N + 1)$, and reconstructing the output samples by the Chinese remainder theorem. When one of the input sequences is fixed, the Chinese remainder operation can be viewed as the inverse operation of the reductions, and the total number of additions for reductions and Chinese remainder operations is $4N$. Thus, all cyclic convolutions of length 2^t are computed recursively from only two short algorithms, corresponding to the convolution of

Table 3.9. Optimum block size and number of operations for digital filters computed by circular convolutions of N terms, $N = 2^t$, and mixed radix nesting of polynomials.

Filter tap length N_1	Multiplications per point $M_2(N, N_1)$	Additions per point $A_2(N, N_1)$	Optimum block size N
2	2.00	4.43	8
4	2.80	9.80	8
8	4.16	16.43	32
16	5.98	23.39	64
32	7.19	31.11	128
64	8.00	43.83	512
128	9.34	51.28	512
256	11.91	62.37	2048
512	12.80	76.09	8192

2 terms and the polynomial product modulo ($z^2 + 1$) given, respectively, in Sects. 3.7.1 and 3.7.2. Using this approach, a circular convolution of 1024 points is computed with 9455 multiplications and 48585 additions.

We give in Table 3.9 the number of operations per output sample for digital filters with fixed tap coefficients computed by the overlap-add algorithm with circular convolutions evaluated by the mixed radix nesting of polynomials. It can be seen, by comparison with Table 3.6 that this method yields a smaller operation count than the split nesting technique for digital filters of more than 64 taps. It should be noted that these computational savings are achieved along with a simpler computational structure, since all the algorithms are designed to support vector lengths which are powers of two, as opposed to the product of relatively prime radices for split nesting. Another advantage of the mixed radix nesting of polynomials is that only two very short algorithms need to be stored, all other algorithms being implemented by a single FFT-type routine, as described in Sect. 3.2.2.

3.6 Distributed Arithmetic

Thus far, we have restricted our discussion of fast convolution algorithms to algebraic methods. These algorithms can be used with any kind of arithmetic and are easily programmed on standard computers with conventional binary arithmetic. We shall see now that significant additional savings can be achieved if the algorithms are reformulated at the bit level. In this reformulation, a B-bit word is represented as a polynomial of B binary coded terms, and the arithmetic is redistributed throughout the algorithm structure [3.10, 11].

With this *distributed arithmetic approach*, an aperiodic scalar convolution is converted into a two-dimensional convolution. To demonstrate this point, we focus initially on the convolution y_l of two length-N data sequences x_m and h_n

$$y_l = \sum_{n=0}^{N-1} h_n x_{l-n}, \qquad l = 0, ..., 2N - 2. \tag{3.115}$$

We assume now that each word of the input sequences is binary coded with B bits

$$h_n = \sum_{b_1=0}^{B-1} h_{n,b_1} 2^{b_1}, \qquad h_{n,b_1} \in \{0, 1\} \tag{3.116}$$

$$x_m = \sum_{b_2=0}^{B-1} x_{m,b_2} 2^{b_2}, \qquad x_{m,b_2} \in \{0, 1\}. \tag{3.117}$$

Substituting (3.116) and (3.117) into (3.115) yields

$$y_l = \sum_{n=0}^{N-1} \sum_{b_1=0}^{B-1} \sum_{b_2=0}^{B-1} h_{n,b_1} x_{l-n,b_2} 2^{b_1+b_2}. \tag{3.118}$$

3.6 Distributed Arithmetic

Changing the coordinates with $b_3 = b_1 + b_2$ gives

$$y_l = \sum_{b_3=0}^{2B-2} \left(\sum_{n=0}^{N-1} \sum_{b_1=0}^{B-1} h_{n,b_1} x_{l-n, b_3-b_1} \right) 2^{b_3} \tag{3.119}$$

which demonstrates that y_l can indeed be considered as the two-dimensional aperiodic convolution of the two binary arrays of $N \times B$ terms, h_{n,b_1} and x_{m,b_2}. Note that in this formulation, each digit $y_{l,b}$ of y_l is no longer binary coded so that the operation defined by (3.119) must be followed by a code conversion if y_l is to be defined in the same format as h_n and x_m.

It can be seen that the computation of y_l by (3.119) uses the well-known fact that a scalar multiplication can be viewed as the convolution of two binary sequences [3.12]. The full potential of this procedure becomes more apparent when the order of operations is modified. By changing the order of operations in (3.118), we obtain

$$y_l = \sum_{b_1=0}^{B-1} \left[\sum_{n=0}^{N-1} h_{n,b_1} \left(\sum_{b_2=0}^{B-1} x_{l-n, b_2} 2^{b_2} \right) \right] 2^{b_1} \tag{3.120}$$

This shows that y_l is obtained by summing along dimension b_1, the B convolutions y_{l,b_1} defined by

$$y_{l,b_1} = \sum_{n=0}^{N-1} h_{n,b_1} \left(\sum_{b_2=0}^{B-1} x_{l-n, b_2} 2^{b_2} \right). \tag{3.121}$$

Each word y_{l,b_1} is a length-N convolution of the N-bit sequence h_{n,b_1} with the N-word sequence $\left\{ \sum_{b_2=0}^{B-1} x_{l-n, b_2} 2^{b_2} = x_{l-n} \right\}$. Thus, each word of y_{l,b_1} is obtained by multiplying x_{l-n} by h_{n,b_1} and summing along dimension n. Since h_{n,b_1} can only be 0 or 1, multiplication by h_{n,b_1} is greatly simplified and can be considered as a simple addition of N words, where some of the words are zero.

When the sequence x_m is fixed, large savings in number of arithmetic operations can be achieved by precomputing and storing all possible combinations of y_{l,b_1}. Since h_{n,b_1} can only be 0 or 1, the number of such combinations is equal to 2^N. Thus, by storing the 2^N possible combinations y_{l,b_1}, the various values of y_{l,b_1} are obtained by a simple table look-up addressed by the N bits h_{n,b_1} and the computation of y_l reduces to a simple shift-addition of B words. This is equivalent in hardware complexity to a conventional binary multiplier, except for the memory required to store the 2^N combinations of y_{l,b_1}.

Thus, the hardware required to implement a short length convolution in distributed arithmetic is not particularly more complex than that corresponding to an ordinary multiplication. In practice, the amount of memory required to store the precomputed partial products y_{l,b_1} can be halved by coding the words h_n with bit values equal to ± 1 instead of 0, 1. In this case, the 2^N partial products y_{l,b_1} divide into 2^{N-1} amplitude values which can take opposite signs, and only the 2^{N-1} amplitude values need to be stored, provided that the sign inversion is implemented in hardware. Using this method, only 4 or 8 coefficients need to be stored for the aperiodic convolution of length-3 or length-4 input sequences. Larger convolutions are usually implemented by segmenting the input sequences in groups of 3 or 4 samples in order to avoid excessive memory requirements. Hence, replacing the direct computation of a convolution by a distributed arithmetic structure can yield a performance improvement factor of 3 to 4, for approximately the same amount of hardware. Then, in turn, the various nesting algorithms used for the evaluation of large convolutions can also be implemented in distributed arithmetic with comparable improvement factors.

Thus, the complementary blending of nesting algorithms with a distributed arithmetic structure provides an attractive combination for the implementation of digital filters in special purpose hardware. In practice, the concept of distributing the arithmetic operations at the bit level throughout a convolution can be applied in a number of different ways. We shall see in Chap. 8 that a particularly interesting case occurs in the computation of circular convolutions via modular arithmetic. In this case, the two-dimensional formulation of scalar convolutions in distributed arithmetic leads to the definition of number theoretic transforms which can greatly simplify the calculation procedure.

3.7 Short Convolution and Polynomial Product Algorithms

A detailed description of frequently used algorithms for short convolutions and polynomial products is given in this section. These algorithms have been designed to minimize the number of multiplications while avoiding an excessive number of additions. The input sequences are labelled x_m and h_n. Since h_n is assumed to be fixed, expressions involving h_n are presumed to be precomputed and stored. The output sequence is y_l. Expressions written between parentheses indicate grouping of additions. Input and output additions must be executed in the index numerical order. Additional polynomial product algorithms are given in Appendix B.

3.7.1 Short Circular Convolution Algorithms

Convolution of 2 Terms
2 multiplications, 4 additions

$a_0 = x_0 + x_1$ $\qquad\qquad b_0 = (h_0 + h_1)/2$
$a_1 = x_0 - x_1$ $\qquad\qquad b_1 = (h_0 - h_1)/2$
$\qquad m_k = a_k b_k, \quad k = 0,1$
$y_0 = m_0 + m_1$
$y_1 = m_0 - m_1$

Convolution of 3 Terms

4 multiplications, 11 additions

$a_0 = x_0 + x_1 + x_2$ $b_0 = (h_0 + h_1 + h_2)/3$
$a_1 = x_0 - x_2$ $b_1 = h_0 - h_2$
$a_2 = x_1 - x_2$ $b_2 = h_1 - h_2$
$a_3 = a_1 + a_2$ $b_3 = (b_1 + b_2)/3$
$\qquad m_k = a_k b_k, \quad k = 0, \ldots, 3$
$u_0 = m_1 - m_3$ $u_1 = m_2 - m_3$

$y_0 = m_0 + u_0$
$y_1 = m_0 - u_0 - u_1$
$y_2 = m_0 + u_1$

Convolution of 4 Terms

5 multiplications, 15 additions

$a_0 = x_0 + x_2$ $b_0 = h_0 + h_2$
$a_1 = x_1 + x_3$ $b_1 = h_1 + h_3$
$a_2 = a_0 + a_1$ $b_2 = (b_0 + b_1)/4$
$a_3 = a_0 - a_1$ $b_3 = (b_0 - b_1)/4$
$a_4 = x_0 - x_2$ $b_4 = [(h_0 - h_2) - (h_1 - h_3)]/2$
$a_5 = x_1 - x_3$ $b_5 = [(h_0 - h_2) + (h_1 - h_3)]/2$
$a_6 = a_4 + a_5$ $b_6 = (h_0 - h_2)/2$
$\qquad m_k = a_{k+2} b_{k+2}, \quad k = 0, \ldots, 4$
$u_0 = m_0 + m_1$ $u_2 = m_4 - m_3$
$u_1 = m_0 - m_1$ $u_3 = m_4 - m_2$

$y_0 = u_0 + u_2$
$y_1 = u_1 + u_3$
$y_2 = u_0 - u_2$
$y_3 = u_1 - u_3$

Convolution of 5 Terms

10 multiplications, 31 additions

$a_0 = x_0 - x_4$ $b_0 = h_0 - h_2 + h_3 - h_4$
$a_1 = x_1 - x_4$ $b_1 = h_1 - h_2 + h_3 - h_4$
$a_2 = a_0 + a_1$ $b_2 = (-2h_0 - 2h_1 + 3h_2 - 2h_3 + 3h_4)/5$
$a_3 = x_2 - x_4$ $b_3 = -h_0 + h_1 - h_2 + h_3$

68   3. Fast Convolution Algorithms

$a_4 = x_3 - x_4$

$a_5 = a_3 + a_4$

$a_6 = a_0 - a_3$

$a_7 = a_1 - a_4$

$a_8 = a_2 - a_5$

$a_9 = x_0 + x_1 + x_2 + x_3 + x_4$

$b_4 = -h_0 + h_1 - h_2 + h_4$

$b_5 = (3h_0 - 2h_1 + 3h_2 - 2h_3 - 2h_4)/5$

$b_6 = -h_2 + h_3$

$b_7 = h_1 - h_2$

$b_8 = (-h_0 - h_1 + 4h_2 - h_3 - h_4)/5$

$b_9 = (h_0 + h_1 + h_2 + h_3 + h_4)/5$

$m_k = a_k b_k, \quad k = 0, \ldots, 9$

$u_0 = m_0 + m_2$

$u_1 = m_1 + m_2$

$u_2 = m_3 + m_5$

$u_3 = m_4 + m_5$

$u_4 = m_6 + m_8$

$u_5 = m_7 + m_8$

$y_0 = u_0 - u_4 + m_9$

$y_1 = -u_0 - u_1 - u_2 - u_3 + m_9$

$y_2 = u_3 + u_5 + m_9$

$y_3 = u_2 + u_4 + m_9$

$y_4 = u_1 - u_5 + m_9$

Convolution of 7 Terms
16 multiplications, 70 additions

$a_0 = x_1 - x_6$

$a_1 = x_2 - x_6$

$a_2 = x_4 - x_6$

$a_3 = x_5 - x_6$

$a_4 = a_0 + a_1$

$a_5 = -a_0 + a_1$

$a_6 = a_2 + a_3$

$a_7 = -a_2 + a_3$

$a_8 = x_0 - x_6$

$a_9 = a_8 + a_4$

$a_{10} = a_8 + a_5$

$a_{11} = a_9 + a_4 + a_4 + a_5$

$a_{12} = a_1$

$a_{13} = x_3 - x_6$

$b_0 = (-h_6 - 2h_5 + 3h_4 - h_3 - 2h_2 + h_1 + 2h_0)/2$

$b_1 = (10h_6 + 3h_5 - 11h_4 + 10h_3 + 3h_2 - 11h_1 - 4h_0)/14$

$b_2 = (-2h_6 + 3h_5 - h_4 - 2h_3 + 3h_2 - h_1)/6$

$b_3 = (-h_6 + h_4 - h_3 + h_1)/6$

$b_4 = 2h_6 - h_5 - 2h_4 + 3h_3 - h_2 - 2h_1 + h_0$

$b_5 = (-2h_6 + h_5 + 2h_4 - h_3 - 2h_2 + 3h_1 - h_0)/2$

3.7 Short Convolution and Polynomial Product Algorithms

$a_{14} = a_{13} + a_6$

$b_6 = (3h_6 - 11h_5 - 4h_4 + 10h_3 + 3h_2 - 11h_1 + 10h_0)/14$

$a_{15} = a_{13} + a_7$

$b_7 = (3h_6 - h_5 - 2h_3 + 3h_2 - h_1 - 2h_0)/6$

$a_{16} = a_{14} + a_6 + a_6 + a_7$ $b_8 = (h_5 - h_3 + h_1 - h_0)/6$

$a_{17} = a_3$

$b_9 = -h_6 - 2h_5 + h_4 + 2h_3 - h_2 - 2h_1 + 3h_0$

$a_{18} = a_{13} - a_8$

$b_{10} = (2h_4 - h_3 - 2h_2 + h_1)/2$

$a_{19} = a_{14} - a_9$

$b_{11} = (-2h_6 - 2h_5 - 2h_4 + 12h_3 + 5h_2 - 9h_1 - 2h_0)/14$

$a_{20} = a_{15} - a_{10}$

$b_{12} = (-2h_3 + 3h_2 - h_1)/6$

$a_{21} = a_{16} - a_{11}$

$b_{13} = (-h_3 + h_1)/6$

$a_{22} = a_{17} - a_{12}$

$b_{14} = 2h_3 - h_2 - 2h_1 + h_0$

$a_{23} = a_{19} + (x_2 + x_1 + x_0) + (x_2 + x_1 + x_0) + x_6$

$b_{15} = (h_6 + h_5 + h_4 + h_3 + h_2 + h_1 + h_0)/7$

$m_k = a_{k+8} b_k, \quad k = 0, \ldots, 15$

$u_0 = m_0 + m_{10}$ $u_{12} = u_0 + u_{11}$

$u_1 = m_1 + m_{11}$ $u_{13} = u_{10} + u_3$

$u_2 = m_2 + m_{12}$ $u_{14} = u_{13} - u_2$

$u_3 = m_3 + m_{13}$ $u_{15} = (u_{13} + u_3 + u_3 + u_4) + u_2$

$u_4 = m_4 + m_{14}$ $u_{16} = -u_{12} - u_{13} - (u_{13} + u_3 + u_3 + u_4)$

$u_5 = m_5 - m_{10}$ $u_{17} = u_6 + u_8$

$u_6 = m_6 - m_{11}$ $u_{18} = u_{17} + u_7$

$u_7 = m_7 - m_{12}$ $u_{19} = u_5 + u_{18}$

$u_8 = m_8 - m_{13}$ $u_{20} = u_{17} + u_8$

$u_9 = m_9 - m_{14}$ $u_{21} = u_{20} - u_7$

$u_{10} = u_1 + u_3$ $u_{22} = (u_{20} + u_8 + u_8 + u_9) + u_7$

$u_{11} = u_{10} + u_2$ $u_{23} = -u_{19} - u_{20} - (u_{20} + u_8 + u_8 + u_9)$

$y_0 = u_{12} + m_{15}$

$y_1 = u_{16} + u_{23} + m_{15}$

$y_2 = u_{22} + m_{15}$

$y_3 = u_{21} + m_{15}$

$y_4 = u_{19} + m_{15}$

$y_5 = u_{15} + m_{15}$

$y_6 = u_{14} + m_{15}$

Convolution of 8 Terms
14 multiplications, 46 additions

$a_0 = x_0 + x_4$ $\qquad$ $b_0 = h_0 + h_4$

$a_1 = x_1 + x_5$ $\qquad$ $b_1 = h_1 + h_5$

$a_2 = x_2 + x_6$ $\qquad$ $b_2 = h_2 + h_6$

$a_3 = x_3 + x_7$ $\qquad$ $b_3 = h_3 + h_7$

$a_4 = a_0 + a_2$ $\qquad$ $b_4 = b_0 + b_2$

$a_5 = a_1 + a_3$ $\qquad$ $b_5 = b_1 + b_3$

$a_6 = x_0 - x_4$ $\qquad$ $b_6 = \{[-(h_0 - h_4) + (h_2 - h_6)] - [(h_1 - h_5) - (h_3 - h_7)]\}/2$

$a_7 = x_1 - x_5$ $\qquad$ $b_7 = \{[-(h_0 - h_4) + (h_2 - h_6)] + [(h_1 - h_5) + (h_3 - h_7)]\}/2$

$a_8 = x_2 - x_6$ $\qquad$ $b_8 = \{[(h_0 - h_4) + (h_2 - h_6)] + [(h_1 - h_5) + (h_3 - h_7)]\}/2$

$a_9 = x_3 - x_7$ $\qquad$ $b_9 = \{[(h_0 - h_4) + (h_2 - h_6)] + [(h_1 - h_5) - (h_3 - h_7)]\}/2$

$a_{10} = a_0 - a_2$ $\qquad$ $b_{10} = [-(b_0 - b_2) + (b_1 - b_3)]/4$

$a_{11} = a_1 - a_3$ $\qquad$ $b_{11} = [(b_0 - b_2) + (b_1 - b_3)]/4$

$a_{12} = a_4 + a_5$ $\qquad$ $b_{12} = (b_4 + b_5)/8$

$a_{13} = a_4 - a_5$ $\qquad$ $b_{13} = (b_4 - b_5)/8$

$a_{14} = a_7 + a_9$ $\qquad$ $b_{14} = [(h_0 - h_4) - (h_3 - h_7)]/2$

$a_{15} = a_6 + a_8$ $\qquad$ $b_{15} = [(h_0 - h_4) + (h_1 - h_5)]/2$

$a_{16} = a_{15} - a_{14}$ $\qquad$ $b_{16} = (h_0 - h_4)/2$

$a_{17} = a_8 - a_9$ $\qquad$ $b_{17} = [(h_0 - h_4) + (h_2 - h_6)]/2$

$a_{18} = a_6 - a_7$ $\qquad$ $b_{18} = [-(h_0 - h_4) + (h_2 - h_6)]/2$

$a_{19} = a_{10} + a_{11}$ $\qquad$ $b_{19} = (b_0 - b_2)/4$

$\qquad m_k = a_{k+6} b_{k+6}, \quad k = 0, \ldots, 13$

$u_0 = m_8 + m_{10}$ $\qquad$ $u_9 = m_6 - m_7$

$u_1 = m_9 - m_{10}$ $\qquad$ $u_{10} = u_0 - u_2$

$u_2 = m_3 + m_{11}$ $\qquad$ $u_{11} = u_6 + u_8$

$u_3 = m_{11} - m_2$ $\qquad$ $u_{12} = u_1 + u_3$

$u_4 = m_1 + m_{12}$ $\qquad$ $u_{13} = u_7 + u_9$

$u_5 = m_0 - m_{12}$ $\qquad$ $u_{14} = u_0 + u_4$

$u_6 = m_{13} - m_5$ $\qquad$ $u_{15} = -u_6 + u_8$

$u_7 = m_{13} + m_4$ $\qquad u_{16} = u_1 + u_5$
$u_8 = m_7 + m_6$ $\qquad u_{17} = -u_7 + u_9$

$y_0 = u_{10} + u_{11}$
$y_1 = u_{12} + u_{13}$
$y_2 = u_{14} + u_{15}$
$y_3 = u_{16} + u_{17}$
$y_4 = -u_{10} + u_{11}$
$y_5 = -u_{12} + u_{13}$
$y_6 = -u_{14} + u_{15}$
$y_7 = -u_{16} + u_{17}$

Convolution of 9 Terms
19 multiplications, 74 additions

$a_0 = x_0 - x_6$ $\qquad b_0 = -h_0 - h_3 + 2h_6$
$a_1 = x_1 - x_7$ $\qquad b_1 = -h_1 - h_4 + 2h_7$
$a_2 = x_2 - x_8$ $\qquad b_2 = -h_2 - h_5 + 2h_8$
$a_3 = x_3 - x_6$ $\qquad b_3 = h_0 - 2h_3 + h_6$
$a_4 = x_4 - x_7$ $\qquad b_4 = h_1 - 2h_4 + h_7$
$a_5 = x_5 - x_8$ $\qquad b_5 = h_2 - 2h_5 + h_8$
$a_6 = x_0 + x_3 + x_6$ $\qquad b_6 = h_0 + h_3 + h_6$
$a_7 = x_1 + x_4 + x_7$ $\qquad b_7 = h_1 + h_4 + h_7$
$a_8 = x_2 + x_5 + x_8$ $\qquad b_8 = h_2 + h_5 + h_8$
$a_9 = a_0 + a_2$
$a_{10} = a_3 + a_5$
$a_{11} = a_6 + a_7 + a_8$ $\qquad b_9 = (b_6 + b_7 + b_8)/9$
$a_{12} = a_{10} + a_4$ $\qquad b_{10} = (b_0 + 3b_1 + 2b_2 - 2b_3 - 3b_4 - b_5)/18$
$a_{13} = a_9 + a_1$ $\qquad b_{11} = (b_0 - b_2 + b_3 + 3b_4 + 2b_5)/18$
$a_{14} = a_{13} - a_{12}$ $\qquad b_{12} = b_{10} + b_{11}$
$a_{15} = a_{10} - a_4$ $\qquad b_{13} = (-b_0 + b_1 - b_4 + b_5)/6$
$a_{16} = a_9 - a_1$ $\qquad b_{14} = (b_0 - b_2 - b_3 + b_4)/6$
$a_{17} = a_{16} - a_{15}$ $\qquad b_{15} = b_{13} + b_{14}$
$a_{18} = a_3$ $\qquad b_{16} = (2b_0 + b_1 - b_2 - 2b_3 + b_5)/3$
$a_{19} = a_0 - a_3$ $\qquad b_{17} = (2b_0 - b_2 + b_4)/3$

$a_{20} = a_0$

$a_{21} = a_5$

$a_{22} = a_2 - a_5$

$a_{23} = a_2$

$a_{24} = -a_{22} + a_0 - a_4$

$a_{25} = a_{19} + a_5 - a_1$

$a_{26} = -a_{25} + a_{24}$

$a_{27} = a_6 - a_8$

$a_{28} = a_7 - a_8$

$a_{29} = a_{27} + a_{28}$

$b_{18} = b_{17} - b_{16}$

$b_{19} = (b_0 - b_1 - 2b_2 + b_4)/3$

$b_{20} = (-b_1 + b_3 - 2b_5)/3$

$b_{21} = b_{20} - b_{19}$

$b_{22} = (b_0 - b_2 - 2b_3 + 2b_5)/9$

$b_{23} = (-b_0 + b_2 - b_3 + b_5)/9$

$b_{24} = b_{23} - b_{22}$

$b_{25} = (b_6 - b_8)/3$

$b_{26} = (b_7 - b_8)/3$

$b_{27} = (b_{25} + b_{26})/3$

$m_k = a_{k+11} \, b_{k+9}, \quad k = 0, \ldots, 18$

$u_0 = m_1 + m_2$

$u_1 = m_4 + m_5$

$u_2 = m_{14} + m_{15}$

$u_3 = u_0 + u_1$

$u_4 = m_1 + m_3$

$u_5 = m_4 + m_6$

$u_6 = m_{13} + m_{15}$

$u_7 = -u_3 + m_7$

$u_8 = u_4 + u_5$

$u_9 = m_{10} - u_6$

$u_{10} = m_8 + u_2 + u_7$

$u_{11} = u_8 + m_{11} + u_9$

$u_{12} = u_4 - u_5 + u_2$

$u_{13} = u_7 + u_8 + m_9 + u_6$

$u_{14} = u_3 + m_{12} + u_9 + u_2$

$u_{15} = u_0 - u_1 + u_6$

$u_{16} = m_{16} - m_{18}$

$u_{17} = m_{17} - m_{18}$

$u_{18} = m_0 + u_{16}$

$u_{19} = m_0 - u_{16} - u_{17}$

$u_{20} = m_0 + u_{17}$

$y_0 = u_{13} - u_{10} + u_{18}$

$y_1 = u_{14} - u_{11} + u_{19}$

$y_2 = u_{15} - u_{12} + u_{20}$

$y_3 = -u_{13} + u_{18}$

$y_4 = -u_{14} + u_{19}$

$y_5 = -u_{15} + u_{20}$

$y_6 = u_{10} + u_{18}$

$y_7 = u_{11} + u_{19}$

$y_8 = u_{12} + u_{20}$

3.7.2 Short Polynomial Product Algorithms

Polynomial Product modulo ($z^2 + 1$)
3 multiplications, 3 additions

$a_0 = x_0 + x_1$ $\qquad$ $b_0 = h_0$
$a_1 = x_1$ $\qquad$ $b_1 = h_0 + h_1$
$a_2 = x_0$ $\qquad$ $b_2 = h_1 - h_0$
$\qquad m_k = a_k b_k \qquad k = 0, 1, 2$
$y_0 = m_0 - m_1$
$y_1 = m_0 + m_2$

Polynomial Product Modulo ($z^3 - 1$)/($z - 1$)
3 multiplications, 3 additions

$a_0 = x_1$ $\qquad$ $b_0 = h_0 - h_1$
$a_1 = x_0 - x_1$ $\qquad$ $b_1 = h_0$
$a_2 = x_0$ $\qquad$ $b_2 = h_1$
$\qquad m_k = a_k b_k \qquad k = 0, 1, 2$
$y_0 = m_0 + m_1$
$y_1 = m_0 + m_2$

Polynomial Product Modulo ($z^4 + 1$)
9 multiplications, 15 additions

$a_0 = (x_1 + x_3)$ $\qquad$ $b_0 = h_0 - h_3$
$a_1 = (x_0 + x_2) - (x_1 + x_3)$ $\qquad$ $b_1 = h_0$
$a_2 = (x_0 + x_2)$ $\qquad$ $b_2 = h_0 + h_1$
$a_3 = x_3$ $\qquad$ $b_3 = h_0 + h_2 + h_1 - h_3$
$a_4 = x_2 - x_3$ $\qquad$ $b_4 = h_0 + h_2$
$a_5 = x_2$ $\qquad$ $b_5 = h_0 + h_2 + h_1 + h_3$
$a_6 = x_1$ $\qquad$ $b_6 = -h_0 + h_2 + h_1 + h_3$
$a_7 = x_0 - x_1$ $\qquad$ $b_7 = -h_0 + h_2$
$a_8 = x_0$ $\qquad$ $b_8 = -h_0 + h_2 - h_1 + h_3$
$\qquad m_k = a_k b_k \qquad k = 0, ..., 8$
$y_0 = (m_0 + m_1) - (m_3 + m_4)$
$y_1 = (m_2 - m_1) + (m_4 - m_5)$
$y_2 = (m_0 + m_1) + (m_6 + m_7)$
$y_3 = (m_2 - m_1) + (m_8 - m_7)$

3. Fast Convolution Algorithms

Polynomial Product Modulo $(z^5 - 1)/(z - 1)$
9 multiplications, 16 additions

$a_0 = x_0$ $\qquad\qquad\qquad$ $b_0 = h_0$
$a_1 = x_1$ $\qquad\qquad\qquad$ $b_1 = h_1$
$a_2 = x_0 - x_1$ $\qquad\qquad$ $b_2 = -h_0 + h_1$
$a_3 = x_2$ $\qquad\qquad\qquad$ $b_3 = h_2$
$a_4 = x_3$ $\qquad\qquad\qquad$ $b_4 = h_3$
$a_5 = x_2 - x_3$ $\qquad\qquad$ $b_5 = -h_2 + h_3$
$a_6 = x_0 - x_2$ $\qquad\qquad$ $b_6 = h_2 - h_0$
$a_7 = x_1 - x_3$ $\qquad\qquad$ $b_7 = h_3 - h_1$
$a_8 = -a_6 + a_7$ $\qquad\qquad$ $b_8 = b_6 - b_7$
$\qquad m_k = a_k b_k \qquad k = 0, \ldots, 8$
$u_0 = m_0 - m_7 \qquad\qquad u_1 = m_2 + m_0$

$y_0 = u_0 - m_1 + m_5$
$y_1 = u_1 - m_3 - m_7$
$y_2 = u_0 - m_4 + m_6$
$y_3 = u_1 + m_5 + m_6 + m_8$

Polynomial Product Modulo $(z^9 - 1)/(z^3 - 1)$
15 multiplications, 39 additions

$a_0 = x_0 + x_2$
$a_1 = x_3 + x_5$
$a_2 = a_1 + x_4$ $\qquad\qquad$ $b_2 = (h_0 + 3h_1 + 2h_2 - 2h_3 - 3h_4 - h_5)/6$
$a_3 = a_0 + x_1$ $\qquad\qquad$ $b_3 = (h_0 - h_2 + h_3 + 3h_4 + 2h_5)/6$
$a_4 = a_3 - a_2$ $\qquad\qquad$ $b_4 = b_2 + b_3$
$a_5 = a_1 - x_4$ $\qquad\qquad$ $b_5 = (-h_0 + h_1 - h_4 + h_5)/2$
$a_6 = a_0 - x_1$ $\qquad\qquad$ $b_6 = (h_0 - h_2 - h_3 + h_4)/2$
$a_7 = a_6 - a_5$ $\qquad\qquad$ $b_7 = b_5 + b_6$
$a_8 = x_3$ $\qquad\qquad\qquad$ $b_8 = 2h_0 + h_1 - h_2 - 2h_3 + h_5$
$a_9 = x_0 - x_3$ $\qquad\qquad$ $b_9 = 2h_0 - h_2 + h_4$
$a_{10} = x_0$ $\qquad\qquad\qquad$ $b_{10} = b_9 - b_8$
$a_{11} = x_5$ $\qquad\qquad\qquad$ $b_{11} = h_0 - h_1 - 2h_2 + h_4$
$a_{12} = x_2 - x_5$ $\qquad\qquad$ $b_{12} = -h_1 + h_3 - 2h_5$
$a_{13} = x_2$ $\qquad\qquad\qquad$ $b_{13} = b_{12} - b_{11}$

$$a_{14} = -a_{12} + x_0 - x_4 \quad b_{14} = (h_0 - h_2 - 2h_3 + 2h_5)/3$$
$$a_{15} = a_9 + x_5 - x_1 \quad b_{15} = (-h_0 + h_2 - h_3 + h_5)/3$$
$$a_{16} = -a_{15} + a_{14} \quad b_{16} = b_{15} - b_{14}$$
$$m_k = a_{k+2} b_{k+2} \quad k = 0, \ldots, 14$$

$$u_0 = m_0 + m_1 \qquad u_7 = -u_3 + m_6$$
$$u_1 = m_3 + m_4 \qquad u_8 = u_4 + u_5$$
$$u_2 = m_{13} + m_{14} \qquad u_9 = m_9 - u_6$$
$$u_3 = u_0 + u_1$$
$$u_4 = m_0 + m_2$$
$$u_5 = m_3 + m_5$$
$$u_6 = m_{12} + m_{14}$$

$$y_0 = m_7 + u_2 + u_7$$
$$y_1 = u_8 + m_{10} + u_9$$
$$y_2 = u_4 - u_5 + u_2$$
$$y_3 = u_7 + u_8 + m_8 + u_6$$
$$y_4 = u_3 + m_{11} + u_9 + u_2$$
$$y_5 = u_0 - u_1 + u_6$$

Polynomial Product Modulo $(z^7 - 1)/(z - 1)$
15 multiplications, 53 additions

$$a_0 = x_0 + x_2 \qquad b_0 = (-2h_5 + 3h_4 - h_3 - 2h_2 + h_1 + 2h_0)/2$$
$$a_1 = a_0 + x_1 \qquad b_1 = (3h_5 - 11h_4 + 10h_3 + 3h_2 - 11h_1 - 4h_0)/14$$
$$a_2 = a_1 + x_2 \qquad b_2 = (3h_5 - h_4 - 2h_3 + 3h_2 - h_1)/6$$
$$a_3 = x_3 + x_5 \qquad b_3 = (h_4 - h_3 + h_1)/6$$
$$a_4 = a_3 + x_4 \qquad b_4 = -h_5 - 2h_4 + 3h_3 - h_2 - 2h_1 + h_0$$
$$a_5 = a_4 + x_5 \qquad b_5 = (h_5 + 2h_4 - h_3 - 2h_2 + 3h_1 - h_0)/2$$
$$a_6 = x_0 \qquad b_6 = (-11h_5 - 4h_4 + 10h_3 + 3h_2 - 11h_1 + 10h_0)/14$$
$$a_7 = a_1 \qquad b_7 = (-h_5 - 2h_3 + 3h_2 - h_1 - 2h_0)/6$$
$$a_8 = a_0 - x_1 \qquad b_8 = (h_5 - h_3 + h_1 - h_0)/6$$
$$a_9 = a_2 + a_5 - x_0 \qquad b_9 = -2h_5 + h_4 + 2h_3 - h_2 - 2h_1 + 3h_0$$
$$a_{10} = x_2 \qquad b_{10} = (2h_4 - h_3 - 2h_2 + h_1)/2$$
$$a_{11} = x_3 \qquad b_{11} = (-2h_5 - 2h_4 + 12h_3 + 5h_2 - 9h_1 - 2h_0)/14$$
$$a_{12} = a_4 \qquad b_{12} = (-2h_3 + 3h_2 - h_1)/6$$

76   3. Fast Convolution Algorithms

$a_{13} = a_3 - x_4$ $b_{13} = (-h_3 + h_1)/6$
$a_{14} = a_5 + a_5 - x_3$ $b_{14} = 2h_3 - h_2 - 2h_1 + h_0$
$a_{15} = x_5$
$a_{16} = a_{11} - a_6$
$a_{17} = a_{12} - a_7$
$a_{18} = a_{13} - a_8$
$a_{19} = a_{14} - a_9$
$a_{20} = a_{15} - a_{10}$

$\qquad m_k = a_{k+6} b_k \qquad k = 0, \ldots, 14$

$u_0 = m_5 + m_0$ $u_{13} = u_{10} + u_3$
$u_1 = m_6 + m_1$ $u_{14} = u_{13} - u_2$
$u_2 = m_7 + m_2$ $u_{15} = u_{13} + u_3 + u_3 + u_4 + u_2$
$u_3 = m_8 + m_3$ $u_{16} = -u_{12} - u_{15} - u_{14}$
$u_4 = m_9 + m_4$ $u_{17} = u_7 - u_8$
$u_5 = m_{10} + m_0$ $u_{18} = u_5 + u_{17} + u_7$
$u_6 = m_{11} + m_1$ $u_{19} = u_{17} - u_6 - u_8$
$u_7 = m_{12} + m_2$ $u_{20} = (u_7 + u_8) + (u_7 + u_8) + u_9$
$u_8 = m_{13} + m_3$ $u_{21} = u_{19} + u_{19}$
$u_9 = m_{14} + m_4$ $u_{22} = u_{21} - u_{18}$
$u_{10} = u_1 + u_3$ $u_{23} = u_{21} - u_{20}$
$u_{11} = u_{10} + u_2$
$u_{12} = u_0 + u_{11}$

$y_0 = u_{18}$ $y_3 = u_{14} + u_{21}$
$y_1 = u_{16} + u_{19}$ $y_4 = u_{12} + u_{22}$
$y_2 = u_{15} + u_{23}$ $y_5 = u_{20}$

Polynomial Product Modulo ($z^8 + 1$)
21 multiplications, 77 additions

$a_0 = x_0 + x_2$
$a_1 = x_1 + x_3$
$a_2 = x_0 - x_2$
$a_3 = x_7 - x_5$
$a_4 = x_4 + x_6$

3.7 Short Convolution and Polynomial Product Algorithms

$a_5 = x_5 + x_7$
$a_6 = x_4 - x_6$
$a_7 = x_1 - x_3$
$a_8 = a_0 + a_1$ $\qquad b_0 = (h_0 + h_1 + h_2 - h_3 + h_4 + h_5 + h_6 + h_7)/4$
$a_9 = a_4 + a_5$ $\qquad b_1 = (-h_0 - h_1 - h_2 - h_3 + h_4 + h_5 + h_6 - h_7)/4$
$a_{10} = a_8 + a_9$ $\qquad b_2 = (h_3 - h_4 - h_5 - h_6)/4$
$a_{11} = a_0 - a_1$ $\qquad b_3 = (5h_0 - 5h_1 + 5h_2 - 7h_3 + 5h_4 - 5h_5 + 5h_6 - h_7)/20$
$a_{12} = a_4 - a_5$ $\qquad b_4 = (-5h_0 + 5h_1 - 5h_2 + h_3 + 5h_4 - 5h_5 + 5h_6 - 7h_7)/20$
$a_{13} = a_{11} + a_{12}$ $\qquad b_5 = (3h_3 - 5h_4 + 5h_5 - 5h_6 + 4h_7)/20$
$a_{14} = a_2 + a_3$ $\qquad b_6 = (h_0 + h_1 - h_2 - h_3 + h_4 - h_5 - h_6 + 3h_7)/4$
$a_{15} = a_6 + a_7$ $\qquad b_7 = (-h_0 + h_1 + h_2 - 3h_3 + h_4 + h_5 - h_6 - h_7)/4$
$a_{16} = a_{14} + a_{15}$ $\qquad b_8 = (-h_1 + 2h_3 - h_4 + h_6 - h_7)/4$
$a_{17} = a_2 - a_3$ $\qquad b_9 = (5h_0 - 5h_1 - 5h_2 + h_3 + 5h_4 + 5h_5 - 5h_6 - 3h_7)/20$
$a_{18} = a_6 - a_7$ $\qquad b_{10} = (-5h_0 - 5h_1 + 5h_2 + 3h_3 + 5h_4 - 5h_5 - 5h_6 + h_7)/20$
$a_{19} = a_{17} + a_{18}$ $\qquad b_{11} = (5h_1 - 2h_3 - 5h_4 + 5h_6 + h_7)/20$
$a_{20} = x_0 + x_4$ $\qquad b_{12} = h_0 - h_3 + h_4$
$a_{21} = x_0$ $\qquad b_{13} = -2h_0 + h_3 - h_7$
$a_{22} = x_4$ $\qquad b_{14} = h_3 - 2h_4 + h_7$
$a_{23} = x_3 + x_7$ $\qquad b_{15} = -h_2 + h_6 - 2h_7$
$a_{24} = x_3$ $\qquad b_{16} = 2h_3 - 2h_6 + 2h_7$
$a_{25} = x_7$ $\qquad b_{17} = 2h_2 - 2h_3 + 2h_7$
$a_{26} = a_{15} - a_9 + x_0 - a_{23}$ $\qquad b_{18} = (-h_3 + h_7)/5$
$a_{27} = a_8 - a_{14} + x_3 + x_4 - x_7$ $\qquad b_{19} = (-h_3 - h_7)/5$
$a_{28} = a_{26} + a_{27}$ $\qquad b_{20} = h_3/5$

$\qquad m_k = a_{k+8}\, b_k \qquad k = 0, \ldots, 20$

$u_0 = m_0 + m_2$ $\qquad u_{16} = u_4 - u_6$
$u_1 = m_1 + m_2$ $\qquad u_{17} = u_5 + u_7$
$u_2 = m_3 + m_5$ $\qquad u_{18} = u_5 - u_7$
$u_3 = m_4 + m_5$ $\qquad u_{19} = m_{18} - m_{19}$
$u_4 = m_6 + m_8$ $\qquad u_{20} = u_9 + m_{14}$

$u_5 = m_7 + m_8$ $\qquad u_{21} = u_{19} + u_{19}$
$u_6 = m_9 + m_{11}$ $\qquad u_{22} = m_{15} - u_{21}$
$u_7 = m_{10} + m_{11}$ $\qquad u_{23} = u_{22} + m_{16}$
$u_8 = m_{20} + m_{19}$ $\qquad u_{24} = -u_{22} - m_{17}$
$u_9 = m_{12} + u_8$ $\qquad u_{25} = u_8 + u_8$
$u_{10} = u_9 + m_{13}$ $\qquad u_{26} = u_{25} + u_{25}$
$u_{11} = u_0 + u_2$ $\qquad u_{27} = u_{19} + u_{25}$
$u_{12} = u_0 - u_2$
$u_{13} = u_1 + u_3$
$u_{14} = u_1 - u_3$
$u_{15} = u_4 + u_6$

$y_0 = u_{13} + u_{17} + u_{20}$
$y_1 = u_{12} - u_{18} + u_{23}$
$y_2 = u_{11} - u_{15} + u_{25}$
$y_3 = u_{12} + u_{18} + u_{27}$
$y_4 = u_{11} + u_{15} + u_{10} + u_{19}$
$y_5 = -u_{14} - u_{16} + u_{24} + u_{26}$
$y_6 = -u_{13} + u_{17} + u_{21} + u_{25}$
$y_7 = -u_{14} + u_{16} + u_{19}$

3.7.3 Short Aperiodic Convolution Algorithms

Aperiodic Convolution of 2 Sequences of Length 2

3 multiplications, 3 additions
(1 input addition, 2 output additions)

$a_0 = x_0$ $\qquad\qquad b_0 = h_0$
$a_1 = x_0 + x_1$ $\qquad b_1 = h_0 + h_1$
$a_2 = x_1$ $\qquad\qquad b_2 = h_1$
$\qquad m_k = a_k b_k, \quad k = 0, ..., 2$
$y_0 = m_0$
$y_1 = m_1 - m_0 - m_2$
$y_2 = m_2$

3.7 Short Convolution and Polynomial Product Algorithms

Aperiodic Convolution of 2 Sequences of Length 3

5 multiplications, 20 additions
(7 input additions, 13 output additions)

$a_0 = x_1 + x_2$

$a_1 = x_2 - x_1$

$a_2 = x_0 \qquad\qquad b_0 = h_0/2$

$a_3 = x_0 + a_0 \qquad b_1 = (h_0 + h_1 + h_2)/2$

$a_4 = x_0 + a_1 \qquad b_2 = (h_0 - h_1 + h_2)/6$

$a_5 = a_0 + a_0 + a_1 + a_3 \quad b_3 = (h_0 + 2h_1 + 4h_2)/6$

$a_6 = x_2 \qquad\qquad b_4 = h_2$

$\qquad m_k = a_{k+2} b_k, \qquad k = 0, \ldots, 4$

$u_0 = m_4 + m_4 \qquad u_3 = m_2 + m_2$

$u_1 = m_1 + m_1 \qquad u_4 = u_0 - m_0 - m_3$

$u_2 = m_0 + m_0 \qquad u_5 = m_1 + m_2$

$y_0 = u_2 \qquad\qquad y_3 = -u_4 - u_5$

$y_1 = u_1 - u_3 + u_4 \qquad y_4 = m_4$

$y_2 = -u_2 + u_3 + u_5 - m_4$

4. The Fast Fourier Transform

The object of this chapter is to briefly summarize the main properties of the discrete Fourier transform (DFT) and to present various fast DFT computation techniques known collectively as the fast Fourier transform (FFT) algorithm. The DFT plays a key role in physics because it can be used as a mathematical tool to describe the relationship between the time domain and frequency domain representation of discrete signals. The use of DFT analysis methods has increased dramatically since the introduction of the FFT in 1965 because the FFT algorithm decreases by several orders of magnitude the number of arithmetic operations required for DFT computations. It has thereby provided a practical solution to many problems that otherwise would have been intractable.

4.1 The Discrete Fourier Transform

The DFT $\bar{X}_k$ of a sequence x_m of N terms is defined by

$$\bar{X}_k = \sum_{m=0}^{N-1} x_m W^{mk}, \qquad k = 0, \ldots, N-1$$

$$W = e^{-j2\pi/N}, \; j = \sqrt{-1} \qquad (4.1)$$

The sequence x_m can be viewed as representing N consecutive samples $x(mT)$ of a continuous signal $x(t)$, while the sequence $\bar{X}_k$ can be considered as representing N consecutive samples $\bar{X}(kf)$ in the frequency domain. Thus, the DFT is an approximation of the continuous Fourier transform of a function. The relationship between the discrete and continuous Fourier transform is well known and can be found in [4.1–4]. In our discussion of the DFT, we shall restrict our attention to some of the properties that are used in various parts of this book.

An important property of the DFT is that x_m and $\bar{X}_k$ are uniquely related by a transform pair, with the direct transform defined by (4.1) and an inverse transform defined by

$$y_l = \frac{1}{N} \sum_{k=0}^{N-1} \bar{X}_k W^{-lk}, \qquad l = 0, \ldots, N-1. \qquad (4.2)$$

It can easily be verified that (4.2) is the inverse of (4.1) by substituting $\bar{X}_k$, given by (4.1), into (4.2). This yields

$$y_l = \sum_{m=0}^{N-1} x_m \frac{1}{N} \sum_{k=0}^{N-1} W^{(m-l)k}. \qquad (4.3)$$

Since $W^N = 1$, $m - l$ is defined modulo N. For $m - l \equiv 0$ modulo N, $S = \sum_{k=0}^{N-1} W^{(m-l)k} = N$. For $m - l \not\equiv 0$ modulo N, we have $S = [W^{(m-l)N} - 1]/(W^{m-l} - 1)$ and since $W^{m-l} \neq 1$, $S = 0$. Therefore, the only nonzero case corresponds to $l \equiv m$, which gives $y_l = x_m$.

The DFT can be used to compute a circular convolution y_l of N terms, with

$$y_l = \sum_{n=0}^{N-1} h_n x_{l-n}, \quad l = 0, \ldots, N - 1. \tag{4.4}$$

This is done by computing the DFTs $\bar{H}_k$ and $\bar{X}_k$ of h_n and x_m, by multiplying $\bar{H}_k$ by $\bar{X}_k$, and by computing the inverse transform c_l of $\bar{H}_k \bar{X}_k$. Hence c_l is given by

$$c_l = \sum_{n=0}^{N-1} \sum_{m=0}^{N-1} h_n x_m \frac{1}{N} \sum_{k=0}^{N-1} W^{(m+n-l)k}. \tag{4.5}$$

Using the same procedure as above, one finds that $S = \sum_{k=0}^{N-1} W^{(m+n-l)k}$ becomes $S \equiv 0$ for $m + n - l \not\equiv 0$ modulo N and $S \equiv N$ for $m + n - l \equiv 0$ modulo N. Thus, $m \equiv l - n$ and $y_l = c_l$. Hence an N-point circular convolution is calculated by three DFTs plus N multiplications. When the DFTs are computed directly, this approach is not of practical value because each DFT is computed with N^2 multiplications whereas the direct computation of the convolution requires N^2 multiplications. We shall see however that the method becomes very efficient when the DFTs are evaluated by a fast algorithm.

4.1.1 Properties of the DFT

In order to present the main properties of the DFT as compactly as possible, we shall use the following notation to represent a DFT relationship between a sequence x_m and its transform $\bar{X}_k$:

$$\{x_m\} \xleftrightarrow{\text{DFT}} \{\bar{X}_k\}. \tag{4.6}$$

Assuming a second sequence h_n and its DFT $\bar{H}_k$, we now establish the following DFT properties.

Linearity

$$\{x_m\} + \{h_n\} \xleftrightarrow{\text{DFT}} \{\bar{X}_k\} + \{\bar{H}_k\} \tag{4.7}$$

$$\{p\, x_m\} \xleftrightarrow{\text{DFT}} \{p\, \bar{X}_k\}. \tag{4.8}$$

These properties follow directly from the definitions (4.1) and (4.2).

Symmetry

$$\{x_{-m}\} \xleftrightarrow{\text{DFT}} \{\bar{X}_{-k}\}. \tag{4.9}$$

This can be seen by noting that the transform $\bar{A}_k$ of x_{-m} is given by

$$\bar{A}_k = \sum_{m=0}^{N-1} x_{-m} W^{mk} = \sum_{m=0}^{N-1} x_m W^{-mk} \tag{4.10}$$

Thus $\bar{A}_k = \bar{X}_{-k}$.

Time Shifting

$$\{x_{m+l}\} \xleftrightarrow{\text{DFT}} \{W^{-lk} \bar{X}_k\} \tag{4.11}$$

This property is established by computing the DFT $\bar{A}_k$ of x_{m+l}

$$\bar{A}_k = \sum_{m=0}^{N-1} x_{m+l} W^{mk} = \sum_{m=0}^{N-1} x_{m+l} W^{(m+l-l)k}. \tag{4.12}$$

Hence

$$\bar{A}_k = W^{-lk} \bar{X}_k. \tag{4.13}$$

Frequency Shifting

$$\{W^{lm} x_m\} \xleftrightarrow{\text{DFT}} \{\bar{X}_{k+l}\}. \tag{4.14}$$

This property follows directly from the proof given for the time shifting property by replacing the direct DFT with an inverse DFT.

DFT of a permuted sequence

$$\{x_{pm}\} \xleftrightarrow{\text{DFT}} \{\bar{X}_{qk}\}. \tag{4.15}$$

We assume that the sequence x_m is permuted, with m replaced by pm modulo N, where p is an integer relatively prime to N. The DFT $\bar{A}_k$ of x_{pm} is given by

$$\bar{A}_k = \sum_{m=0}^{N-1} x_{pm} W^{mk} \tag{4.16}$$

Since $(p, N) = 1$, we can find an integer q such that $qp \equiv 1$ modulo N. Equation (4.16) is not changed if m is replaced by qm modulo N. We then have

$$\bar{A}_k = \sum_{m=0}^{N-1} x_{pqm} W^{mqk} = \bar{X}_{qk}. \tag{4.17}$$

Correlation of real sequences

$$\left\{\sum_{n=0}^{N-1} h_n x_{l+n}\right\} \xleftrightarrow{\text{DFT}} \{\bar{H}_k^* \bar{X}_k\}. \tag{4.18}$$

Since a correlation is derived from a convolution by inverting one of the input sequences, the DFT convolution property implies that the transform of the correlation of the two sequences h_n and x_m is obtained by evaluating the DFT of the convolution of h_{-n} by x_m. Hence

$$\left\{\sum_{n=0}^{N-1} h_n \, x_{l+n}\right\} \xrightarrow{\text{DFT}} \{\bar{H}_{-k} \, \bar{X}_k\} \tag{4.19}$$

with

$$\bar{H}_{-k} = \sum_{n=0}^{N-1} h_n \, e^{j2\pi nk/N}. \tag{4.20}$$

Since h_n is real, (4.20) therefore implies that $\bar{H}_{-k} = \bar{H}_k^*$, where $\bar{H}_k^*$ is the complex conjugate of $\bar{H}_k$.

Parseval's theorem

$$\sum_{m=0}^{N-1} x_m^2 = \left\{\frac{1}{N} \sum_{k=0}^{N-1} |\bar{X}_k|^2\right\}. \tag{4.21}$$

The Parseval theorem is a direct consequence of the correlation property because $\sum_{m=0}^{N-1} x_m^2$ is the first term of the autocorrelation of x_m. Thus,

$$\sum_{m=0}^{N-1} x_m^2 = \sum_{m=0}^{N-1} x_m x_m = \frac{1}{N} \sum_{k=0}^{N-1} \bar{X}_k^* X_k = \frac{1}{N} \sum_{k=0}^{N-1} |X_k|^2, \tag{4.22}$$

where $|X_k|$ is the magnitude of X_k

4.1.2 DFTs of Real Sequences

In many practical applications, the input sequence x_m is real. In this case, the DFT $\bar{X}_k$ of x_m has special properties that can be found by rewriting (4.1) as

$$\bar{X}_k = \sum_{m=0}^{N-1} x_m \cos(2\pi mk/N) - j \sum_{m=0}^{N-1} x_m \sin(2\pi mk/N). \tag{4.23}$$

Since x_m is real, (4.23) implies that the real part $\text{Re}\{\bar{X}_k\}$ of $\bar{X}_k$ is even and that the imaginary part $\text{Im}\{\bar{X}_k\}$ of $\bar{X}_k$ is odd

$$\text{Re}\{\bar{X}_k\} = \text{Re}\{\bar{X}_{-k}\} \tag{4.24}$$

$$\text{Im}\{\bar{X}_k\} = -\text{Im}\{\bar{X}_{-k}\} \tag{4.25}$$

$$|\bar{X}_k| = |\bar{X}_{-k}|. \tag{4.26}$$

Similarly, when x_m is a pure imaginary sequence, we have

$$\text{Re}\{\bar{X}_k\} = -\text{Re}\{\bar{X}_{-k}\} \tag{4.27}$$

$$\text{Im}\{\bar{X}_k\} = \text{Im}\{\bar{X}_{-k}\} \tag{4.28}$$

$$|\bar{X}_k| = |\bar{X}_{-k}|. \tag{4.29}$$

These properties can be used to compute simultaneously the transforms $\bar{X}_k$ and $\bar{X}_k^1$ of two real N-point sequences x_m and x_m^1 with a single, complex DFT. This is done by evaluating the DFT $\bar{Y}_k$ of the sequence $x_m + jx_m^1$, with

$$\bar{Y}_k = \sum_{m=0}^{N-1} (x_m + jx_m^1) W^{mk} \tag{4.30}$$

$$\bar{Y}_k = \bar{X}_k + j\bar{X}_k^1. \tag{4.31}$$

Hence

$$\text{Re}\{\bar{Y}_k\} = \text{Re}\{\bar{X}_k\} - \text{Im}\{\bar{X}_k^1\} \tag{4.32}$$

$$\text{Im}\{\bar{Y}_k\} = \text{Im}\{\bar{X}_k\} + \text{Re}\{\bar{X}_k^1\}. \tag{4.33}$$

Then, by using the symmetry property of pure real and pure imaginary sequences

$$\text{Re}\{\bar{X}_k\} = (\text{Re}\{\bar{Y}_k\} + \text{Re}\{\bar{Y}_{-k}\})/2 \tag{4.34}$$

$$\text{Im}\{\bar{X}_k\} = (\text{Im}\{\bar{Y}_k\} - \text{Im}\{\bar{Y}_{-k}\})/2 \tag{4.35}$$

$$\text{Re}\{\bar{X}_k^1\} = (\text{Im}\{\bar{Y}_k\} + \text{Im}\{\bar{Y}_{-k}\})/2 \tag{4.36}$$

$$\text{Im}\{\bar{X}_k^1\} = (\text{Re}\{\bar{Y}_{-k}\} - \text{Re}\{\bar{Y}_k\})/2. \tag{4.37}$$

4.1.3 DFTs of Odd and Even Sequence

If x_m is a real, even sequence, $x_m = x_{-m}$. In this case, we have

$$\sum_{m=0}^{N-1} x_m \sin(2\pi mk/N) = \sum_{m=0}^{N/2-1} (x_m - x_{-m}) \sin(2\pi mk/N) = 0. \tag{4.38}$$

This implies that the DFT $\bar{X}_k$ of x is even and real, since

$$\bar{X}_k = \sum_{m=0}^{N-1} x_m \cos(2\pi mk/N). \tag{4.39}$$

Similarly, if x_m is an odd, real sequence, with $x_m = -x_{-m}$, the DFT $\bar{X}_k$ is odd and imaginary.

An immediate consequence of these properties is that, if x_m is a conjugate symmetric sequence defined by $x_m = x_{-m}^*$, the DFT $\bar{X}_k$ of x_m is real. Similarly, if

x_m is a conjugate antisymmetric sequence defined by $x_m = -x^*_{-m}$, the DFT $\bar{X}_k$ of x_m is imaginary. These properties can be used to compute the DFTs of two complex even sequences in one transform step [4.5]. This is done by constructing an auxiliary sequence y_m which is derived from the two complex even sequences x_m and x_m^1 by

$$y_m = x_m W^m + x_m^1. \tag{4.40}$$

Then, from the shift theorem, the DFT $\bar{Y}_k$ of y_m is given by

$$\bar{Y}_k = \bar{X}_{k+1} + \bar{X}_k^1. \tag{4.41}$$

Since x_m and x_m^1 are complex even sequences, their DFTs $\bar{X}_k$ and $\bar{X}_k^1$ are such that $\bar{X}_{k+1} = \bar{X}_{-k-1}$ and $\bar{X}_k^1 = \bar{X}_{-k}^1$. This implies that

$$\bar{Y}_{-k-1} = \bar{X}_k + \bar{X}_{k+1}^1. \tag{4.42}$$

$\bar{X}_0$ and $\bar{X}_0^1$ are computed directly with

$$\bar{X}_0 = \sum_{m=0}^{N-1} x_m \tag{4.43}$$

$$\bar{X}_0^1 = \sum_{m=0}^{N-1} x_m^1. \tag{4.44}$$

Then, all the other values of $\bar{X}_k$ and $\bar{X}_k^1$ are obtained recursively by

$$\bar{X}_{k+1} = \bar{Y}_k - \bar{X}_k^1 \tag{4.45}$$

$$\bar{X}_{k+1}^1 = \bar{Y}_{-k-1} - \bar{X}_k. \tag{4.46}$$

Similar techniques can be used to compute the DFTs of two complex odd sequences or of four real even or odd sequences in one transform step.

4.2 The Fast Fourier Transform Algorithm

We consider a DFT $\bar{X}_k$ of dimension N, where N is composite, with

$$\bar{X}_k = \sum_{m=0}^{N-1} x_m W^{mk}, \quad k = 0, ..., N-1$$

$$W = e^{-j2\pi/N}. \tag{4.47}$$

If N is the product of two factors, with $N = N_1 N_2$, we can redefine the indices m and k by

$$m = N_1 m_2 + m_1, \quad m_1, k_1 = 0, ..., N_1 - 1 \tag{4.48}$$

4. The Fast Fourier Transform

$$k = N_2 k_1 + k_2, \qquad m_2, k_2 = 0, \ldots, N_2 - 1. \tag{4.49}$$

Substituting (4.48) and (4.49) into (4.47) yields

$$\bar{X}_{N_2 k_1 + k_2} = \sum_{m_1=0}^{N_1-1} W^{N_2 m_1 k_1} W^{m_1 k_2} \sum_{m_2=0}^{N_2-1} x_{N_1 m_2 + m_1} W^{N_1 m_2 k_2} \tag{4.50}$$

or, with $W^{N_2} = e^{-j2\pi/N_1} = W_1$ and $W^{N_1} = e^{-j2\pi/N_2} = W_2$,

$$\bar{X}_{N_2 k_1 + k_2} = \sum_{m_1=0}^{N_1-1} W_1^{m_1 k_1} W^{m_1 k_2} \sum_{m_2=0}^{N_2-1} x_{N_1 m_2 + m_1} W_2^{m_2 k_2}, \tag{4.51}$$

which shows that the DFT of length $N_1 N_2$ can be viewed as a DFT of size $N_1 \times N_2$, except for the introduction of the *twiddle factors* $W^{m_1 k_2}$. Thus, the computation of $\bar{X}_k$ by (4.51) is done in three steps, with the first step corresponding to the evaluation of the N_1 DFTs $\bar{Y}_{m_1, k_2}$ corresponding to the N_1 distinct values of m_1

$$\bar{Y}_{m_1, k_2} = \sum_{m_2=0}^{N_2-1} x_{N_1 m_2 + m_1} W_2^{m_2 k_2}. \tag{4.52}$$

$\bar{Y}_{m_1, k_2}$ is then multiplied by the twiddle factors $W^{m_1 k_2}$ and $\bar{X}_k$ is obtained by calculating N_2 DFTs of N_1 points on the N_2 input sequences $\bar{Y}_{m_1, k_2} W^{m_1 k_2}$, with

$$\bar{X}_{N_2 k_1 + k_2} = \sum_{m_1=0}^{N_1-1} \bar{Y}_{m_1, k_2} W^{m_1 k_2} W_1^{m_1 k_1}. \tag{4.53}$$

Note that the computation procedure could have been organized in reverse order, with the multiplications by the twiddle factors preceding the evaluation of the first DFTs instead of being done after the calculation of these DFTs. In this case,

$$\bar{X}_{N_2 k_1 + k_2} = \sum_{m_2=0}^{N_2-1} W_2^{m_2 k_2} \sum_{m_1=0}^{N_1-1} (x_{N_1 m_2 + m_1} W^{m_1 k_2}) W_1^{m_1 k_1}. \tag{4.54}$$

Hence there are generally two different forms for the FFT algorithm, each being equivalent in terms of computational complexity. It should be noted that, in both procedures, the order of the input and output row-column indices are permuted. Thus, while the input sequence can be viewed as N_2 polynomials of N_1 terms, the output sequence is organized as N_1 polynomials of N_2 terms. This implies that a permutation step must be added at the end of the three basic steps described above to complete the FFT procedure.

The FFT algorithm derives its efficiency by replacing the computation of one large DFT with that of several smaller DFTs. Since the number of operations required to directly compute an N-point DFT is proportional to N^2, the number of operations decreases rapidly when the computation structure is partitioned into that of many small DFTs. In the simple case of a DFT of length $N_1 N_2$, the direct computation would require $N_1^2 N_2^2$ multiplications. If we now evaluate this DFT by the FFT algorithm corresponding to (4.52) and (4.53), the computation breaks down into that of N_1 DFTs of N_2 terms, N_2 DFTs of N_1 terms plus $N_1 N_2$ multiplications by twiddle factors. Thus, the number M of multiplications required to evaluate the DFT of $N_1 N_2$ points with this simple two-factor algorithm reduces to

$$M = N_1 N_2^2 + N_2 N_1^2 + N_1 N_2 = N_1 N_2 (N_1 + N_2 + 1), \qquad (4.55)$$

which is obviously less than $N_1^2 N_2^2$. In practice, the FFT algorithm is extremely powerful because the procedure can be used iteratively when N is highly composite. In such cases, and with the two-factor decomposition discussed above, N_1 and N_2 are composite and the DFTs of lengths N_1 and N_2 are again computed by an FFT procedure. With this approach, each stage provides an additional reduction in number of operations so that the algorithm is the most efficient when N is highly composite. This feature, together with the need for a regular computational structure, motivates application of the FFT algorithm to DFT lengths which are the power of an integer. In most cases, N is chosen to be a power of two, and this was the original form of the FFT algorithm [4.6].

4.2.1 The Radix-2 FFT Algorithm

We now consider the DFT $\bar{X}_k$ of an N-point sequence x_m, with $N = 2^t$. In this case, the first stage of the FFT can be defined by choosing $N_1 = 2$ and $N_2 = 2^{t-1}$. This is equivalent to splitting the N-point input sequence x_m into two $(N/2)$-point sequences x_{2m} and x_{2m+1} corresponding, respectively, to the even and odd samples of x_m. Under these conditions, $\bar{X}_k$ becomes

$$\bar{X}_k = \sum_{m=0}^{N/2-1} x_{2m} W^{2mk} + W^k \sum_{m=0}^{N/2-1} x_{2m+1} W^{2mk} \qquad (4.56)$$

and, since $W^{N/2} = -1$,

$$\bar{X}_{k+N/2} = \sum_{m=0}^{N/2-1} x_{2m} W^{2mk} - W^k \sum_{m=0}^{N/2-1} x_{2m+1} W^{2mk}$$

$$k = 0, \ldots, N/2 - 1. \qquad (4.57)$$

With this approach, called *decimation in time*, the computation of an N-point DFT is replaced by that of two DFTs of length $N/2$ plus N additions and $N/2$ multiplications by W^k. The same procedure can be applied again to replace the two DFTs of length $N/2$ by 4 DFTs of length $N/4$ at the cost of N additions and $N/2$ multiplications. A systematic application of this method computes the DFT of length 2^t in $t = \log_2 N$ stages, each stage converting 2^i DFTs of length 2^{t-i} into 2^{i+1} DFTs of length 2^{t-i-1} at the cost of N additions and $N/2$ multiplications.

4. The Fast Fourier Transform

Consequently, the number of complex multiplications M and complex additions A required to compute a DFT of length N by the radix-2 FFT algorithm is

$$M = (N/2)\log_2 N \qquad (4.58)$$

$$A = N\log_2 N \qquad (4.59)$$

The decimation in time approach is illustrated in Fig. 4.1 for an 8-point DFT. In

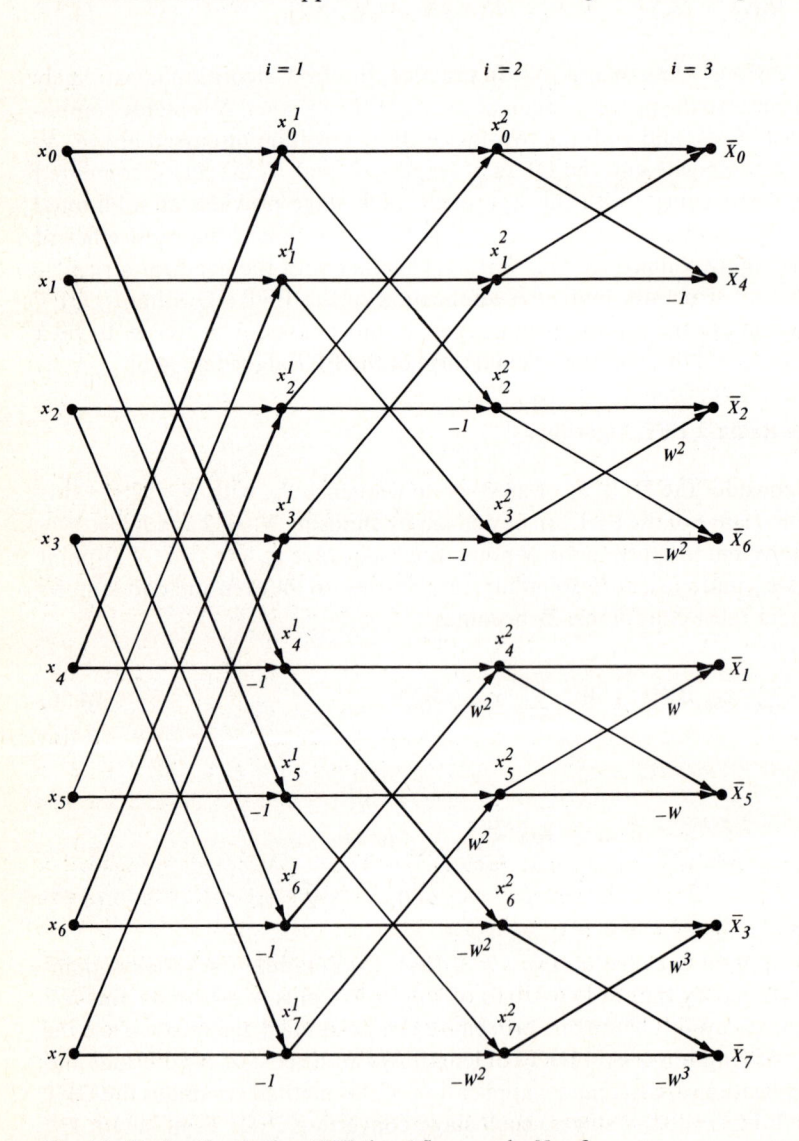

Fig. 4.1. Decimation in time FFT signal flow graph. $N = 8$.

this signal flow graph, each node represents a variable and each arrow terminating at a node represents the additive contribution of the variable at the originating node of the arrow. Multiplications by a constant are represented by the constant written near the arrowhead.

A second form of the FFT algorithm can be obtained by simply splitting the N-point input sequence x_m into two $(N/2)$-point sequences x_m and $x_{m+N/2}$ corresponding, respectively, to the $N/2$ first samples and to the $N/2$ last samples of x_m. With this approach, called *decimation in frequency*, $\bar{X}_k$ becomes

$$\bar{X}_k = \sum_{m=0}^{N/2-1} (x_m + W^{Nk/2} x_{m+N/2}) W^{mk}. \tag{4.60}$$

We now compute the even-and odd-numbered samples of $\bar{X}_k$ separately. For k even, replacing k by $2k$, we obtain

$$\bar{X}_{2k} = \sum_{m=0}^{N/2-1} (x_m + x_{m+N/2}) W^{2mk}, \qquad k = 0, ..., N/2 - 1. \tag{4.61}$$

Replacing k by $2k + 1$ for k odd, we get

$$\bar{X}_{2k+1} = \sum_{m=0}^{N/2-1} (x_m - x_{m+N/2}) W^m W^{2mk}, \qquad k = 0, ..., N/2 - 1. \tag{4.62}$$

Thus $\bar{X}_k$ is computed by (4.61) and (4.62) in terms of two DFTs of length $N/2$, but with a premultiplication by W^m of the input sequence in (4.62). Therefore, the computation of a DFT of N terms is replaced by that of two DFTs of $N/2$ terms at the cost of N complex additions and $N/2$ complex multiplications. As with the decimation in time algorithm, the same procedure can be used recursively to compute the DFT in $\log_2 N$ stages, each stage converting 2^i DFT of length 2^{t-i} into 2^{i+1} DFTs of length 2^{t-i-1} at the cost of N additions and $N/2$ multiplications. This means that the decimation in frequency algorithm requires the same number of operations as the decimation in time algorithm. The computation structure for decimation in frequency is shown in Fig. 4.2 for $N = 8$. It can be seen that the flow graph has the same geometry as the decimation in time flow graph, but different coefficients.

Since the FFT algorithm computes a DFT with $N \log N$ operations instead of N^2 for the direct approach, practical reduction of the computation load can be very large. In the case of a 1024-point DFT, for instance, we have $N = 2^{10}$ and the direct computation requires 2^{20} complex multiplications. On the other hand, the FFT algorithm computes the same DFT with only $5 \cdot 2^{10}$ complex multiplications or about 200 times fewer multiplications. Significant additional reduction can be obtained by noting that a number of the multiplications are trivial multiplications by ± 1 or $\pm j$.

In the case of a decimation in time algorithm, the twiddle factors in the first stage are given by $W^{kN/2} = (-1)^k$. Thus all multiplications in the first stage are

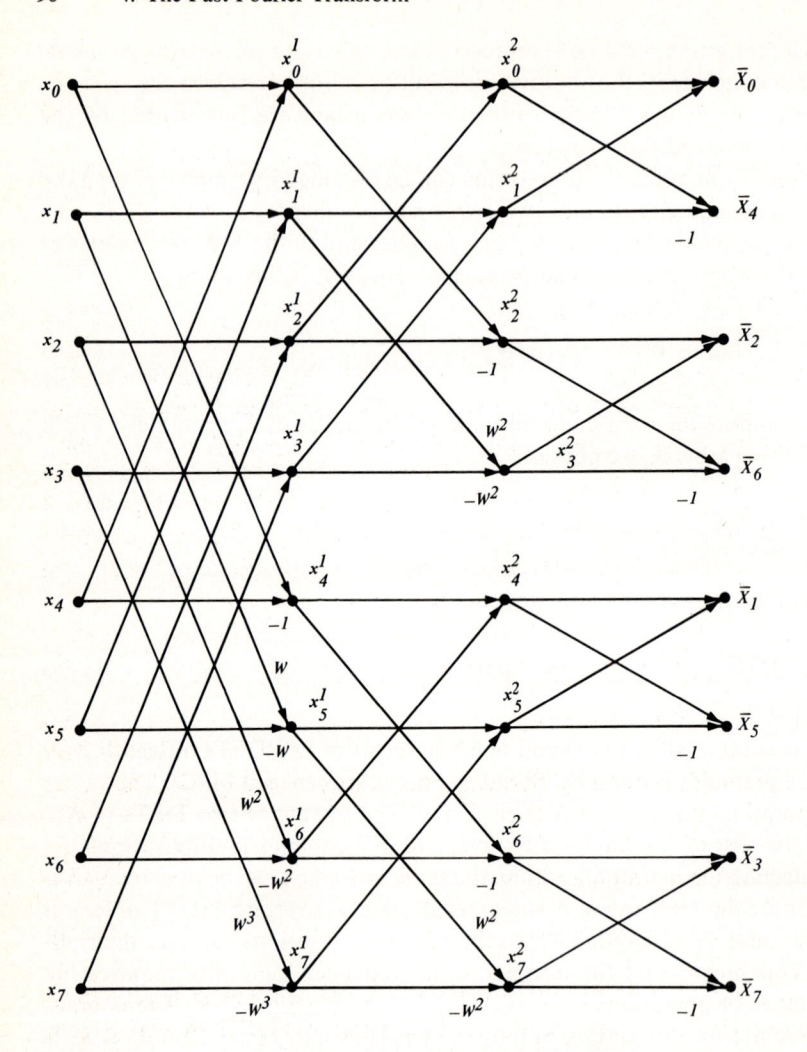

Fig. 4.2. Decimation in frequency signal flow graph. $N = 8$.

trivial. The multiplications in the second stage are also trivial because the twiddle factors are then defined by $W^{kN/4} = (-j)^k$. In the following stages, the twiddle factors are given by $W^{kN/8}$, $W^{kN/16}$, ... and the number of trivial multiplications is $N/4$, $N/8$, ... Under these conditions, the number of nontrivial complex multiplications becomes $(N/2)(-3 + \log_2 N) + 2$ and, if the complex multiplications are implemented with 4 real multiplications and 2 real additions, the numbers of real multiplications M and real additions A required to implement the radix-2 FFT algorithm are

$$M = 2N(-3 + \log_2 N) + 8 \tag{4.63}$$

$$A = 3N(-1 + \log_2 N) + 4. \tag{4.64}$$

If the complex multiplications are implemented with 3 multiplications and 3 additions (Sect. 3.7.2), M and A become

$$M = (3N/2)(-3 + \log_2 N) + 6 \tag{4.65}$$

$$A = (N/2)(-9 + 7\log_2 N) + 6. \tag{4.66}$$

It can also be noted that, since $W^{N/8} = (1 - j)/\sqrt{2}$, the multiplications by $W^{kN/8}$ can be implemented with 2 real multiplications and 2 real additions. Since the stages of order 3, 4, 5, ... use, respectively, $N/4$, $N/8$, ... such multiplications, we have a total of $N/2 - 2$ multiplications by $W^{kN/8}$ and the total number of multiplications for the DFT reduces to $(N/2)(-4 + \log_2 N) + 4$ complex multiplications plus $N/2 - 2$ multiplications by $W^{kN/8}$. Thus, when the complex multiplications are implemented with 4 real multiplications and 2 real additions, the number M and A of real operations is

$$M = N(-7 + 2\log_2 N) + 12 \tag{4.67}$$

$$A = 3N(-1 + \log_2 N) + 4 \tag{4.68}$$

and, for the complex multiplication algorithm using 3 multiplications and 3 additions,

$$M = (N/2)(-10 + 3\log_2 N) + 8 \tag{4.69}$$

$$A = (N/2)(-10 + 7\log_2 N) + 8. \tag{4.70}$$

Hence significant additional reduction is obtained when full use is made of the symmetries in sine and cosine functions. In the case of a DFT of 1024 points, for instance, the straightforward computation by (4.56) would require $20 \cdot 2^{10}$ real multiplications, as opposed to about $10 \cdot 2^{10}$ real multiplications when the FFT is computed by the approach corresponding to (4.69).

4.2.2 The Radix-4 FFT Algorithm

We now turn our attention to a DFT of dimension $N = 2^t$ with t even. In this case, the first stage of the FFT can be defined by choosing $N_1 = 4$ and $N_2 = 2^{t-2}$. This is equivalent to splitting the N-point input sequence x_m into the 4 sequences of $N/4$ points corresponding to x_{4m}, x_{4m+1}, x_{4m+2}, and x_{4m+3} for $m = 0, ..., N/4 - 1$. For this partition, $\bar{X}_k$ becomes

92   4. The Fast Fourier Transform

$$\bar{X}_k = \sum_{l=0}^{3} W^{lk} \sum_{m=0}^{N/4-1} x_{4m+l} W^{4mk} \tag{4.71}$$

and, since $W^{N/4} = -j$,

$$\bar{X}_{k+N/4} = \sum_{l=0}^{3} (-j)^l W^{lk} \sum_{m=0}^{N/4-1} x_{4m+l} W^{4mk} \tag{4.72}$$

$$\bar{X}_{k+N/2} = \sum_{l=0}^{3} (-1)^l W^{lk} \sum_{m=0}^{N/4-1} x_{4m+l} W^{4mk} \tag{4.73}$$

$$\bar{X}_{k+3N/4} = \sum_{l=0}^{3} j^l W^{lk} \sum_{m=0}^{N/4-1} x_{4m+l} W^{4mk}$$

$$k = 0, \ldots, N/4 - 1. \tag{4.74}$$

Hence this radix-4 decimation in time algorithm [4.7, 8] converts an N-point DFT into 4 DFTs of length $N/4$ at the cost of N complex multiplications by the twiddle factors W^{lk} and $3N$ complex additions for recombining the output samples of the DFTs of $N/4$ points. The same procedure can be applied recursively in $t/2$ stages, each stage reducing the dimensions of the DFTs by a factor of four. Thus, the DFT is computed by the radix-4 algorithm with $(N/2)\log_2 N$ complex multiplications and $(3N/2)\log_2 N$ complex additions. This number of operations is higher than with the radix-2 FFT algorithm. However, we now show that the computational complexity can be drastically reduced by exploiting the symmetries of the sine and cosine functions.

We observe first that, denoting the $N/4$ point DFTs by $\bar{X}_{l,k}$, we can decrease the number of additions per stage to $2N$ instead of $3N$, by the following computation procedure:

$$\bar{X}_{l,k} = \sum_{m=0}^{N/4-1} x_{4m+l} W^{4mk}, \quad l = 0, \ldots, 3 \tag{4.75}$$

$$Y_{l,k} = W^{lk} \bar{X}_{l,k} \tag{4.76}$$

$$\bar{X}_k = (Y_{0,k} + Y_{2,k}) + (Y_{1,k} + Y_{3,k}) \tag{4.77}$$

$$\bar{X}_{k+N/4} = (Y_{0,k} - Y_{2,k}) - j(Y_{1,k} - Y_{3,k}) \tag{4.78}$$

$$\bar{X}_{k+N/2} = (Y_{0,k} + Y_{2,k}) - (Y_{1,k} + Y_{3,k}) \tag{4.79}$$

$$\bar{X}_{k+3N/4} = (Y_{0,k} - Y_{2,k}) + j(Y_{1,k} - Y_{3,k}). \tag{4.80}$$

With regard to multiplications, we note that the twiddle factors in the successive stages are given by W^{lk}, W^{4lk}, W^{16lk}, ... Thus, the twiddle factors take the values W^{lk4^i} for $i = 0, 1, \ldots$. Each stage i splits the computation of 4^i DFTs of length $N/4^i$ into that of 4^{i+1} DFTs of length $N/4^{i+1}$, with $k = 0, \ldots, N/4^{i+1} - 1$. Since

4.2 The Fast Fourier Transform Algorithm

the total number of multiplications by twiddle factors per stage is N, each stage divides into 4^i groups of twiddle factors W^{lk4^i}, with $k = 0, \ldots, N/4^{i+1} - 1$ and $l = 0, \ldots, 3$. For the last stage, the twiddle factors correspond to W_4^{lk} and are computed by trivial multiplications by $\pm 1, \pm j$. For the other stages and $l = 1, 3$, the only simple multiplications correspond to $k = 0$ and $k = (N/2)4^{i+1}$. These cases correspond, respectively, to a multiplication by 1 and a multiplication by $W_8^l = [(1-j)/\sqrt{2}]^l$. For $l = 0$, we have $W^{lk} = 1$. For $l = 2$, the multiplications by W^{lk4^i} are implemented with 2 trivial multiplications, 2 multiplications by an odd power of W_8, and $(N/4^{i+1}) - 4$ complex multiplications. Since we have 4^i groups per stage, this corresponds to $(3N/4) - 8 \cdot 4^i$ complex multiplications and 4^{i+1} multiplications by odd powers of W_8 per stage. Moreover, $N = 2^t = 4^{t/2}$. We must therefore sum these numbers of multiplications over $(t/2) - 1$ stages. Thus, the number M_1 of nontrivial complex multiplications is given by

$$M_1 = (3N/8)\log_2 N - (17/12)N + 8/3. \tag{4.81}$$

Similarly, the number M_2 of multiplications by odd powers of W_8 is given by

$$M_2 = (N - 4)/3. \tag{4.82}$$

Under these conditions, if the complex multiplications are implemented with 4 real multiplications and 2 real additions, the number of real multiplications M and real additions A is

$$M = (3N/2)\log_2 N - 5N + 8 \tag{4.83}$$

$$A = (11N/4)\log_2 N - (13N/6) + (8/3) \tag{4.84}$$

and, when the complex multiplications are implemented with 3 multiplications and 3 additions,

$$M = (9N/8)\log_2 N - (43N/12) + (16/3) \tag{4.85}$$

$$A = (25N/8)\log_2 N - (43N/12) + (16/3). \tag{4.86}$$

Thus, the radix-4 algorithm significantly reduces the number of operations in comparison with the radix-2 algorithm. This is shown in Table 4.1 which gives the number of real operations for DFTs computed by radix-2 and radix-4 FFT algorithms corresponding to (4.69, 70) and (4.85, 86), respectively. It can be seen that the radix-4 algorithm reduces the number of multiplications to a level about 25% below that of the radix-2 algorithm, while the number of additions is approximately the same. Slight additional improvement can also be obtained by using radix-8 or radix-16 algorithms [4.7, 8]. When N is not a power of a single radix, one is prompted to use a mixed-radix approach. A DFT of 32 points, for example, could be computed by a two-stage radix-4 decomposition followed by

4. The Fast Fourier Transform

Table 4.1. Number of nontrivial real operations for radix-2 and radix-4 FFTs where the complex multiplications are implemented with 3 real multiplications and 3 real additions and where the symmetries of the trigonometric functions are fully used

DFT size N	Radix-2 FFT		Radix-4 FFT	
	Number of multiplications	Number of additions	Number of multiplications	Number of additions
4	0	16	0	16
16	24	152	20	148
64	264	1032	208	976
256	1800	5896	1392	5488
1024	10248	30728	7856	28336
4096	53256	151560	40624	138928

a one-stage radix-2 FFT. When properly designed, such mixed radix methods can be optimum from the standpoint of the number of arithmetic operations, but the additional computational savings are achieved at the expense of a somewhat more complex implementation.

4.2.3 Implementation of FFT Algorithms

The FFT algorithm may be organized in a variety of different ways as a function of the order in which data are accessed and stored and the implementation of the twiddle factors in the computation structure. In order to illustrate the implementation of an actual FFT, we consider here the radix-2 decimation in time algorithm depicted previously in Fig. 4.1.

It is apparent that the computation proceeds in t stages, denoted by i, with $i = 1, 2, ..., t$, and that each stage must compute the $N/2$ *butterfly* operations

$$x_l^i = x_l^{i-1} + W^d\, x_{l+N/2}^{i-1} \tag{4.87}$$

$$x_{l+N/2^i}^i = x_l^{i-1} - W^d\, x_{l+N/2^i}^{i-1}, \tag{4.88}$$

where x_l^{i-1} and x_l^i are, respectively, the input and output data samples corresponding to the ith stage. Since the input and output samples in (4.87, 88) have the same indices, the computation may be executed *in place*, by writing the output results over the input data. Thus, the FFT may be implemented with only N complex storage locations, plus auxiliary storage registers to support the butterfly computation.

The complex value of W^d as a function of index l and stage i can be determined by using a *bit-reversal* method. This is done by writing l as a t-bit binary number, scaling this number by $t-i$ bits to the right, and reversing the order of the bits. Thus, if we consider, for instance, the node x_5^2 corresponding to the second stage of the 8-point DFT illustrated in Fig. 4.1, we have $l = 5$ and $i = 2$. In this case, l is 1 0 1 in binary notation. Scaling by $t - i = 1$ bit to the right yields 0 1 0. Finally d is obtained by reversing 0 1 0, which gives also 0 1 0 or integer 2, and we have $x_5^2 = x_5^1 + W^2 x_7^1$.

The bit-reversal process can also be implemented very simply by counting in bit-reversed notation. For an 8-point DFT, a conventional 3-bit counter yields the successive integers 0, 1, 2, 3, 4, 5, 6, 7. If the counter bit positions are reversed, we have 0, 4, 2, 6, 1, 5, 3, 7, which gives the one-to-one correspondence between the natural order sequence and the bit-reversed order sequence.

The coefficients W^d may also be computed, in each stage, via recursion formula with

$$W^d = WW^{d-1}. \tag{4.89}$$

These coefficients may be precomputed and stored for each stage in order to save computation time at the expense of increased memory.

The algorithm illustrated in Fig. 4.1 produces the DFT output samples $\bar{X}_k$ in bit-reversed order. Thus, these samples must usually be reordered at the end of the computation by performing a bit-reversal operation on the indices k. We shall see in Sect. 4.6, however, that this operation is unnecessary when the DFTs are used to compute convolutions.

The foregoing considerations also apply generally to algorithms using radices greater than 2, and a Fortran program for these FFT forms can be found in [4.8]. In practice, there are many variations of the basic FFT algorithm which correspond to different trade-offs between speed of execution and memory requirements. It is possible, for instance, to devise schemes with identical geometry from stage to stage or with input data and output data in natural order. When the FFT is programmed in a high-level language with sophisticated functions for the manipulation of arrays such as APL [4.9], the implementation can be strikingly simple. This is well illustrated in the radix-2 FFT program designed by McAuliffe and reprinted in Fig. 4.3 with the kind permission of the author.

```
      ∇ Z←TF33 A;K;M;W;O;P;Q;R;S;V;N
[1]     W← 2 1 ○.○○(,⌽(2,P)ρ(PρV←0),-O-V[-O-2×⍳P])⍴P,0ρO←⍳1,0ρS←2,N,0ρR←
        (M+1)ρ2,0ρZ←A[;V←,(⌽⍳M)⌽((K+M+1+2●P←0.5×N)ρ2)ρ⍳N+ ̄1↑ρA]
[2]     →(0<K←K-1)/2,0ρW←W[;,⌽(2,P)ρ⍳P]+0,0ρZ←Sρ(-/[O] W×Z),+/[O] W×⊖Z←S
        ρ((O+K),((-K)⌽0,Mρ1)/⍳M+1)⍉Rρ(,+/[K+O] Z),,-/[K+O] Z←RρZ
      ∇
```

Fig. 4.3. Radix-2 FFT program written in APL.

This APL program uses just two instructions, the first one for generating coefficient values and the second for performing the actual data computation. In this program, the N-point DFT is computed by executing

$$Z \leftarrow \text{TF33 } A, \tag{4.90}$$

where TF33 is the name of the FFT subroutine and A is an array of 2 lines and N columns, the first line representing the real part of the input sequence and the second line representing the imaginary part of the input sequence. The input sequence is given by the array Z which has the same structure as the input data array A. The reader is cautioned, however, to note that this program actually computes an inverse DFT rather than the direct DFT as defined by (4.1).

We give the execution times for various DFT lengths computed with this program on an IBM 370/168 computer operating on APL under VM370 in Table 4.2. These figures can be compared with the execution times for direct computation of the same DFTs in APL with the same system. It can be seen that the reduction in arithmetic load made possible by the FFT algorithm does translate into a comparable reduction in execution time. This is quite apparent for large DFTs and, for instance, a 1024-point DFT is calculated in only 791 ms vis the FFT program, as opposed to 165335 ms for direct computation.

Table 4.2. Comparative execution times in milliseconds for DFTs computed by the FFT program of Fig. 4.3 and by direct computation. IBM 370/168—APL VM370

DFT size N	Execution times (CPU time) [ms]	
	Radix-2 FFT	DFT
4	17	16
8	24	32
16	32	80
32	43	234
64	63	776
128	109	2840
256	188	10765
512	368	41708
1024	791	165335

4.2.4 Quantization Effects in the FFT

Since the FFT is implemented with finite precision arithmetic, the results of the computation are affected by the roundoff noise incurred in the butterfly calculations, the scaling of the data, and the approximate representation of the coefficients W^d. These effects have been studied for fixed point and floating point computations [4.10–12]. We shall restrict our discussion here solely to fixed point radix-2 FFT algorithms.

Consider first the impact of scaling. At each stage, we must compute the butterflies,

$$x_l^i = x_l^{i-1} + W^d x_{l+N/2^i}^{i-1} \tag{4.91}$$

$$x^i_{l+N/2^i} = x^{i-1}_l - W^d x^{i-1}_{l+N/2^i} \qquad (4.92)$$

Thus, the magnitude of the signal samples tends to increase at each stage, the upper bounds on the modulus of x^i_l being given by

$$\text{Max } |x^i_l| \leq 2 \text{ Max } (|x^{i-1}_l|, |x^{i-1}_{l+N/2^i}|). \qquad (4.93)$$

Hence the signal magnitude increases by a maximum of one bit at each stage and a scaling procedure is needed to avoid overflow. An especially efficient scaling procedure would be to compute each stage without scaling, then to scale the entire sequence by one bit, only if an overflow is detected. Alternatively, a simpler, but less efficient method based upon systematic scaling by one bit at each stage can also be employed. In this case, the implementation is simple, but suboptimum. Nevertheless, an evaluation of the quantization effects using this simple scheme provides an upper bound on quantization noise. Thus, in the following analytical development, we shall assume that the data is scaled by one bit at each stage.

It is well known [4.1] that if the product of two B-bit numbers is rounded to B bits, the error variance is given by

$$\sigma^2 = 2^{-2B}/12. \qquad (4.94)$$

Moreover, when two B-bit numbers are added together, the sum may be a $(B+1)$-bit number. Thus, when there is an overflow, the sum must be scaled by $1/2$ and one bit is lost. The variance of the corresponding error is

$$\sigma_1^2 = 2^{-2B}/2 = 6\sigma^2. \qquad (4.95)$$

We shall now assume that errors are uncorrelated and that an overflow occurs at each stage. Since the data input at the first stage of the transform is scaled by $1/2$, the variance $V(x_n)$ of x_n is given by

$$V(x_n) = 6\sigma^2. \qquad (4.96)$$

The first stage computes a set of N data samples with multiplications by ± 1 and additions. Furthermore, the output samples x^1_n from this first stage must be scaled by $1/2$. Hence

$$V(x^1_n) = 2V(x_n) + 4 \cdot 6\sigma^2 = 36\sigma^2, \qquad (4.97)$$

where the factor of 4 accounts for the fact that the error caused by scaling at the first stage is twice the error at the zeroth stage. Similarly, the second stage implements multiplications by only $\pm 1, \pm j$ and we have

$$V(x^2_n) = 2V(x^1_n) + 4^2 \, 6\sigma^2 \qquad (4.98)$$

$$V(x_n^2) = 2^2 \, 6 \, \sigma^2 + 2 \cdot 4 \cdot 6 \sigma^2 + 4^2 \, 6 \sigma^2. \tag{4.99}$$

In the third stage, half the butterfly operations are nontrivial. For these, we have

$$\operatorname{Re}\{x_n^3\} = \operatorname{Re}\{x_n^2\} + \operatorname{Re}\{x_{n+N/8}^2\} \operatorname{Re}\{W^d\} - \operatorname{Im}\{x_{n+N/8}^2\} \operatorname{Im}\{W^d\}, \tag{4.100}$$

which yields

$$V(x_n^3) = V(x_n^2) + (\overline{\operatorname{Re}^2\{x_n^2\}} + \overline{\operatorname{Im}^2\{x_n^2\}})V(W^d)$$
$$+ (\operatorname{Re}^2\{W^d\} + \operatorname{Im}^2\{W^d\})V(x_n^2) + 4^3 \, \sigma^2 + 4^3 \, 6 \sigma^2, \tag{4.101}$$

where the bars over the symbols represent here an average over the sequence. Thus,

$$V(x_n^3) = V(x_n^2) + \overline{|x_n^2|}^2 \, \sigma^2 + V(x_n^2) + 4^3 \, \sigma^2 + 4^3 \, 6 \, \sigma^2, \tag{4.102}$$

where the first term in (4.102) is the variance of the first term in (4.100) and the two next terms in (4.102) correspond to the complex multiplication. The terms $4^3 \, \sigma^2$ derive from the rounding after addition and $4^3 \, 6 \, \sigma^2$ corresponds to rescaling. We now define λ as the average squared modulus of the input sequence

$$\lambda = \overline{(x_n)}^2. \tag{4.103}$$

Since λ increases by a factor of two at each stage, we have

$$V(x_n^3) = 2V(x_n^2) + 2^2 \lambda \sigma^2 + 4^3 \sigma^2 + 4^3 6 \sigma^2. \tag{4.104}$$

However, since the second and third terms in (4.104) appear only when the multiplications are nontrivial and, since half the multiplications in the third stage are trivial, $V(x_n^3)$ reduces to

$$V(x_n^3) = 2V(x_n^2) + 2\lambda \sigma^2 + 4^3 \sigma^2/2 + 4^3 6 \sigma^2 \tag{4.105}$$

$$V(x_n^3) = 2^3 6 \sigma^2 + 2^2 4 \cdot 6 \sigma^2 + 2 \cdot 4^2 6 \sigma^2 + 2 \lambda \sigma^2 + 4^3 \sigma^2/2 + 4^3 6 \sigma^2 \tag{4.106}$$

and, assuming a similar computation procedure for all other stages, we have, for the last stage,

$$V(X_k) \simeq 2^{2t+3} \, \sigma^2 + [M - (5/2)] 2^{t-1} \lambda \sigma^2 + 2^{t+2} \sigma^2. \tag{4.107}$$

Since the mean square of the absolute values of the output sequence X_k (we delete here our usual bar sign on transforms in order to avoid confusion with averaging) is $2^t \lambda$, the ratio of rms noise output to rms signal output is, for large DFTs,

$$\frac{\text{rms (error)}}{\text{rms (signal)}} \simeq \frac{\sqrt{N}\, 2^{-B}\, (0.3)\sqrt{8}}{\text{rms (input)}}, \tag{4.108}$$

which demonstrates that the error-to-signal ratio of the FFT process increases as $\sqrt{N}$ or 1/2 bits per stage.

Another source of error is due to the use of truncated coefficients. Weinstein [4.12] has shown, by a simplified statistical analysis, that this effect translates into an error-to-signal ratio which increases very slowly with N. Experimental results have tended to confirm this analysis result.

4.3 The Rader-Brenner FFT

The evaluation of DFTs by the conventional FFT algorithm requires complex multiplications. We shall show now that a simple modification of the FFT algorithm replaces these complex multiplications by multiplications of a complex number by either a pure real or a pure imaginary number [4.13]. This is realized by computing an N-point DFT, with $N = 2^t$

$$\bar{X}_k = \sum_{m=0}^{N-1} x_m W^{mk}, \quad k = 0, \ldots, N-1 \tag{4.109}$$

via a decimation in frequency radix-2 FFT form, which for k even, yields

$$\bar{X}_{2k} = \sum_{m=0}^{N/2-1} (x_m + x_{m+N/2}) W^{2mk}, \quad k = 0, \ldots, N/2 - 1, \tag{4.110}$$

where k is replaced by $2k$.
For k odd, replacing k by $2k + 1$, yields

$$\bar{X}_{2k+1} = \sum_{m=0}^{N/2-1} (x_m - x_{m+N/2}) W^m W^{2mk}, \quad k = 0, \ldots, N/2 - 1. \tag{4.111}$$

Thus, the first stage of the decimation in frequency FFT decomposition replaces one DFT of length N by two DFTs of length $N/2$ at the cost of N complex additions and $N/2$ complex multiplications. In order to simplify the calculation of the DFT $\bar{X}_{2k+1}$, we define the $(N/2)$-point auxiliary sequence a_m by

$$\begin{cases} a_m = (x_m - x_{m+N/2})/[2 \cos(2\pi m/N)], & m \neq 0, N/4. \\ a_0 = 0 \quad a_{N/4} = 0. \end{cases} \tag{4.112}$$

We then compute the $(N/2)$-point DFT $\bar{A}_k$ of a_m

$$\bar{A}_k = \sum_{m=0}^{N/2-1} a_m W^{2mk}, \quad k = 0, \ldots, N/2 - 1. \tag{4.113}$$

4. The Fast Fourier Transform

$\bar{X}_{2k+1}$ can be recovered from $\bar{A}_k$ by noting that

$$\bar{A}_k + \bar{A}_{k+1} = \sum_{m=0}^{N/2-1} a_m(1 + W^{2m})W^{2mk} \qquad (4.114)$$

or

$$\bar{A}_k + \bar{A}_{k+1} = \sum_{m=0}^{N/2-1} [2a_m \cos(2\pi m/N)]W^m W^{2mk}. \qquad (4.115)$$

And, since $W^{N/4} = -j$,

$$\begin{cases} \bar{X}_{2k+1} = \bar{A}_k + \bar{A}_{k+1} + v_0 & \text{for } k \text{ even} \\ \bar{X}_{2k+1} = \bar{A}_k + \bar{A}_{k+1} + v_1 & \text{for } k \text{ odd} \end{cases} \qquad (4.116)$$

with

$$v_0 = x_0 - x_{N/2} - j(x_{N/4} - x_{3N/4}) \qquad (4.117)$$

$$v_1 = x_0 - x_{N/2} + j(x_{N/4} - x_{3N/4}). \qquad (4.118)$$

Under these conditions, the $N/2$ complex multiplications by the twiddle factors W^m in the first stage are replaced with $(N/2) - 2$ multiplications by the pure real numbers $1/[2 \cos(2\pi m/N)]$. Note here that the contributions of $x_0 - x_{N/2}$ and $X_{N/4} - x_{3N/4}$ must be treated separately, because $\cos(2\pi m/N) = 0$ for $m = N/4$. The same method is used recursively to compute the $(N/2)$-point transforms $\bar{X}_{2k}$ and $\bar{A}_k$, and then the transforms of dimensions $N/4$, $N/8$ ... until complete decomposition is achieved.

Since the multiplication of a complex number by a scalar value is implemented with two real multiplications, each stage is computed with $N-4$ nontrivial real multiplications. We need also N complex additions for evaluating $x_m + x_{m+N/2}$ and $x_m - x_{m+N/2}$ plus $N + 2$ complex additions for calculating (4.116–118). However, two complex additions are saved in the computation of $\bar{A}_k$, because $a_0 = 0$ and $a_{N/4} = 0$. Thus, for each stage, the number of real multiplications M and real additions A become

$$M = N - 4 \qquad (4.119)$$

$$A = 4N. \qquad (4.120)$$

The two last stages of the decomposition correspond to transforms of dimensions 4 and 2 which are computed by the conventional FFT methods with trivial multiplications by ± 1 and $\pm j$. Moreover, the two preceding stages, corresponding, respectively, to DFTs of lengths 16 and 8, are also computed more efficiently by conventional methods such as a radix-4 algorithm (Sect. 4.2.2) or the Winograd algorithm [4.14]. Under these conditions, the number of real operations

Table 4.3. Number of nontrivial real operations for complex DFTs computed by the Rader-Brenner method

DFT size N	Number of real multiplications	Number of real additions	Multiplications per point	Additions per point
8	4	52	0.50	6.50
16	20	148	1.25	9.25
32	68	424	2.12	13.25
64	196	1104	3.06	17.25
128	516	2720	4.03	21.25
256	1284	6464	5.02	25.25
512	3076	14976	6.01	29.25
1024	7172	34048	7.00	33.25
2048	16388	76288	8.00	37.25

for the DFTs of complex input sequences evaluated by the Rader-Brenner method is given in Table 4.3. It can be seen, by comparison with Table 4.1, that the Rader-Brenner technique reduces the number of multiplications over the radix-2 and radix-4 FFT algorithms, while requiring about 10% more additions.

The same method may also be implemented in a decimation in time arrangement [4.13]. In this case, the premultiplications by $1/[2\cos(2\pi m/N)]$ are replaced by postmultiplications by $1/[2\cos(2\pi m/N)]$ and the computational complexity is the same as with the decimation in frequency approach. It should be noted that, for large transforms, $\cos(2\pi m/N)$ becomes very small for some values of m. Then, the multiplications by $1/[2\cos(2\pi m/N)]$ for these values introduce large errors. Cho and Temes [4.15], however, have proposed a modification of the basic Rader-Brenner algorithm to overcome this limitation.

In many instances, one needs only the odd output terms of a DFT. These terms are generated by (4.111) and can be viewed as the modified DFT

Table 4.4. Number of nontrivial real operations for complex reduced DFTs computed by the Rader-Brenner algorithm

Reduced DFT size	Number of real multiplications	Number of real additions
8	16	64
16	48	212
32	128	552
64	320	1360
128	768	3232
256	1792	7488
512	4096	17024
1024	9216	38144

$$\bar{Y}_k = \sum_{m=0}^{N/2-1} y_m W^m W^{2mk}, \qquad k = 0, \ldots, N/2 - 1$$

$$W = e^{-j2\pi/N}. \tag{4.121}$$

Such a modified DFT, which occurs naturally in the first stage of a decimation in frequency FFT algorithm, is often called a *reduced* DFT [4.16] or an *odd* DFT [4.17] and is used, for instance, in the computation of multidimensional DFTs by polynomial transforms (Chap. 7). The Rader-Brenner algorithm applies directly to the calculation of such reduced DFTs and we give, in Table 4.4, the number of nontrivial real operations for reduced DFTs computed via this method.

4.4 Multidimensional FFTs

We consider first a two-dimensional DFT of size $N_1 \times N_2$, with

$$\bar{X}_{k_1,k_2} = \sum_{m_1=0}^{N_1-1} \sum_{m_2=0}^{N_2-1} x_{m_1,m_2} W_1^{m_1 k_1} W_2^{m_2 k_2}$$

$$W_1 = e^{-j2\pi/N_1}, \; W_2 = e^{-j2\pi/N_2}$$

$$k_1 = 0, \ldots, N_1 - 1, \qquad k_2 = 0, \ldots, N_2 - 1. \tag{4.122}$$

In order to evaluate this DFT, we first rewrite (4.122) as

$$\bar{X}_{k_1,k_2} = \sum_{m_1=0}^{N_1-1} W_1^{m_1 k_1} \sum_{m_2=0}^{N_2-1} x_{m_1,m_2} W_2^{m_2 k_2}. \tag{4.123}$$

As a first step, we evaluate the N_1 DFTs $\bar{Y}_{m_1,k_2}$ of length N_2 which correspond to the N_1 distinct values of m_1

$$\bar{Y}_{m_1,k_2} = \sum_{m_2=0}^{N_2-1} x_{m_1,m_2} W_2^{m_2 k_2}. \tag{4.124}$$

$\bar{X}_{k_1,k_2}$ is then obtained by calculating N_2 DFTs of length N_1 on the N_2 sequences $\bar{Y}_{m_1,k_2}$ corresponding to the N_2 distinct values of k_2

$$\bar{X}_{k_1,k_2} = \sum_{m_1=0}^{N_1-1} \bar{Y}_{m_1,k_2} W_1^{m_1 k_1}. \tag{4.125}$$

This approach is often called the *row-column* method because it can be viewed as equivalent to organizing the input data into sets of row and column vectors in an array of size $N_1 \times N_2$ and computing, in sequence, first the DFTs of the columns and then the DFTs of the rows. With this technique, the two-dimensional DFT is mapped, respectively, into N_1 DFTs of N_2 terms plus N_2 DFTs of N_1 terms. If N_1 and N_2 are powers of two, the one-dimensional DFTs of lengths N_1 and N_2 can be evaluated by a FFT-type algorithm. In the case of a simple radix-2 decomposition, the number of complex multiplications M becomes

$$M = N_1 N_2 \, [\log_2 (N_1 N_2)]/2 \tag{4.126}$$

and, for a DFT of size $N \times N$,

$$M = 2N \, M_1, \tag{4.127}$$

where M_1 is the number of multiplications required to compute a DFT of length N. The same method also applies to more than two dimensions, and a d-dimensional DFT of size $N \times N \times N \times \ldots$ is calculated with dN^{d-1} DFTs of length N so that the number of multiplications becomes

$$M = dN^{d-1} \, M_1 \tag{4.128}$$

and, in particular,

$$M = (dN^d \log_2 N)/2 \tag{4.129}$$

when the DFTs of N terms are evaluated with a simple radix-2 FFT-type algorithm. We shall see in Chap. 7 that the multidimensional to one-dimensional DFT mapping obtained with the row-column method is suboptimal and that better methods can be devised by using polynomial transforms. In order to support a quantitative comparison of the computational complexities for the two methods, we present in Table 4.5 the number of real operations for various complex two-dimensional DFTs calculated by the row-column method and the Rader-Brenner algorithm.

Table 4.5. Number of nontrivial real operations for complex DFTs of size N × N computed by the row-column method and the Rader-Brenner algorithm

N	Number of real multiplications	Number of real additions	Multiplications per point	Additions per point
8	64	832	1.00	13.00
16	640	4736	2.50	18.50
32	4352	27136	4.25	26.50
64	25088	141312	6.12	34.50
128	132096	696320	8.06	42.50
256	657408	3309568	10.03	50.50
512	3149824	15335424	12.02	58.50
1024	14688256	69730304	14.01	66.50

4.5 The Bruun Algorithm

We shall now discuss an algorithm introduced by Bruun [4.18] which has both theoretical and practical significance. The practical value of this algorithm relates to the fact that the DFT of real data can be computed almost entirely with real arithmetic, thereby simplifying the implementation of DFTs for real data. We shall present here a modified version of the original algorithm which will allow us to introduce a polynomial definition of the DFTs that will be used in later parts of this book.

We consider again an N-point DFT, with $N = 2^t$

$$\bar{X}_k = \sum_{m=0}^{N-1} x_m W^{mk}, \quad k = 0, ..., N-1$$

$$W = e^{-j2\pi/N}, \quad j = \sqrt{-1}. \tag{4.130}$$

In order to develop the algorithm, we replace (4.130) with a polynomial representation of the DFT defined by the two following equations:

$$X(z) \equiv \sum_{m=0}^{N-1} x_m z^m \quad \text{modulo } (z^N - 1) \tag{4.131}$$

$$\bar{X}_k \equiv X(z) \quad \text{modulo } (z - W^k). \tag{4.132}$$

Equations (4.131) and (4.132) are equivalent to (4.130) because the definition of (4.132) modulo $(z - W^k)$ means that we can replace z by W^k in (4.131). At this point, the definition of (4.131) modulo $(z^N - 1)$ is unnecessary. However, this definition is valid because $z^N \equiv W^{kN} = 1$. We note that the N roots of $z^N - 1$ are given by W^k for $k = 0, ..., N-1$, with

$$z^N - 1 = \prod_{k=0}^{N-1} (z - W^k). \tag{4.133}$$

Moreover, since $N = 2^t$, we can express $z^N - 1$ as the product of two polynomials of $N/2$ terms in z, with

$$z^N - 1 = (z^{N/2} - 1)(z^{N/2} + 1) \tag{4.134}$$

and

$$z^{N/2} - 1 = \prod_{k_1=0}^{N/2-1} (z - W^{2k_1}) \tag{4.135}$$

$$z^{N+2} + 1 = \prod_{k_1=0}^{N/2-1} (z - W^{2k_1+1}), \quad k_1 = 0, ..., N/2 - 1. \tag{4.136}$$

4.5 The Bruun Algorithm

Hence, for k even, all the values of W^k correspond to the polynomial $z^{N/2} - 1$ and we can replace (4.131) and (4.132) with

$$X_1(z) = \sum_{m=0}^{N/2-1} (x_m + x_{m+N/2})z^m \equiv X(z) \text{ modulo } (z^{N/2} - 1) \tag{4.137}$$

$$\bar{X}_k(z) \equiv X_1(z) \text{ modulo } (z - W^k), \quad k \text{ even.} \tag{4.138}$$

Similarly, for k odd, all the values of W^k correspond to the polynomial $z^{N/2} + 1$ and $\bar{X}_k(z)$ is computed by

$$X_2(z) = \sum_{m=0}^{N/2-1} (x_m - x_{m+N/2})z^m \equiv X(z) \text{ modulo } (z^{N/2} + 1) \tag{4.139}$$

$$\bar{X}_k(z) \equiv X_2(z) \text{ modulo } (z - W^k), \quad k \text{ odd.} \tag{4.140}$$

The form (4.137–140) can be easily recognized as equivalent to the first stage of a decimation in frequency FFT decomposition, since (4.137, 138) represent a DFT of $N/2$ terms while (4.139,140) represent an odd DFT of $N/2$ terms. At this stage, we depart from the conventional FFT decomposition by noting that any polynomial of the form $z^{4q} + az^{2q} + 1$ factors into two real polynomials,

$$z^{4q} + az^{2q} + 1 = (z^{2q} + \sqrt{2-a}\, z^q + 1)(z^{2q} - \sqrt{2-a}\, z^q + 1). \tag{4.141}$$

This implies, therefore,

$$z^{N/2} + 1 = (z^{N/4} + \sqrt{2}\, z^{N/8} + 1)(z^{N/4} - \sqrt{2}\, z^{N/8} + 1) \tag{4.142}$$

with

$$z^{N/4} + \sqrt{2}\, z^{N/8} + 1 = \prod_{k_1 \in B_1} (z - W^{2k_1+1}) \tag{4.143}$$

$$z^{N/4} - \sqrt{2}\, z^{N/8} + 1 = \prod_{k_1 \in B_2} (z - W^{2k_1+1}), \tag{4.144}$$

where B_1 is the set of $N/4$ values of k_1 such that W^{2k_1+1} is a root of $z^{N/4} + \sqrt{2}\, z^{N/8} + 1$ and B_2 is the set of the $N/4$ other values of k_1. Under these conditions, the odd DFT represented by (4.139,140) can be replaced by

$$X_3(z) \equiv X_2(z) \text{ modulo } (z^{N/4} + \sqrt{2}\, z^{N/8} + 1) \tag{4.145}$$

$$\bar{X}_{2k_1+1} \equiv X_3(z) \text{ modulo } (z - W^{2k_1+1}), \quad k_1 \in B_1 \tag{4.146}$$

and

$$X_4(z) \equiv X_2(z) \text{ modulo } (z^{N/4} - \sqrt{2}\, z^{N/8} + 1) \tag{4.147}$$

106   4. The Fast Fourier Transform

$$\bar{X}_{2k_1+1} \equiv X_4(z) \quad \text{modulo } (z - W^{2k_1+1}), \quad k_1 \in B_2. \tag{4.148}$$

The same decomposition process can then be repeated by systematically expressing the polynomials of the form $z^{4q} + az^{2q} + 1$ as the products of two real polynomials of degree $2q$, until the polynomials are reduced to degree 2. At this point, further decomposition as the product of two real polynomials is no longer possible and two DFT output terms are obtained for each degree 2 polynomial by replacing z with the two complex roots of the polynomial. This process is summarized in Fig. 4.4 for a DFT of 8 terms. In this diagram, each box represents a reduction modulo the polynomial indicated in the box.

We note that the first stage, which corresponds to the reductions modulo $(z^{N/2} - 1)$ and modulo $(z^{N/2} + 1)$ is computed by (4.137) and (4.139) with N com-

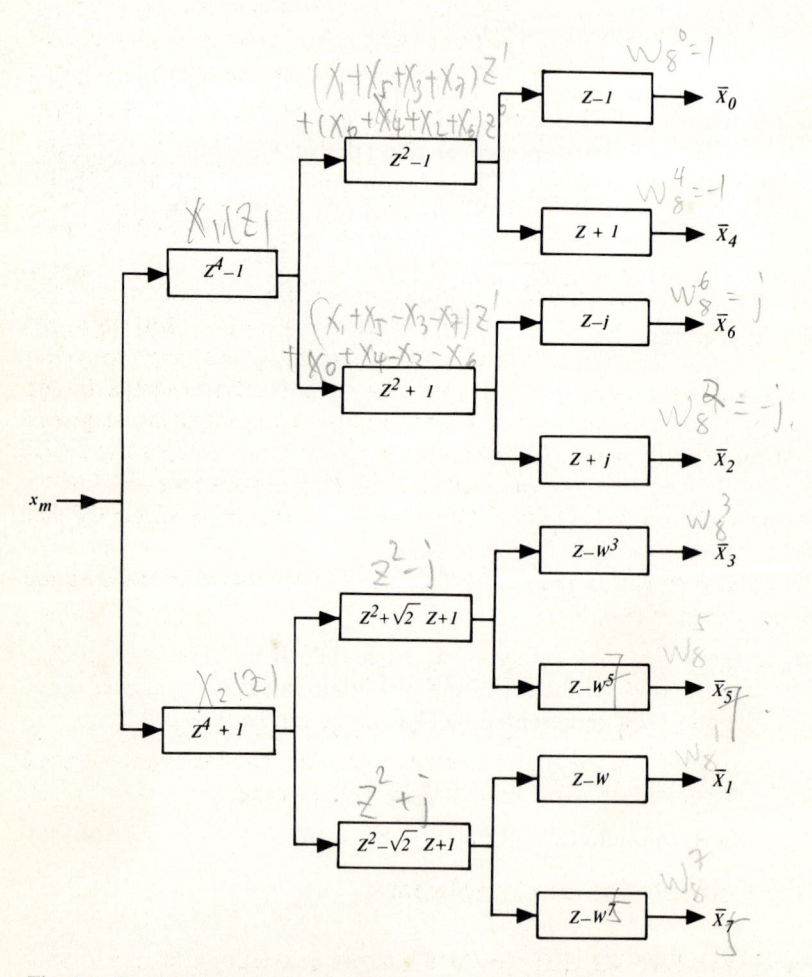

Fig. 4.4. Computation of a 8-point DFT by Bruun's algorithm.

plex additions. In the second stage, the reductions modulo $(z^{N/4} - 1)$ and modulo $(z^{N/4} + 1)$ are computed with $N/2$ complex additions. For the reductions modulo $(z^{N/4} + \sqrt{2}\, z^{N/8} + 1)$, we have $z^{N/4} \equiv -\sqrt{2}\, z^{N/8} - 1$ and $z^{3N/8} \equiv z^{N/8} + \sqrt{2}$, and for the reductions modulo $(z^{N/4} - \sqrt{2}\, z^{N/8} + 1)$, $z^{N/4} \equiv \sqrt{2}\, z^{N/8} - 1$ and $z^{3N/8} \equiv z^{N/8} - \sqrt{2}$. Since $\sqrt{2}$ is real, the complex multiplications are implemented with two real multiplications and the two reductions are implemented with $N/2$ real multiplications and $3N/2$ additions. The second stage corresponds to $a = 0$ in (4.141). In the following stages, a takes successively the values $\pm\sqrt{2}, \pm\sqrt{2 \pm \sqrt{2}}$ ... and the reductions proceed similarly, with multiplications by the real factors a and $\sqrt{2-a}$, the only difference with the second stage being that the multiplications by a are no longer trivial.

In the last stage, we have two reductions with trivial multiplications by ± 1, $\pm j$ and two reductions with multiplications by powers of $W^{N/8}$ which require 2 real multiplications for each reduction. The $N/2 - 4$ other reductions correspond to multiplications by W^k, W^{-k} which are implemented with 4 real multiplications and 8 real additions. Hence, with the exception of the last stage, all multiplications are done with real factors. In practice, the original algorithm proposed by Bruun uses aperiodic convolutions instead of reductions, and this original approach requires slightly fewer arithmetic operations than the method described here.

The principal use of the Bruun algorithm is in the calculation of the DFT for real data sequences. In this case, since the coefficients in the $t - 1$ first stages are real, these stages are implemented in real arithmetic. Moreover, since the reductions in the last stage correspond to multiplications of a real data sample by the complex conjugate coefficients W^k and W^{-k}, the operations in the last stage can also be viewed as implemented in real arithmetic, with 2 real multiplications and 1 real addition for each reduction modulo $(z - W^k)$ and modulo $(z - W^{-k})$. Thus, the Bruun algorithm provides a convenient way of computing the DFT of a real data vector using only real arithmetic.

It should also be noted that the Bruun algorithm is closely related to the Rader-Brenner algorithm, since

$$(z - W^k)(z - W^{-k}) = z^2 - 2z \cos(2\pi k/N) + 1. \tag{4.149}$$

Hence the various coefficients used in the Bruun algorithm are identical to corresponding coefficients in the Rader-Brenner algorithm, and the main difference relates to multiplications by W^k and W^{-k} in the last stage.

4.6 FFT Computation of Convolutions

We have seen in Sect. 4.1 that the DFT has the convolution property. This means that the circular convolution y_l of two sequences h_n and x_m can be computed by evaluating the DFTs $\bar{H}_k$ and $\bar{X}_k$ of h_n and x_m, by multiplying, term by term, $\bar{H}_k$ by $\bar{X}_k$, and by computing the inverse DFT of $\bar{H}_k \bar{X}_k$. Hence we have

$$y_l = \text{DFT}^{-1}\{[\text{DFT}(h_n)][\text{DFT}(x_m)]\} \tag{4.150}$$

for the convolution

$$y_l = \sum_{n=0}^{N-1} h_n x_{l-n}, \quad l = 0, \ldots, N-1. \tag{4.151}$$

Since a DFT can be computed by the FFT algorithm, this method requires a number of operations proportional to $N \log N$ and, therefore, requires considerably less computation than the direct method. More precisely, if the DFTs are calculated via simple radix-2 algorithm with one of the input sequences fixed, the circular convolution of length N, with $N = 2^t$, requires the computation of two FFTs and N complex multiplications. Consequently, the number of complex multiplications M required to evaluate the convolution is

$$M = N(1 + \log_2 N). \tag{4.152}$$

For large convolutions, this is considerably less than the N^2 multiplications required for the direct computation of (4.151).

Frequently, one may wish to evaluate a real convolution by the FFT method. This can be done by computing the convolutions of two successive blocks simultaneously. Assuming that h_n is fixed, we compute the convolution of h_n with the two real N-point sequences x_m and x_{m+N} by first constructing the auxiliary sequence $x_m + jx_{m+N}$. The complex convolution of h_n with $x_m + jx_{m+N}$ is then computed by DFTs to yield the complex convolution $y_l + jy_{l+N}$. Thus, the convolution of the first block with h_n is defined by the real part of the complex convolution, and the convolution corresponding to the second block by the imaginary part. With this method, the number of operations required to compute a real convolution is half that of a complex convolution.

Tables 4.6 and 4.7 list the number of real operations corresponding to the calculation of real one-dimensional and two-dimensional convolutions by FFTs, using the Rader-Brenner algorithm. In these tables, we have assumed that one of the sequences is fixed, that two real convolutions are computed for each complex DFT, and that complex multiplication is implemented with 3 real multiplications and 3 real additions. It can be verified easily that the FFT approach reduces drastically the number of operations: for example, a real circular convolution of 1024 points computed by FFTs requires only 8708 multiplications, as opposed to 1048576 multiplications for the direct method, or about 100 times fewer multiplications.

When real convolutions are calculated by dedicated special purpose FFT hardware, it is often desirable to compute the DFT and the inverse DFT in a single transform step with the same hardware, rather than evaluating two real

Table 4.6. Number of real operations for real circular convolutions computed by the Rader-Brenner algorithm (2 real convolutions per DFT; one input sequence fixed)

Convolution size N	Number of real multiplications	Number of real additions	Multiplications per point	Additions per point
8	16	64	2.00	8.00
16	44	172	2.75	10.75
32	116	472	3.62	14.75
64	292	1200	4.56	18.75
128	708	2912	5.53	22.75
256	1668	6848	6.52	26.75
512	3844	15744	7.51	30.75
1024	8708	35584	8.50	34.75
2048	19460	79360	9.50	38.75

Table 4.7. Number of real operations for real circular convolutions of size $N \times N$ computed by the Rader-Brenner algorithm. (2 real convolutions per DFT; one input sequence fixed)

N	Number of real multiplications	Number of real additions	Multiplications per point	Additions per point
8	160	928	2.50	14.50
16	1024	5120	4.00	20.00
32	5888	28672	5.75	28.00
64	31232	147456	7.62	36.00
128	156672	720896	9.56	44.00
256	755712	3407872	11.53	52.00
512	3543040	15728640	13.52	60.00
1024	16261120	71303168	15.51	68.00

convolutions simultaneously with separate hardware for the DFT and the inverse DFT. This can be accommodated using an approach, proposed by McAuliffe [4.19], which is based on the computation of the DFTs of two real sequences in a single complex DFT step.

We have already seen in Sect. 4.1 that the DFTs $\bar{X}_k$ and $\bar{X}_k^1$ of two real N-point sequences x_m and x_m^1 can be evaluated as a single complex DFT by computing the DFT $\bar{Y}_k$ of the auxiliary sequence $x_m + jx_m^1$. The sequences $\bar{X}_k$ and $\bar{X}_k^1$ are then deduced from $\bar{Y}_k$ by

$$\bar{X}_k = (\bar{Y}_k + \bar{Y}_{-k}^*)/2 \qquad (4.153)$$

$$\bar{X}_k^1 = (\bar{Y}_k - \bar{Y}_{-k}^*)/2j, \qquad (4.154)$$

where $\bar{Y}_{-k}^*$ is the complex conjugate of $\bar{Y}_{-k}$. Following this procedure, the convolution y_l of the two real sequences x_m and h_n is computed as shown in Fig. 4.5.

110   4. The Fast Fourier Transform

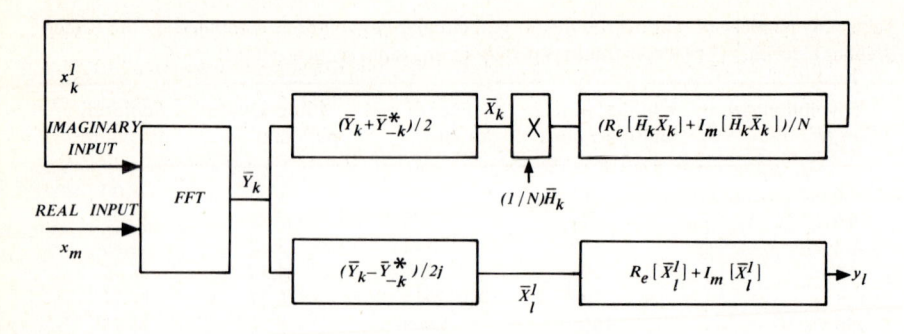

Fig. 4.5. Computation of a real convolution in a single FFT step.

The transform $\bar{X}_k$ of x_m is derived from $\bar{Y}_k$ by (4.153). $\bar{X}_k$ is then multiplied with $\bar{H}_k/N$, where $\bar{H}_k$ is the DFT of h_n, and the real parts and imaginary part of $\bar{H}_k \bar{X}_k/N$ are added, thus yielding the sequence x_k^1,

$$x_k^1 = \frac{1}{2N} \sum_{n=0}^{N-1} \sum_{m=0}^{N-1} h_n x_m [W^{(m+n)k} + W^{-(m+n)k} - jW^{(m+n)k} + jW^{-(m+n)k}]. \tag{4.155}$$

The sequence x_k^1 is then used as the imaginary input to the FFT $\bar{Y}_k$, and the transform $\bar{X}_l^1$ of x_k^1 is obtained by (4.154). Hence

$$\bar{X}_l^1 = \frac{1}{2N} \sum_{n=0}^{N-1} \sum_{m=0}^{N-1} h_n x_m \sum_{k=0}^{N-1} [W^{(m+n+l)k} + W^{-(m+n-l)k} - jW^{(m+n+l)k} + jW^{-(m+n-l)k}]. \tag{4.156}$$

The terms in the summation over W are different from zero only for $m + n + l \equiv 0$ modulo N and $m + n - l \equiv 0$ modulo N. Thus, we have

$$\bar{X}_l^1 = \sum_{n=0}^{N-1} (h_n x_{-n-l} + h_n x_{l-n} - jh_n x_{-n-l} + jh_n x_{l-n})/2, \tag{4.157}$$

where a summation of the real and imaginary parts of $\bar{X}_k^1$ obviously yields the convolution y_l. Clearly, one must account for the FFT computation delay in the process and, in practice, the imaginary input to the FFT hardware usually corresponds to the block x_{m-N}, while the real input corresponds to the block x_m. Hence real convolutions of dimension N can be computed with a single N-point FFT hardware structure.

It should also be noted that some simplifications of the FFT process are possible when used to compute convolutions. In particular, when a DFT is computed by an FFT algorithm, since either the input sequence or the output sequence must be in bit-reversed order, some amount of computation is required to reorder the sequence. However, when the FFT method is used to compute convolutions, this requirement can be ignored because it is always possible to organize the two direct FFTs to provide outputs in the bit-reversed order which is a compatible input to an inverse FFT that produces its output sequence in natural order.

5. Linear Filtering Computation of Discrete Fourier Transforms

The FFT algorithm reduces drastically the number of arithmetic operations required to compute discrete Fourier transforms and is easily implemented on most existing computers. Thus, it is usually advantageous to compute linear filtering processes via the circular convolution property of the DFT with the FFT algorithm. Under these conditions, it would seem paradoxical to develop linear filtering algorithms for the computation of the DFT. This may explain why some algorithms which have been introduced in 1968 by Bluestein [5.1, 2] and Rader [5.3] have long been regarded as a curiosity.

However, a number of recent developments have given an increased importance to the use of such linear filtering algorithms for the computation of the DFTs. For real time execution, new devices, such as charge coupled devices (CCD) or acoustic surface wave devices (ASW) have been developed to implement fairly complex filters on a single chip and can be used as basic building blocks in the computation of DFTs. Moreover, new results in complexity theory have shown that some convolutions can indeed be computed more efficiently with linear filtering methods. This point has been clearly demonstrated by Winograd [5.4], who has introduced a fast DFT algorithm based on the nesting of small DFTs computed as convolutions. This algorithm, which is fundamentally different from the FFT is, for a variety of vector lengths, more efficient than the FFT. In this chapter, we shall first discuss several basic algorithms that can be used to convert DFTs into convolutions: namely, the chirp z-transform algorithm and Rader's algorithm. We shall then show how large DFTs can be computed from a set of small DFTs by the Good prime factor technique and the Winograd Fourier transform algorithm.

5.1 The Chirp z-Transform Algorithm

Consider the DFT of a sequence x_n

$$\bar{X}_k = \sum_{n=0}^{N-1} x_n W^{nk}, \quad k = 0, \ldots, N-1$$

$$W = e^{-j2\pi/N}, \quad j = \sqrt{-1} \tag{5.1}$$

We now rearrange the exponents n and k by noting that

$$nk = -(k-n)^2/2 + n^2/2 + k^2/2. \tag{5.2}$$

Thus, (5.1) becomes

$$\bar{X}_k = W^{k^2/2} \sum_{n=0}^{N-1} x_n W^{n^2/2} W^{-(k-n)^2/2}, \qquad (5.3)$$

which shows that $\bar{X}_k$ may be computed by convolving the sequence $x_n W^{n^2/2}$ with the sequence $W^{-n^2/2}$, and postmultiplying by $W^{k^2/2}$ as indicated on Fig. 5.1. With this method, the DFT is computed by N complex premultiplications, N complex postmultiplications, and one complex finite impulse response (FIR) filter. The impulse response of the FIR filter is that of a chirp filter, well known in radar signal processing; hence the name *chirp z-transform* given to this DFT computation technique [5.1–3, 5].

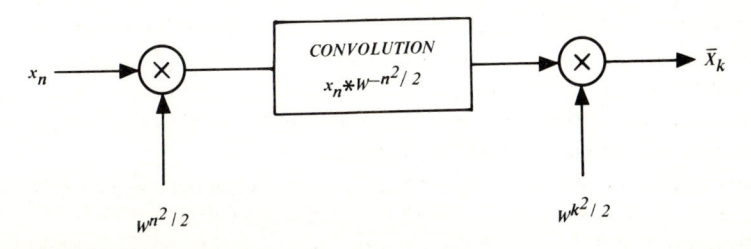

Fig. 5.1. DFT computation using chirp filtering

Since we are evaluating an N-point DFT, we only need to compute N output terms of the chirp filter. Moreover, the indices n and k are defined modulo N and, for N even, $W^{-(n+N)^2/2} = W^{-n^2/2}$. Thus, for N even, the chirp filtering process can be regarded as an N-point circular convolution of complex sequences. For N odd, $W^{-(n+N)^2/2} = -W^{-n^2/2}$, so that the chirp filtering process corresponds to a circular convolution of size $2N$ where the N-point input sequence $x_n W^{n^2/2}$ is augmented by appending N zeros and where only the first N output samples are computed.

One of the significant points demonstrated by the chirp z-transform algorithm is that the DFT may always be computed with a number of operations proportional to $N \log N$, even if N is not highly composite. This can be seen by considering, for instance, the case of N even. Here, the circular convolution of N points can always be computed as a circular convolution of length d, with $d \geqslant 2N - 1$, by using the overlap-add technique. If d is chosen to be a power of 2, this augmented circular convolution can in turn be evaluated by FFTs with a number of operations proportional to $N \log N$.

5.1.1 Real Time Computation of Convolutions and DFTs Using the Chirp z-Transform

Relatively complex FIR filters may be implemented very efficiently on a single chip either with CCD or with ASW devices. In these filters, the filter coefficient tap values are determined by the geometry of electrodes photoengraved on the chip. Thus, these devices, which operate at very high speed on sampled analog signals, are well adapted to the implementation of filters with fixed tap values. When CCD or ASW devices are used to compute the DFT, the chirp-filter structure shown in Fig. 5.1 is generally employed, with complex multiplications and complex convolutions implemented with real multiplications and real convolutions arranged in the conventional butterfly configuration. The filters are integrated on one or several chips with off-chip multipliers for the premultiplications and postmultiplications [5.6].

For filtering applications, the direct implementation of filters by CCD or ASW devices is often unattractive because it can require a new chip design for each filter design and it is not readily applicable to time-variant filters. Thus, it is generally preferable to build digital filters with CCD or ASW implemented Fourier transform circuits. In this case, some of the premultiplication and postmultiplication circuits can be eliminated by combining the postmultiplication in one of the direct transforms with the premultiplication in the inverse transform. This may be seen more precisely as follows.

We want to evaluate the circular convolution y_l of two sequences x_n and h_m

$$y_l = \sum_{n=0}^{N-1} x_n h_{l-n}. \tag{5.4}$$

This is done via the chirp z-transform $\bar{X}_k$ of x_n by (5.3) and the chirp z-transform $\bar{H}_k$ of h_m by

$$\bar{H}_k = W^{k^2/2} \sum_{m=0}^{N-1} h_m W^{m^2/2} W^{-(k-m)^2/2}. \tag{5.5}$$

The convolution product y_l is obtained by calculating $Y_k = \bar{H}_k \bar{X}_k$ and computing the inverse chirp z-transform of Y_k

$$y_l = (1/N) W^{-l^2/2} \sum_{k=0}^{N-1} Y_k W^{-k^2/2} W^{(l-k)^2/2}. \tag{5.6}$$

Note that in (5.6), Y_k is multiplied by $W^{-k^2/2}$ while the postmultiplication of $\bar{X}_k$ in (5.3) is equivalent to multiplying Y_k by $W^{k^2/2}$. Thus, the postmultiplication in $\bar{X}_k$ and the premultiplication in (5.6) cancel and can be dropped.

5.1.2 Recursive Computation of the Chirp z-Transform

When a DFT is evaluated by the chirp z-transform technique, most of the computation occurs in the chirp filtering process. The z-transform, $H(z)$ of the impulse response of the chirp filter is given by

$$H(z) = \sum_{n=0}^{2N-1} W^{-n^2/2} z^{-n}. \qquad (5.7)$$

We assume now that N is a perfect square, with $N = N_1^2$, and we change the index n with

$$n = N_1 n_2 + n_1, \qquad n_2 = 0, \ldots, 2N_1 - 1$$
$$n_1 = 0, \ldots, N_1 - 1. \qquad (5.8)$$

Hence,

$$H(z) = \sum_{n_1=0}^{N_1-1} W^{-n_1^2/2} z^{-n_1} \sum_{n_2=0}^{2N_1-1} W^{-N_1 n_1 n_2}(-1)^{n_2} z^{-N_1 n_2}, \qquad (5.9)$$

which, in turn, implies

$$H(z) = \sum_{n_1=0}^{N_1-1} W^{-n_1^2/2} z^{-n_1}[(1 - z^{-2N})/(1 + W^{-N_1 n_1} z^{-N_1})]. \qquad (5.10)$$

Thus, $H(z)$ may be implemented with a bank of N_1 filters corresponding to the N_1 distinct values of n_1, each being implemented with a premultiplication by $W^{-n_1^2/2}$, a delay of n_1 samples, and the recursive filter $(1 - z^{-2N})/(1 + W^{-N_1 n_1} z^{-N_1})$.

5.1.3 Factorizations in the Chirp Filter

We return now to the transversal filter form of the chirp z-transform shown in Fig. 5.1. The tap coefficients of this filter are given by $W^{-n^2/2}$ for $n = 0, \ldots, N - 1$. These N tap values cannot be all distinct because $-n^2/2$ is defined modulo N, and the congruence $-n^2/2 \equiv a$ modulo N has no solution for certain values of a. Thus, the chirp filter can be implemented with less than N distinct multipliers by adding, prior to multiplication, the data samples which correspond to the same tap value. We note that the number of distinct taps is given by the number of distinct quadratic residues modulo N [5.2, 7]. It is therefore possible to use the results of Sect. 2.1.4 to find the number of distinct multipliers required to implement a given chirp filter.

Consider first the case corresponding to N, an odd prime, with $N = p$. Then, we know by theorem 2.9 that the number of distinct quadratic residues is given by

$$Q(p) = 1 + (p - 1)/2. \qquad (5.11)$$

If we eliminate the trivial zero solution which corresponds to multiplication by 1, the number of nontrivial multipliers M reduces to

$$M = (p - 1)/2. \tag{5.12}$$

For N composite, the two following theorems can be used to find the number of distinct quadratic residues:

Theorem 5.1: If N is composite, with $N = N_1 N_2 \ldots N_k$ and $N_i = p_i^{c_i}$, with the p_i being distinct primes, the number $Q(N)$ of quadratic residues modulo N is given by

$$Q(N) = Q(N_1) Q(N_2) \ldots Q(N_k). \tag{5.13}$$

This theorem is proved by using the Chinese remainder theorem. If a is a quadratic residue modulo N, it must also be quadratic residue modulo the mutually prime factors of N. Since the representation given by the Chinese remainder theorem is unique, two distinct quadratic residues a and b must necessarily differ in at least one of their residues a_i, b_i. If $N = N_1 N_2$, we have $Q(N_1)$ distinct quadratic residues modulo N_1 and $Q(N_2)$ distinct quadratic residues modulo N_2. Therefore, we have $Q(N_1)Q(N_2)$ distinct quadratic residues modulo $N_1 N_2$ and (5.13) follows by induction.

Theorem 5.2: For $N = p^c$, the number of distinct quadratic residues is given by

$$Q(p^c) = 1 + (p^{c+1} - p^d)/2(p + 1) \tag{5.14}$$

if p is an odd prime, and

$$Q(2^c) = 2 + (2^{c-1} - 2^d)/3 \tag{5.15}$$

if $p = 2$. In (5.14) and (5.15), we have $d = 0$ if c is odd and $d = 1$ if c is even.

Proof of this theorem can be found in [5.8].

Thus, combining data samples prior to multiplying them can significantly reduce the number of multiplications required to process the chirp filter. In the case of a DFT of 16 points, for instance, $Q(16) = 4$ so that the number of multiplications is reduced by a factor of 4 when direct computation is replaced by a factorization.

5.2 Rader's Algorithm

We have seen that any DFT can be converted into a convolution by the chirp z-transform algorithm at the cost of $2N$ complex multiplications performed on the input and output data samples. We shall see now that DFTs can also be converted into circular convolutions by an entirely different method initially introduced by Rader [5.3]. This method is, in some cases, computationally more efficient than the chirp z-transform algorithm because it replaces the premultitiplications and postmultiplications in the chirp z-transform algorithm by a simple rearrangement of input and output data samples.

We consider first the simple case of a DFT of size $N = p$, p being an odd prime

$$\bar{X}_k = \sum_{n=0}^{p-1} x_n W^{nk}, \quad k = 0, \ldots, p-1$$

$$W = e^{-j2\pi/p}, \quad j = \sqrt{-1}; \quad (5.16)$$

for $k = 0$, $\bar{X}_k$ is computed by a simple summation

$$\bar{X}_0 = \sum_{n=0}^{p-1} x_n; \quad (5.17)$$

for $k \neq 0$, we have

$$\bar{X}_k = x_0 + \sum_{n=1}^{p-1} x_n W^{nk}. \quad (5.18)$$

The indices n and k are defined modulo p. We have seen in Sect. 2.1.3 that, if u is the set of integers $0, 1, \ldots, p-2$, there are always primitive roots g defined modulo p such that g^u modulo p takes once and only once all the values $1, 2, \ldots, p-1$ when u takes successively the values $0, 1, \ldots, p-2$. Thus, for $n, k \neq 0$, we can replace n and k by u and v defined by

$$n \equiv g^u \text{ modulo } p$$
$$k \equiv g^v \text{ modulo } p \quad , u, v = 0, \ldots, p-2. \quad (5.19)$$

Under these conditions, (5.18) becomes

$$\bar{X}_{g^v} = x_0 + \sum_{u=0}^{p-2} x_{g^u} W^{g^{u+v}}, \quad (5.20)$$

which shows that $\bar{X}_{g^v} - x_0$ is computed as a circular correlation of the permuted data sequence x_{g^u} with W^{g^u}, or equivalently as a $(p-1)$-point circular convolution of the data sequence $x_{g^{p-1-u}} = x_{g^{-u}}$ with W^{g^u}. Thus, for N an odd prime, most of the computation required to evaluate a DFT of N points reduces to a circular convolution of $N-1$ points. The process is shown schematically in Fig. 5.2.

An obvious implication of Rader's algorithm is that a DFT of size p, where p is an odd prime, can be computed with a number of operations proportional to $p \log p$ if the circular convolution is calculated by an FFT algorithm. We shall see, however, in Sects. 5.3 and 5.4 that the major significance of Rader's algorithm is that it allows one to compute large DFTs very efficiently, when it is combined with other techniques.

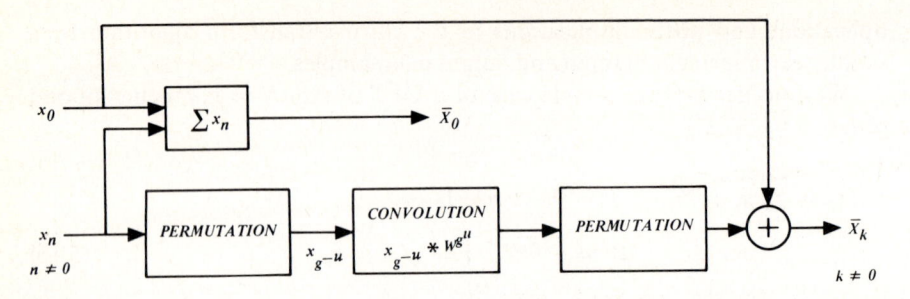

5. Linear Filtering Computation of Discrete Fourier Transforms

Fig. 5.2. Computation of a p-point DFT by Rader's algorithm. p is an odd prime

We shall now extend Rader's algorithm to accommodate composite dimensions [5.4, 9].

5.2.1 Composite Algorithms

Let us now consider DFTs of size $N = p^c$, where p is an odd prime. We have seen in Sect. 2.1.3 that primitive roots g modulo p^c always exist and that these primitive roots are of order $p^{c-1}(p - 1)$. Thus, we can expect to convert a DFT of dimension p^c into a circular convolution of length $p^{c-1}(p - 1)$ plus some additional terms. To demonstrate this point, we first define a change of index

$$k = p\,k_1 + k_2, \quad k_1 = 0, \ldots, p^{c-1} - 1$$
$$k_2 = 0, \ldots, p - 1. \tag{5.21}$$

Subsequently, for $k_2 = 0$, we have $k \equiv 0$ modulo p and $\bar{X}_k$ becomes

$$\bar{X}_{pk_1} = \sum_{n=0}^{p^c-1} x_n\,W^{pnk_1}. \tag{5.22}$$

Since W^{pnk_1} defines n modulo p^{c-1}, we change index n to

$$n = p^{c-1} n_1 + n_2, \quad n_1 = 0, \ldots, p - 1$$
$$n_2 = 0, \ldots, p^{c-1} - 1. \tag{5.23}$$

Thus, for $k_2 = 0$, $\bar{X}_k$ becomes a DFT of p^{c-1} points

$$\bar{X}_{pk_1} = \sum_{n_2=0}^{p^{c-1}-1} \left(\sum_{n_1=0}^{p-1} x_{p^{c-1}n_1+n_2}\right) W^{pk_1 n_2}. \tag{5.24}$$

Next, for $k \not\equiv 0$ modulo p, we compute separately the terms corresponding to $n \equiv 0$ modulo p and to $n \not\equiv 0$ modulo p,

$$\bar{X}_k = \bar{A}_k + \bar{B}_k \tag{5.25}$$

$$\bar{A}_k = \sum_{n=0} x_n W^{nk}, \qquad k \not\equiv 0 \text{ modulo } p \qquad (5.26)$$

$$\bar{B}_k = \sum_{n \neq 0} x_n W^{nk}, \qquad k \not\equiv 0 \text{ modulo } p \qquad (5.27)$$

for $n \equiv 0$ modulo p,

$$n = pn_1, \qquad n_1 = 0, \ldots, p^{c-1} - 1. \qquad (5.28)$$

Hence, by reordering index k, we have

$$\begin{aligned} k &= p^{c-1} k_1 + k_2, & k_1 &= 0, \ldots, p - 1 \\ k_2 &\not\equiv 0 \text{ modulo } p, & k_2 &= 1, \ldots, p^{c-1} - 1 \end{aligned} \qquad (5.29)$$

and

$$\bar{A}_{p^{c-1}k_1+k_2} = \sum_{n_1=0}^{p^{c-1}-1} x_{pn_1} W^{pn_1 k_2}. \qquad (5.30)$$

Note that the right-hand side of (5.30) is independent of k_1. Thus, $\bar{A}_k$ is a DFT of size p^{c-1} in which the output terms corresponding to $k_2 \equiv 0$ are not computed.

We turn now our attention to $\bar{B}_k$. Since $n, k \not\equiv 0$ modulo p, $\bar{B}_k$ is of length $p^{c-1}(p-1)$ and $nk \not\equiv 0$ modulo p. Thus, the indices n and k can be generated by a primitive root g defined modulo p^c with

$$\begin{aligned} n &\equiv g^u \text{ modulo } p^c \\ k &\equiv g^v \text{ modulo } p^c \qquad , u, v = 0, \ldots, [p^{c-1}(p-1) - 1] \end{aligned} \qquad (5.31)$$

and, by substituting the indices defined by (5.31) into (5.27), we obtain the correlation of dimension $p^{c-1}(p-1)$

$$\bar{B}_{g^v} = \sum_{u=0}^{p^{c-1}(p-1)-1} x_{g^u} W^{g^{u+v}}. \qquad (5.32)$$

Thus, the DFT of size p^c has been partitioned into two DFTs of size p^{c-1} and one correlation of length $p^{c-1}(p-1)$. The same method can be used recursively to convert the DFTs of size p^{c-1} into correlations. With this approach, a 9-point DFT is evaluated with a 3-point DFT and a 6-point convolution, plus a 3-point DFT where the first output term is not computed. When the 3-point DFTs are also reduced to correlations, the 9-point DFT is computed with 1 multiplication by W^0, 2 convolutions of 2 terms and one convolution of 6 terms.

When N is a power of two, the N-point DFT is partitioned into DFTs of size $N/2$ by the same method, and the DFT terms corresponding to n and k odd are computed as a correlation. However, there are no primitive roots for $N > 4$.

Thus, for $N > 4$, one uses a product of roots $(-1)^{n_1} 3^{n_2}$, with $n_1 = 0, 1$ and $n_2 = 0, ..., (N/4 - 1)$. These roots generate a two-dimensional correlation of size $2 \times (N/4)$.

5.2.2 Polynomial Formulation of Rader's Algorithm

Reducing a DFT into a set of convolutions may become very complex when N is composite. We shall now introduce a polynomial representation of the DFT [5.10] which greatly simplifies the formulation of Rader's algorithm. We begin once again with the N-point DFT

$$\bar{X}_k = \sum_{n=0}^{N-1} x_n W^{nk}, \qquad k = 0, ..., N-1$$

$$W = e^{-j2\pi/N}, \qquad j = \sqrt{-1} \tag{5.33}$$

In order to introduce a polynomial notation, we organize the N-point input sequence x_n as a polynomial $X(z)$ of N terms, defined modulo $(z^N - 1)$,

$$X(z) \equiv \sum_{n=0}^{N-1} x_n z^n \text{ modulo } (z^N - 1). \tag{5.34}$$

Then, (5.33) is replaced by

$$\bar{X}_k \equiv X(z) \quad \text{modulo } (z - W^k). \tag{5.35}$$

Note that (5.34) does not need to be defined modulo $(z^N - 1)$. This representation is therefore superfluous at this stage. However, it is valid, since n is defined modulo N. Equation (5.35) implies that $\bar{X}_k$ is obtained by substituting W^k for z in $X(z)$. A simple inspection shows that (5.34, 35) are a valid alternate representation of (5.33).

We suppose now that N is an odd prime, with $N = p$. Since the only divisors of p are 1 and p, $z^p - 1$ factors into two cyclotomic polynomials, with

$$z^p - 1 = (z - 1)P(z) \tag{5.36}$$

$$P(z) = z^{p-1} + z^{p-2} + ... + 1. \tag{5.37}$$

Thus, the polynomial $X(z)$ is completely determined by its residues modulo $(z - 1)$ and modulo $P(z)$

$$X_1(z) \equiv X(z) \text{ modulo}(z - 1) \tag{5.38}$$

$$X_2(z) \equiv X(z) \text{ modulo} P(z). \tag{5.39}$$

Note also that the roots of $z^p - 1$ are given by $z = W^k$ for $k = 0, ..., p-1$.

Moreover, $z = W^0 = 1$ is the root of $z - 1$ and the $p - 1$ roots of $P(z)$ are given by $z = W^k$ for $k \neq 0$. Thus, we can compute $\bar{X}_k$ by

$$\bar{X}_0 = X_1(z) \tag{5.40}$$

$$\bar{X}_k \equiv X_2(z) \quad \text{modulo } (z - W^k) \quad , k \neq 0. \tag{5.41}$$

$X_2(z)$ is a polynomial of degree $p - 2$, since it is defined modulo $P(z)$. Hence, $X_2(z)$ can be expressed as

$$X_2(z) = \sum_{n=0}^{p-2} a_n z^n \tag{5.42}$$

with

$$a_n = x_n - x_{p-1}. \tag{5.43}$$

Thus, for $k \neq 0$, $\bar{X}_k$ is a *reduced* DFT of p terms in which the last input sample is zero and the first output sample is not computed

$$\bar{X}_k = \sum_{n=0}^{p-2} a_n W^{nk}, \quad k \neq 0. \tag{5.44}$$

The final result will not be changed if $X(z)$ is multiplied by z^{p-1} modulo $(z^p - 1)$ and $X_2(z)$ is multiplied by z modulo $P(z)$. In this case, $X_2(z)$ becomes

$$X_2(z) = \sum_{n=0}^{p-2} b_n z^n \tag{5.45}$$

with

$$b_n = x_{n+1} - x_0 \tag{5.46}$$

and $\bar{X}_k$ reduces to

$$\bar{X}_k \equiv zX_2(z) \text{ modulo}(z - W^k) = \sum_{n=1}^{p-1} b_{n-1} W^{nk}. \tag{5.47}$$

Since $n, k \neq 0$ in (5.47), this expression defines a $(p - 1)$-point convolution if the indices n and k are expressed as powers of a primitive root. Thus, for N a prime, Rader's algorithm is represented in polynomial notation as shown in Fig. 5.3. Note that the boxes shown in this figure for polynomial ordering and multiplications by z^{p-1} and z are given for illustrative purpose, but do not usually correspond to any processing since they merely indicate the origin of the data index. Thus, we shall usually delete such boxes in subsequent polynomial representations of DFTs.

The main contribution of the polynomial representation is that it greatly

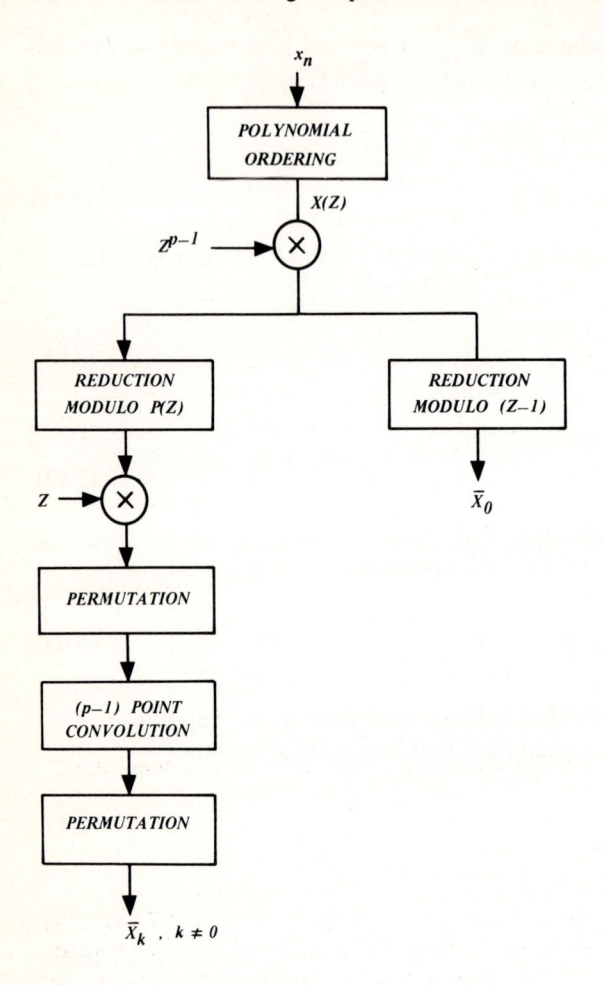

5. Linear Filtering Computation of Discrete Fourier Transforms

Fig. 5.3. Polynomial representation of Rader's algorithm for a p-point DFT, p odd prime

simplifies the decomposition of composite DFTs into convolutions. If we consider, for instance, a 9-point DFT, we know that $z^9 - 1$ factors into 3 cyclotomic polynomials, since the only divisors of 9 are 1, 3, and 9. These polynomials are given by $P_1(z) = z - 1$, $P_2(z) = (z^3 - 1)/(z - 1) = z^2 + z + 1$, and $P_3(z) = (z^9 - 1)/(z^3 - 1) = z^6 + z^3 + 1$. Thus, the 9-point DFT can be computed as shown in Fig. 5.4 by successive reductions modulo $P_1(z)$, $P_2(z)$, and $P_3(z)$. The reduction modulo $(z^3 - 1)$ yields a 3-point DFT which can in turn be calculated as a 2-point convolution plus one multiplication by the approach of Fig. 5.3. The reduction modulo $(z^6 + z^3 + 1)$ yields a reduced DFT which computes $\bar{X}_k$ for $k \not\equiv 0$ modulo 3. This reduced DFT is evaluated with one 2-point convolution plus one 6-point convolution by (5.30) and (5.32), respectively.

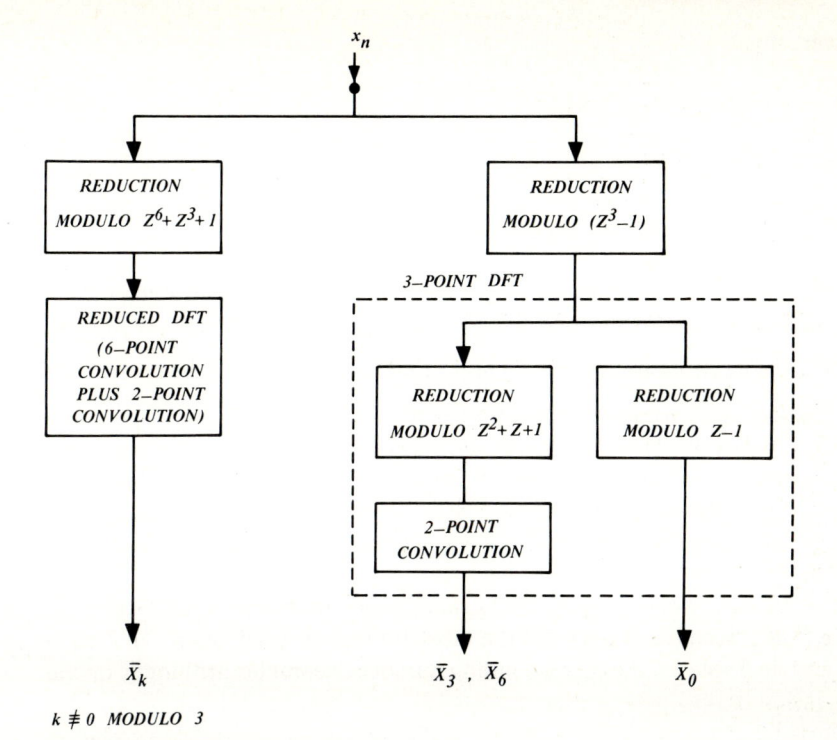

Fig. 5.4. Computation of a 9-point DFT by Rader's algorithm. Polynomial representation

5.2.3 Short DFT Algorithms

We have seen in Chap. 3 that short convolutions can be computed very efficiently by interpolation techniques. Thus, Rader's algorithm yields efficient implementations for small DFTs. In practice, we shall not have to use Rader's algorithm for large DFTs because there are several other methods, to be discussed in the following sections, which allow one to construct a large DFT from a limited set of small DFTs. Thus, we shall be concerned here only with the efficient implementation of Rader's algorithm for small DFTs.

In practice, the convolutions derived from Rader's method are computed by using the same techniques as those described in Chap. 3. However, some additional simplification is possible because here the sequence of coefficients, W^{g^n}, has special properties.

Consider first the case of a p-point DFT, where p is an odd prime. Then, the convolution is of length $d = p - 1$, with d even. Since g^n modulo p generates a cyclic group of order $p - 1$, we have $g^{p-1} \equiv 1$ modulo p. Therefore, we have $g^{(p-1)/2} \equiv -1$ modulo p and

$$W^{g^n + (p-1)/2} \equiv W^{-g^n}. \tag{5.48}$$

Moreover, since

$$W^{g^n} = \cos(2\pi g^n/p) - j\sin(2\pi g^n/p), \tag{5.49}$$

we have

$$W^{g^{n+(p-1)/2}} = \cos(2\pi g^n/p) + j\sin(2\pi g^n/p), \tag{5.50}$$

which suggests an even symmetry about midpoint for real coefficients and an odd symmetry for imaginary coefficients. Thus, when the coefficient polynomial $\sum_{n=0}^{d-1} W^{g^n} z^n$ is reduced modulo($z^{d/2} - 1$) and modulo($z^{d/2} + 1$), all the coefficients in the reduced polynomials become pure real numbers and pure imaginary numbers respectively. This means that all complex multiplications reduce to the multiplication of a complex number by either a pure real of a pure imaginary number and are therefore implemented with only two real multiplications. This feature is common to all convolutions derived by partitioning a DFT via Rader's algorithm.

When $N = p^e$, where p is an odd prime, some additional simplification is possible [5.10]. We give in Sect. 5.5 the most frequently used small DFT algorithms and, in Table 5.1 the corresponding number of complex arithmetic operations.

Table 5.1. Number of complex operations for short DFTs computed by Rader's algorithm. Trivial multiplications by ± 1, $\pm j$ are given between parentheses. The number of real operations is twice the number of operations given in this table

DFT size N	Number of multiplications	Number of additions
2	2 (2)	2
3	3 (1)	6
4	4 (4)	8
5	6 (1)	17
7	9 (1)	36
8	8 (6)	26
9	11 (1)	44
16	18 (8)	74

In many applications, it is acceptable to have the DFT output multiplied by a constant integer l. In particular, when the DFT method is used to compute circular convolutions of a fixed sequence with many data sequences, the transform of the fixed sequence is usually precomputed and can be premultiplied by $1/l^3$. In such cases, it is possible to design improved short DFT algorithms in which the number of nontrivial multiplications is minimized. For $N = p$, with p

an odd prime, this is done by using the property $\sum_{n=0}^{p-2} W^{g^n} = -1$ and this gives an algorithm with a scaling factor equal to $p - 1$ and with two trivial multiplications by ± 1 instead of one. The corresponding number of operations are given in Sect. 5.5.

5.3 The Prime Factor FFT

For large DFTs, the derivation of Rader's algorithm becomes cumbersome and computationally inefficient. In this section, we shall discuss an alternative computation technique which allows one to compute a large DFT of size N by combining several small DFTs of sizes $N_1, N_2, ..., N_e$ which are relative prime factors of N. This technique, which is known today as the *prime factor* FFT, was proposed by Good [5.11, 12] prior to the introduction of the FFT and has both theoretical and practical significance. Its main theoretical contribution is in showing how a one-dimensional DFT can be mapped by simple permutations into a multidimensional DFT. This approach has also been shown recently [5.13, 14] to be of practical interest when it is combined with Rader's algorithm. Furthermore, Good's algorithm provides one of the foundations on which the very efficient Winograd Fourier transform algorithm is based [5.4].

5.3.1 Multidimensional Mapping of One-Dimensional DFTs

We first consider the simple case of a DFT $\bar{X}_k$ of size N, where N is the product of two mutually prime factors N_1 and N_2

$$\bar{X}_k = \sum_{n=0}^{N-1} x_n W^{nk}, \quad k = 0, ..., N-1$$

$$W = e^{-j2\pi/N}, j = \sqrt{-1} \tag{5.51}$$

$$N = N_1 N_2, \quad (N_1, N_2) = 1. \tag{5.52}$$

Our objective is to convert this one-dimensional DFT into a two-dimensional DFT of size $N_1 \times N_2$. In order to do this, one must covert each of the indices n and k, defined modulo N, into two sets of indices n_1, k_1, and n_2, k_2, defined, respectively, modulo N_1 and modulo N_2. We have seen already, in Sect. 2.1.2, two different methods of doing this, one based on the Chinese remainder theorem, and the other on the use of simple permutations. We shall initially employ this simpler method by defining the index transformation

$$\begin{aligned} n &\equiv N_1 n_2 + N_2 n_1 \text{ modulo } N, & n_1, k_1 &= 0, ..., N_1 - 1 \\ k &\equiv N_1 k_2 + N_2 k_1 \text{ modulo } N, & n_2, k_2 &= 0, ..., N_2 - 1. \end{aligned} \tag{5.53}$$

Note that this definition is valid only for $(N_1, N_2) = 1$. Now, since $N_1 N_2 \equiv 0$ modulo N, substituting n and k defined by (5.53) into (5.51) yields

$$\bar{X}_{N_1 k_2 + N_2 k_1} = \sum_{n_1=0}^{N_1-1} \sum_{n_2=0}^{N_2-1} x_{N_1 n_2 + N_2 n_1} W_1^{N_2 n_1 k_1} W_2^{N_1 n_2 k_2} \tag{5.54}$$

with

$$W_1 = e^{-j2\pi/N_1}, \qquad W_2 = e^{-j2\pi/N_2}. \tag{5.55}$$

We note that (5.54) is a two-dimensional DFT of size $N_1 \times N_2$, but with the exponents $n_1 k_1$ and $n_2 k_2$ permuted, respectively, by N_2 and N_1. Thus, in order to obtain the two-dimensional DFT in the conventional lexicographic order, it is convenient to replace k_1 and k_2 by their permuted values $t_2 k_1$ and $t_1 k_2$ such that $N_2 t_2 \equiv 1$ modulo N_1 and $N_1 t_1 \equiv 1$ modulo N_2. This is equivalent to replacing the mapping of k given by (5.53) with its Chinese remainder equivalent

$$k \equiv N_1 t_1 k_2 + N_2 t_2 k_1 \text{ modulo } N. \tag{5.56}$$

Then, $\bar{X}_k$ reduces to

$$\bar{X}_{N_1 t_1 k_2 + N_2 t_2 k_1} = \sum_{n_1=0}^{N_1-1} \sum_{n_2=0}^{N_2-1} x_{N_1 n_2 + N_2 n_1} W_1^{n_1 k_1} W_2^{n_2 k_2}, \tag{5.57}$$

which is the usual representation of a DFT of size $N_1 \times N_2$. Thus, by using for n the permutation defined by (5.53), and for k the Chinese remainder correspondence defined by (5.56) (or vice versa), we are able to map a one-dimensional convolution of length $N_1 N_2$ into a two-dimensional convolution of size $N_1 \times N_2$.

The same method can be used recursively to define a one to many multidimensional mapping. More precisely, if N is the product of d mutually prime factors N_i, with

$$N = \prod_{i=1}^{d} N_i, \tag{5.58}$$

then, the one-dimensional DFT of length N is converted into a d-dimensional DFT of size $N_1 \times N_2 \ldots \times N_d$ by the change of indices

$$n \equiv \sum_{i=1}^{d} N n_i / N_i \text{ modulo } N, \qquad n_i = 0, \ldots, N_i - 1 \tag{5.59}$$

$$k \equiv \sum_{i=1}^{d} N t_i k_i / N_i \text{ modulo } N, \qquad k_i = 0, \ldots, N_i - 1, \tag{5.60}$$

where t_i is given by

$$N t_i / N_i \equiv 1 \text{ modulo } N_i. \tag{5.61}$$

It can be verified easily that, in the product nk modulo N, with n and k defined by (5.59) and (5.60), all cross-products $n_i k_u$ for $i \neq u$ cancel, so that

$$nk \equiv \sum_{i=1}^{d} N n_i k_i / N_i \text{ modulo } N, \tag{5.62}$$

which demonstrates that the multidimensional representation indeed has the desired format.

Once the one-dimensional DFT has been converted into a multidimensional DFT, several different strategies can be used to compute the multidimensional DFT. We shall see, in Sect. 5.4, that one possible approach consists of nesting various N_i-point DFT algorithms. However, we shall first present here the original method described by Good, which is based on the conventional row-column approach to multidimensional DFT computation (Sect. 4.4) [5.11–13].

5.3.2 The Prime Factor Algorithm

We now consider a two-dimensional DFT $\bar{X}_{k_1,k_2}$ of size $N_1 \times N_2$, with $(N_1, N_2) = 1$

$$\bar{X}_{k_1,k_2} = \sum_{n_1=0}^{N_1-1} \sum_{n_2=0}^{N_2-1} x_{n_1,n_2} W_1^{n_1 k_1} W_2^{n_2 k_2}$$

$$W_1 = e^{-j2\pi/N_1}, \; W_2 = e^{-j2\pi/N_2}, \quad k_1 = 0, \ldots, N_1 - 1$$
$$k_2 = 0, \ldots, N_2 - 1. \tag{5.63}$$

This DFT is either a genuine two-dimensional DFT, or is derived from a one-dimensional DFT by the mapping defined by (5.53, 56, 57). Equation (5.63) can be rewritten as

$$\bar{X}_{k_1,k_2} = \sum_{n_2=0}^{N_2-1} W_2^{n_2 k_2} \sum_{n_1=0}^{N_1-1} x_{n_1,n_2} W_1^{n_1 k_1}. \tag{5.64}$$

This illustrates that $\bar{X}_{k_1,k_2}$ can be evaluated by first computing one DFT of N_1 terms for each value of n_2. This gives N_1 sets of N_2 points $\bar{X}_{n_2,k_1}$ which are the input sequences to N_1 DFTs of N_2 points. Thus, with this method, $\bar{X}_{k_1,k_2}$ is calculated with N_2 DFTs of length N_1 plus N_1 DFTs of length N_2. A detailed representation of the computation process is shown in Fig. 5.5 for a 12-point DFT using the 3-point and 4-point DFT algorithms of Sects. 5.5.2 and 5.5.3.

In order to evaluate the number of multiplications, M, and additions, A, which are necessary to compute a DFT by the prime factor algorithm, we assume that M_1, M_2 and A_1, A_2 are the number of multiplications and additions required the calculate the DFTs of lengths N_1 and N_2, respectively. Then, we have obviously

$$M = N_1 M_2 + N_2 M_1 \tag{5.65}$$

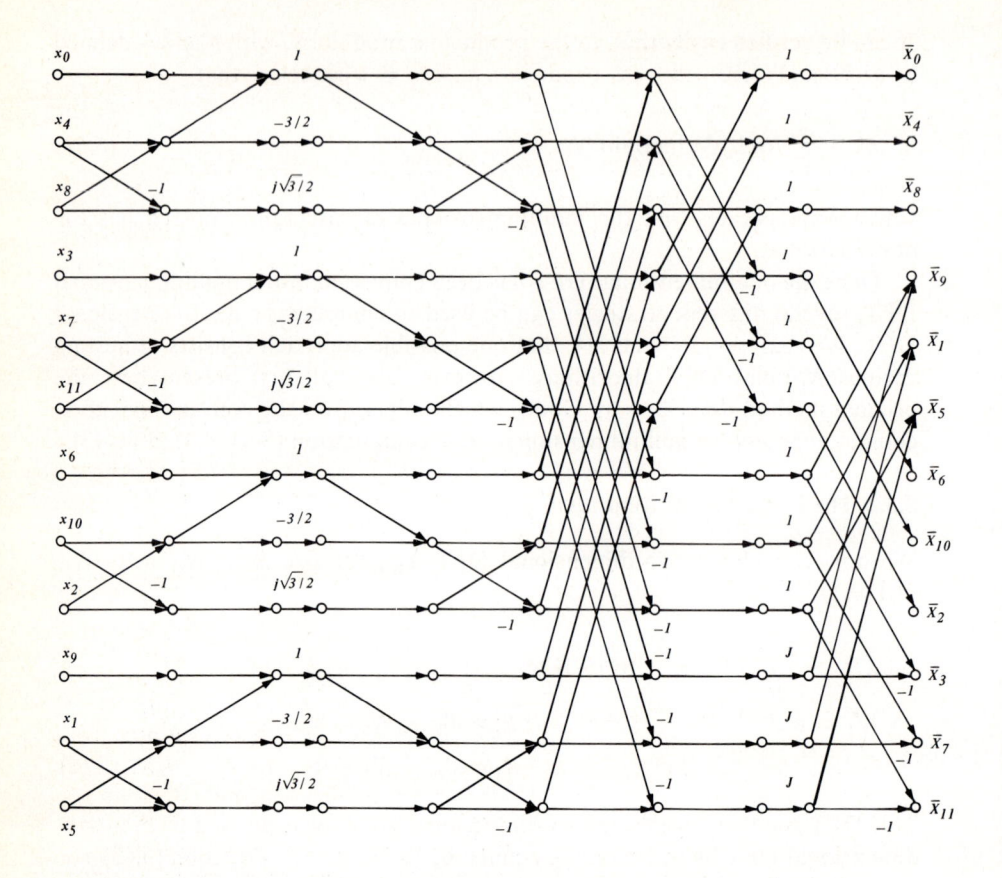

128   5. Linear Filtering Computation of Discrete Fourier Transforms

Fig. 5.5. Flow graph of a 12-point DFT computed by the prime factor algorithm

$$A = N_1 A_2 + N_2 A_1. \tag{5.66}$$

The same method can be extended recursively to cover the case of more than two factors. Thus, for a d-dimensional DFT, we have

$$N = \prod_{i=1}^{d} N_i, \quad (N_i, N_l) = 1 \text{ for } i \neq l \tag{5.67}$$

and

$$M = \sum_{i=1}^{d} NM_i/N_i \tag{5.68}$$

$$A = \sum_{i=1}^{d} NA_i/N_i, \tag{5.69}$$

where M_i and A_i are, respectively, the number of multiplications and additions for a DFT of size N_i.

Thus, the computation of a large DFT is reduced to that of a set of small DFTs of lengths $N_1, N_2, \ldots, N_d$. These small DFTs can be computed with any algorithm, but it is extremely attractive to use Rader's algorithm for this application because this particular algorithm is very efficient for small DFTs.

Table 5.2 lists the number of nontrivial real arithmetic operations for various DFTs computed by the prime factor and Rader algorithms (Table 5.1). It can be seen, by comparison with Table 4.3 that this approach compares favorably with the FFT method.

Table 5.2. Number of nontrivial real operations for DFTs computed by the prime factor and Rader algorithms

DFT size N	Number of real multiplications M	Number of real additions A	Multiplications per point M/N	Additions per point A/N
30	100	384	3.33	12.80
48	124	636	2.58	13.25
60	200	888	3.33	14.80
120	460	2076	3.83	17.30
168	692	3492	4.12	20.79
240	1100	4812	4.58	20.05
504	2524	13388	5.01	26.56
840	5140	23172	6.12	27.59
1008	5804	29548	5.76	29.31
2520	17660	84076	7.01	33.36

Note that the data given in Table 5.2 apply to multidimensional DFTs as well as to one-dimensional DFTs. For example, it may be seen that 100 nontrivial multiplications are required to compute a DFT of size 30. The number of multiplications would be the same for DFTs of sizes $2 \times 3 \times 5$, 6×5, 10×3, or 2×15, since the only difference between the members of this group is the index mapping. We shall see, however, in Chap. 7, that it is possible to devise even more efficient computation techniques for multidimensional DFTs. Therefore, the main utility of the prime factor algorithm resides in the calculation of one-dimensional DFTs.

5.3.3 The Split Prime Factor Algorithm

We shall now show that the efficiency of the prime factor algorithm can be improved by splitting the calculations [5.10]. This can be seen by considering again a two-dimensional DFT of size $N_1 \times N_2$

130   5. Linear Filtering Computation of Discrete Fourier Transforms

$$\bar{X}_{k_1,k_2} = \sum_{n_2=0}^{N_2-1} W_2^{n_2 k_2} \sum_{n_1=0}^{N_1-1} x_{n_1,n_2} W_1^{n_1 k_1}$$

$$W_1 = e^{-j2\pi/N_1}, \quad W_2 = e^{-j2\pi/N_2} \quad k_1 = 0, \ldots, N_1 - 1$$
$$k_2 = 0, \ldots, N_2 - 1. \tag{5.70}$$

In order to simplify the discussion, we shall assume that N_1 and N_2 are both odd primes. In this case, Rader's algorithm reduces each of the DFTs of size N_1 or N_2 to one multiplication plus one correlation of size $N_1 - 1$ or $N_2 - 1$. Therefore, $\bar{X}_{k_1,k_2}$ is evaluated via the prime factor algorithm as one DFT of N_1 points, one correlation of $N_2 - 1$ points and one correlation of $(N_2 - 1) \times (N_1 - 1)$ points, with

$$\bar{X}_{k_1,0} = \sum_{n_1=0}^{N_1-1} \left(\sum_{n_2=0}^{N_2-1} x_{n_1,n_2} \right) W_1^{n_1 k_1} \tag{5.71}$$

$$\bar{X}_{0,g^{v_2}} = \sum_{u_2=0}^{N_2-2} \left[\sum_{n_1=0}^{N_1-1} (x_{n_1,g^{u_2}} - x_{n_1,0}) \right] W_2^{g^{v_2+u_2}} \tag{5.72}$$

$$\bar{X}_{h^{v_1},g^{v_2}} = \sum_{u_2=0}^{N_2-2} W_2^{g^{v_2+u_2}} \sum_{u_1=0}^{N_1-2} (x_{h^{u_1},g^{u_2}} - x_{h^{u_1},0} - x_{0,g^{u_2}} + x_{0,0}) W_1^{h^{v_1+u_1}} \tag{5.73}$$

where h and g are primitive roots modulo N_1 and N_2 and

$$k_1 \equiv h^{v_1} \quad \text{modulo } N_1, \quad n_1 \equiv h^{u_1} \quad \text{modulo } N_1$$
$$k_2 \equiv g^{v_2} \quad \text{modulo } N_2, \quad n_2 \equiv g^{u_2} \quad \text{modulo } N_2$$
$$u_1, v_1 = 0, \ldots, N_1 - 2$$
$$u_2, v_2 = 0, \ldots, N_2 - 2. \tag{5.74}$$

We note that the two-dimensional correlation defined by (5.73) is half separable. Hence we can compute this correlation by the row-column method as $N_2 - 1$ correlations of $N_1 - 1$ points plus $N_1 - 1$ correlations of $N_2 - 1$ points. If M_1 and M_2 are the number of complex multiplications required to compute the DFTs of lengths N_1 and N_2, the correlations of lengths $N_1 - 1$ and $N_2 - 1$ are computed, respectively, with $M_1 - 1$ and $M_2 - 1$ complex multiplications, because, for N prime, Rader's algorithm reduces a DFT of N points into one multiplication plus one correlation of $N - 1$ points. Under these conditions, the total number of complex multiplications required to compute $\bar{X}_{k_1,k_2}$ reduces to

$$M = N_1 M_2 + N_2 M_1 - N_1 - N_2 + 1. \tag{5.75}$$

Since the conventional prime factor algorithm would have required $N_1 M_2 + N_2 M_1$ multiplications, splitting the computation eliminates $N_1 + N_2 - 1$ complex multiplications. When the two-dimensional convolution is reduced modulo cyclotomic polynomials, the various terms remain half separable and additional savings can be realized. This can be seen more precisely by representing the DFT $\bar{X}_{k_1,k_2}$ defined by (5.70) in polynomial notation, and employing an approach similar to that described in Sect. 5.2.2

$$X(z_1, z_2) \equiv \sum_{n_1=0}^{N_1-1} \sum_{n_2=0}^{N_2-1} x_{n_1,n_2} z_1^{n_1} z_2^{n_2} \text{ modulo } (z_1^{N_1} - 1), (z_2^{N_2} - 1) \tag{5.76}$$

$$\bar{X}_{k_1,k_2} \equiv X(z_1, z_2) \text{ modulo } (z_1 - W_1^{k_1}), (z_2 - W_2^{k_2}). \tag{5.77}$$

Ignoring the permutations and the multiplications by z_1, $z_1^{N_1-1}$, z_2, and $z_2^{N_2-1}$, we can use this polynomial formulation to represent the split prime factor algorithm very simply, as indicated by the diagram in Fig. 5.6, which corresponds to a DFT of size 5×7. With this method, the main part of the computation corresponds to the evaluation of a correlation of dimension 4×6 which can be regarded as a polynomial product modulo $(z_1^4 - 1)$, $(z_2^6 - 1)$. Since both $z_1^4 - 1$ and $z_2^6 - 1$ are composite, the computation of the correlation of dimension 4×6 can be split into that of the cyclotomic polynomials which are factors of $z_1^4 - 1$ and $z_1^6 - 1$ and given by

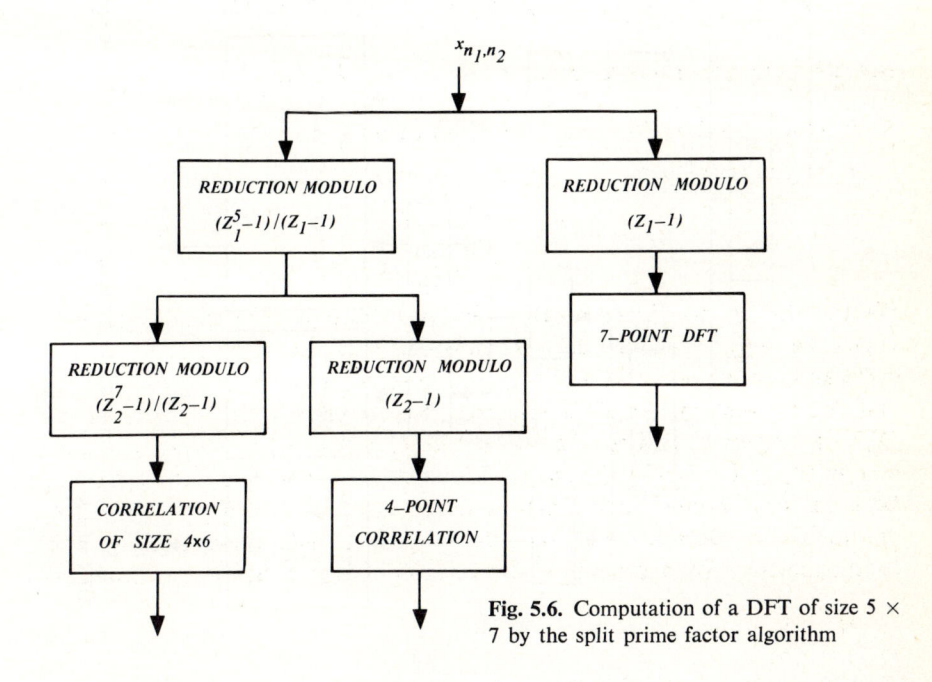

Fig. 5.6. Computation of a DFT of size 5×7 by the split prime factor algorithm

$$z_1^4 - 1 = (z_1 - 1)(z_1 + 1)(z_1^2 + 1) \tag{5.78}$$

$$z_2^6 - 1 = (z_2 - 1)(z_2 + 1)(z_2^2 + z_2 + 1)(z_2^2 - z_2 + 1). \tag{5.79}$$

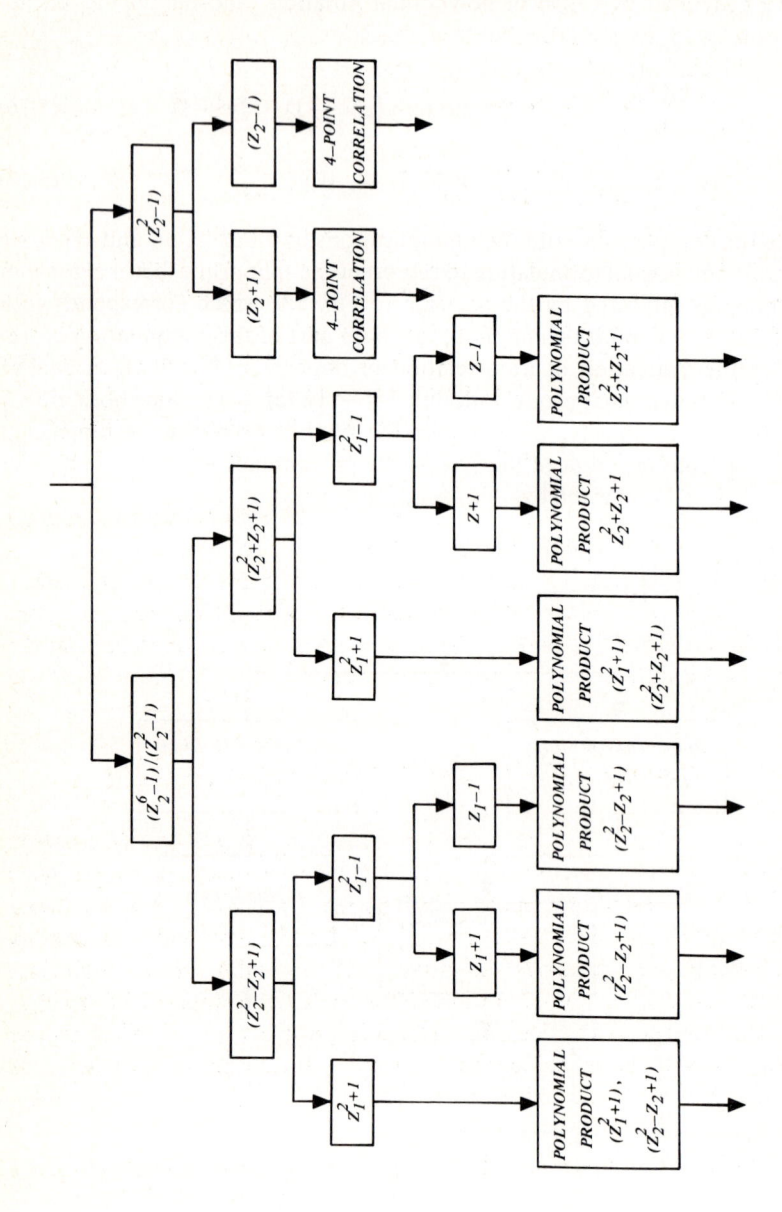

Fig. 5.7. Calculation of the correlation of 4×6 points in the complete split prime factor evaluation of a DFT of size 5×7

The complete method is illustrated in Fig. 5.7 with the various reductions modulo cyclotomic polynomials. Since all the expressions remain half separable throughout the decomposition, the two-dimensional polynomial products are computed by the row-column method. Thus, for instance, the two-dimensional polynomial product modulo $(z_1^2 + 1), (z_2^2 + z_2 + 1)$ is calculated as 2 polynomial products modulo $(z_2^2 + z_2 + 1)$ plus 2 polynomial products modulo $(z_1^2 + 1)$.

With split-prime factorization, a DFT of size 5×7 is evaluated with 76 complex multiplications and 381 additions if the correlation of size 4×6 is computed directly by the row-column method. If the computation of the correlation of size 4×6 is reduced to that of polynomial products, as shown in Fig. 5.7, this correlation is calculated with only 46 multiplications and 150 additions instead of 62 multiplications and 226 additions with the row-column method. Thus, the complete split-prime factor computation reduces the total number of operations to 60 complex multiplications and 305 complex additions. By comparison, the conventional prime factor algorithm requires 87 complex multiplications and 299 additions. Thus, splitting the computations saves, in this case, about 30% of the multiplications.

The same split computation technique can also be applied to sequence lengths with more than two factors as well as those with composite factors. It should also be noted that the computational savings provided by the method increases as a function of the DFT size. Thus, for large DFTs, the split-prime factor method reduces significantly the number of arithmetic operations at the expense of requiring a more complex implementation.

5.4 The Winograd Fourier Transform Algorithm (WFTA)

We have seen in the preceding section that a composite DFT of size N, where N is the product of d prime factors $N_1, N_2, \ldots, N_d$, can be mapped, by simple index permutations, into a multidimensional DFT of size $N_1 \times N_2 \times \ldots \times N_d$. When this multidimensional DFT is evaluated by the conventional row-column method, the algorithm becomes the prime factor algorithm. In the following, we shall discuss another way of evaluating the multidimensional DFT which is based on a nesting algorithm introduced by Winograd [5.4, 15, 16]. This method is particularly effective in reducing the number of multiplications when it is combined with Rader's algorithm.

5.4.1 Derivation of the Algorithm

We commence with a DFT of size $N_1 \times N_2$

$$\bar{X}_{k_1, k_2} = \sum_{n_2=0}^{N_2-1} W_2^{n_2 k_2} \sum_{n_1=0}^{N_1-1} x_{n_1, n_2} W_1^{n_1 k_1}$$

$$k_1 = 0, \ldots, N_1 - 1, \quad k_2 = 0, \ldots, N_2 - 1$$
$$W_1 = e^{-j2\pi/N_1}, \quad W_2 = e^{-j2\pi/N_2}. \tag{5.80}$$

This DFT may either be a genuine two-dimensional DFT or a one-dimensional DFT of length $N = N_1 N_2$, with $(N_1, N_2) = 1$, which has been mapped into a two-dimensional form using the index mapping scheme of Good's algorithm. We note that the two-dimensional DFT defined by (5.80) can be regarded as a one-dimensional DFT of length N_2 where each scalar is replaced by a vector of N_1 terms, and each multiplication is replaced by a DFT of length N_1. More precisely, the DFT defined by (5.80) can be expressed as

$$\bar{X}_{k_1,k_2} = \sum_{n_2=0}^{N_2-1} W_2^{n_2 k_2} A X_{n_2}, \tag{5.81}$$

where X_{n_2} is an N_1 element column vector of the input data x_{n_1,n_2} and A is an $N_1 \times N_1$ matrix of the complex exponentials $W_1^{n_1 k_1}$

$$X_{n_2} = \begin{bmatrix} x_{0,n_2} \\ x_{1,n_2} \\ \vdots \\ \vdots \\ x_{N_1-1,n_2} \end{bmatrix} \tag{5.82}$$

$$A = \begin{bmatrix} W_1^0 & W_1^0 & \cdots & W_1^0 \\ W_1^0 & W_1^1 & \cdots & W_1^{(N_1-1)} \\ \vdots & \vdots & & \vdots \\ W_1^0 & W_1^{(N_1-1)} & \cdots & W_1^{(N_1-1)^2} \end{bmatrix}. \tag{5.83}$$

Thus, the polynomial DFT defined by (5.81) is a DFT of length N_2 where each multiplication by $W_2^{n_2 k_2}$ is replaced with a multiplication by $W_2^{n_2 k_2} A$. This last operation itself is equivalent to a DFT of length N_1 in which each multiplication by $W_1^{n_1 k_1}$ is replaced with a multiplication by $W_2^{n_2 k_2} W_1^{n_1 k_1}$.

It can be seen that the Winograd algorithm breaks the computation of a DFT of size $N_1 N_2$ or $N_1 \times N_2$ into the evaluation of small DFTs of length N_1 and N_2 in a manner which is fundamentally different from that corresponding to the prime factor algorithm. In fact, the method used here is essentially similar to the nesting method described by Agarwal and Cooley for convolutions (Sect. 3.3.1).

The Winograd method is particularly interesting when the small DFTs are evaluated via Rader's algorithm. In this case, the small DFTs are calculated with A^1 input additions, M complex multiplications, and A^2 output additions. Thus, the Winograd algorithm for a DFT of size $N_1 \times N_2$ can be represented as shown in Fig. 5.8. In this case, if M_2, A_2^1 and A_2^2 are the number of complex multiplications and input, output additions for the DFT of length N_2, the total number of multiplications M and additions A for the DFT of size $N_1 \times N_2$ becomes

5.4 The Winograd Fourier Transform Algorithm (WFTA)

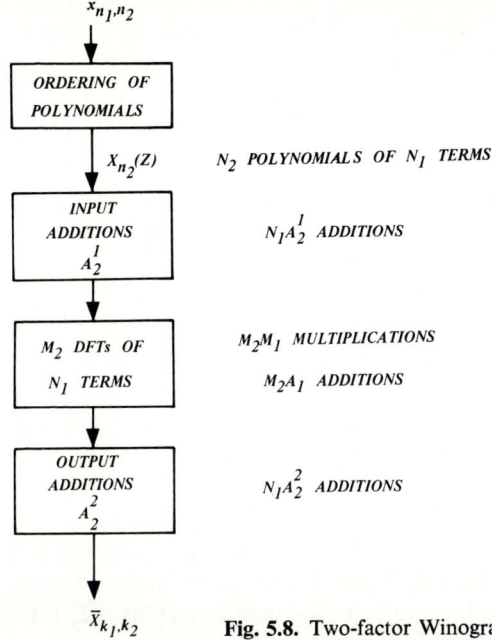

Fig. 5.8. Two-factor Winograd Fourier transform algorithm

$$M = M_1 M_2 \tag{5.84}$$

$$A = N_1(A_2^1 + A_2^2) + M_2 A_1, \tag{5.85}$$

where M_1 and A_1 are, respectively, the number of multiplications and additions for the N_1-point DFT. Since the total number of additions A_2 for the N_2-point DFT is given by $A_2 = A_2^1 + A_2^2$, (5.85) reduces to

$$A = N_1 A_2 + M_2 A_1. \tag{5.86}$$

The same method can be applied recursively to cover the case of more than two factors. Hence a multidimensional DFT of size $N_1 \times N_2 \ldots \times N_d$ or a one-dimensional DFT of length $N_1 N_2 \ldots N_d$, where $(N_i, N_k) = 1$ for $i \neq k$, is computed with M multiplications and A additions, where M and A are given by

$$M = \prod_{i=1}^{d} M_i \tag{5.87}$$

$$A = A_1 N_2 \ldots N_d + M_1 A_2 N_3 \ldots N_d + \ldots + M_1 M_2 \ldots M_{d-1} A_d, \tag{5.88}$$

where M_i and A_i are the number of complex multiplications and additions for an N_i-point DFT.

Note that the number of additions depends upon the order in which the operations are executed. If we take, for instance, the two-factor algorithm, computing the DFT of $N_1 \times N_2$ points as a DFT of N_1 points in which all multiplications are replaced by a DFT of size N_2 would give a number of additions

$$A = N_2 A_1 + M_1 A_2. \tag{5.89}$$

The first nesting method will require less additions than the second nesting method if $N_1 A_2 + M_2 A_1 < N_2 A_1 + M_1 A_2$ or

$$(M_2 - N_2)/A_2 < (M_1 - N_1)/A_1. \tag{5.90}$$

Thus, the values $(M_i - N_i)/A_i$ characterize the order in which the various short algorithms must be nested in order to minimize the number of additions.

It should be noted that, in (5.84, 86–88), M and A are, respectively, the total number of complex multiplications and additions. However, when the small DFTs are computed by Rader's algorithms, all complex multiplications reduce to multiplications of a complex number by a pure real or a pure imaginary number and are implemented with only two real additions. Moreover, some of the multiplications in the small DFT algorithms are trivial multiplications by ± 1, $\pm j$. Thus, if the number of such complex trivial multiplications is L_i for a N_i point DFT, then the number of nontrivial real multiplications becomes

$$M = (2 \prod_{i=1}^{d} M_i) - (2 \prod_{i=1}^{d} L_i). \tag{5.91}$$

We illustrate the Winograd method in Fig. 5.9 by giving the flow diagram corresponding to a 12-point DFT using the 3-point and 4-point DFT algorithms of Sects. 5.5.2 and 5.5.3.

Table 5.3. Number of nontrivial real operations for one-dimensional DFTs computed by the Winograd Fourier transform algorithm

DFT size N	Number of real multiplications M	Number of real additions A	Multiplications per point M/N	Additions per point A/N
30	68	384	2.27	12.80
48	92	636	1.92	13.25
60	136	888	2.27	14.80
120	276	2076	2.30	17.30
168	420	3492	2.50	20.79
240	632	5016	2.63	20.90
420	1288	11352	3.07	27.03
504	1572	14540	3.12	28.85
840	2580	24804	3.07	29.53
1008	3548	34668	3.52	34.39
2520	9492	99628	3.77	39.53

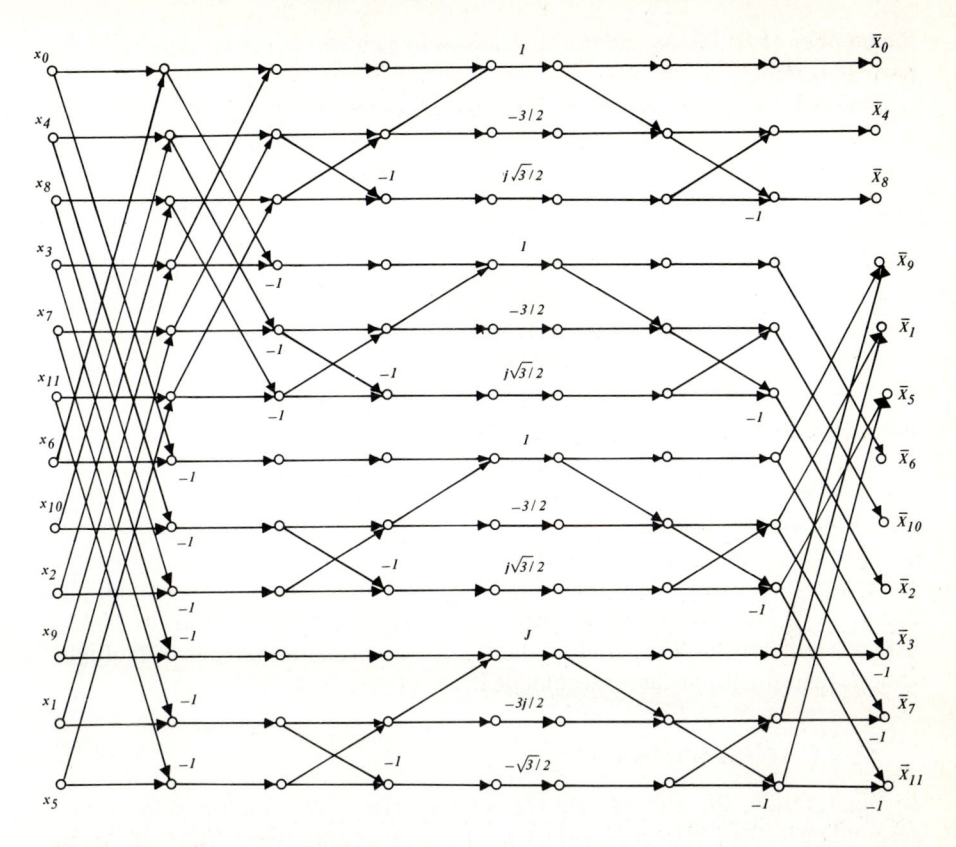

Fig. 5.9. Flow graph of a 12-point DFT computed by the Winograd Fourier transform algorithem

Table 5.3 lists the number of nontrivial real operations for various DFTs computed by the Winograd Fourier transform algorithm, with the small DFTs evaluated by the algorithms given in Sect. 5.5 and calculated with the number of operations summarized in Table 5.1. It can be seen, by comparison with the prime factor technique (Table 5.2), that the Winograd Fourier transform algorithm reduces the number of multiplications by about a factor of two for DFTs of length 840 to 2520, while requiring a slightly larger number of additions. If we now compare with the conventional FFT method, using, for instance, the Rader Brenner algorithm (Table 4.3), we see that the Winograd Fourier transform algorithms reduce the number of multiplications by a factor of 2 to 3, with a number of additions which is only slightly larger. These results show that the principal contribution of the Winograd Fourier transform algorithm concerns a reduction in number of multiplications. It should be noted, however, that the short DFT algorithms can also be redesigned in order to minimize the number of additions at the expense of a larger number of multiplications. Thus, the Winograd Fourier transform approach is very flexible and allows one to adjust the number of additions and multiplications in order to fit the requirement of a particular implementation.

The WFTA is particularly well suited to computing the DFT of real data sequences. In this case, all input additions and all multiplications are real, while only some of the output additions are complex. Thus, contrary to other fast DFT algorithms, the WFTA computes a DFT of real data with nearly half the number of operations required for complex data, without any need for processing two sets of real data simultaneously. Thus, the WFTA is an attractive approach for processing of real data when storage must be conserved and in real time processing applications where the delay required to process simultaneously two consecutive blocks of real data cannot be tolerated.

5.4.2 Hybrid Algorithms

We have seen that a DFT of size $N_1 \times N_2 \ldots \times N_d$ can be computed by either the prime factor algorithm or the Winograd Fourier transform algorithm. In order to compare these two methods more explicitly, we consider here a simple DFT of size $N_1 \times N_2$. Using the prime factor technique, the number of multiplications is $N_1 M_2 + N_2 M_1$, while for the Winograd method it is equal to $M_1 M_2$. This means that the Winograd method requires a smaller number of multiplications than the prime factor technique if

$$N_1/M_1 + N_2/M_2 > 1. \tag{5.92}$$

In this formula, M_1 and M_2 are the number of complex multiplications corresponding to the DFTs of N_1 points and N_2 points, respectively. Thus, $M_1 \geqslant N_1$ and $M_2 \geqslant N_2$. However, since the Rader algorithms are very efficient, M_1 and M_2 are only slightly larger than N_1 and N_2. Moreover, N_i/M_i decreases only slowly with N_i, so that the condition defined by (5.92) is almost always met, except for very large DFTs. Thus, the Winograd algorithm generally yields a smaller number of multiplications than the prime factor method. This can be seen more clearly by considering a DFT of length 2520. In this case, $N/M = 0.53$, which implies $N_1/M_1 + N_2/M_2 = 1.06$ for a DFT of size 2520×2520. Thus, even for such a large DFT, replacing the Winograd method by the split-prime factor technique only marginally reduces the number of multiplications.

If we now consider the number of additions, the situation is completely reversed. This is due to the fact that the prime factor algorithm computes a DFT of $N_1 \times N_2$ points with $N_1 A_2 + N_2 A_1$ additions, while the Winograd method requires $N_1 A_2 + M_2 A_1$ additions. Since $M_2 \geqslant N_2$, the Winograd algorithm always requires more additions than the prime factor technique, except for $M_2 = N_2$.

A quick comparison of Tables 5.2 and 5.3 indicates that, for $N \leqslant 168$, the Winograd method requires fewer multiplications than the prime factor technique, and exactly the same number of additions. For larger DFTs, however, the prime factor method is better than the Winograd method from the standpoint of the number of additions. Thus, for large DFTs, it may be advantageous to combine the two methods when the relative cost of multiplications and additions is about the same [5.13]. For example, a 1008-point DFT could be computed by calculating a DFT of size 16×63 via the prime factor technique and calculating the DFTs of 63 terms by the Winograd algorithm. In this case, the DFT would be computed with 4396 real multiplications and 31852 real additions, as opposed to 3548 real multiplications and 34668 additions for the Winograd method and 5804 multiplications and 29548 additions for the prime factor technique. Thus, a combination of the two methods allows one to achieve a better balance between the number of additions and the number of multiplications.

5.4.3 Split Nesting Algorithms

We have seen in Sect. 5.3.3 that a multidimensional DFT can be converted into a set of one-dimensional and multidimensional convolutions by a sequence of reductions if the small DFTs are computed by Rader's algorithm. In particular, if N_1 and N_2 are odd primes, a DFT of size $N_1 \times N_2$ can be partitioned into a DFT of length N_1 plus one convolution of length $N_2 - 1$ and another of size $(N_1 - 1) \times (N_2 - 1)$. This is shown in Fig. 5.6 for a DFT of size 7×5. Thus, the Winograd algorithm can be regarded as equivalent to converting a DFT into a set of one-dimensional and multidimensional convolutions, and computing the multidimensional convolutions by a nesting algorithm (Sect. 3.3). Consequently, it can be inferred from Sect. 3.3.2 that a further reduction in the number of additions could be obtained by replacing the conventional nesting of convolutions by a split nesting technique. With such an approach, the short DFTs are reduced to convolutions by the Rader algorithm discussed in Sect. 5.2 and the convolutions are in turn reduced into polynomial products, defined modulo cyclotomic polynomials.

In practice, however, this method cannot be directly applied without minor modifications. Alert readers will notice that the number of additions corresponding to some of the short DFT algorithms in Sect. 5.5 does not tally with the number of operations derived directly from the reduction into convolutions discussed in Sect. 5.2. A 7-point DFT, for instance, is computed by the algorithm described in Sect. 5.5.5 with 9 multiplications and 36 additions. The same DFT, evaluated by Rader's algorithm according to Fig. 5.3, however, requires 12 additions for the reductions modulo $(z - 1)$ and $(z^7 - 1)/(z - 1)$ and 34 additions for the 6-point convolution, a total of 46 additions. This difference is due to the fact that the reductions can be partly embedded in the calculation of the convolutions. In the case of a N-point DFT, with N an odd prime, this procedure reduces the number of operations to one convolution of length $N - 1$ plus 2 additions instead of one $(N - 1)$-point convolution plus $2(N - 1)$ additions.

140   5. Linear Filtering Computation of Discrete Fourier Transforms

Thus, direct application of the split nesting algorithm is not attractive because it reduces the number of additions in algorithms which already have an inflated number of additions.

In order to overcome this difficulty, one approach consists in expressing the polynomial products modulo irreducible cyclotomic polynomials of degree higher than 1 in the optimum short DFT algorithms of Sect. 5.5. This is done in Sects. 5.5.4, 5, 7, and 8, respectively, for DFTs of lengths 5, 7, 9, and 16. With this procedure, a 5-point DFT breaks down (Fig. 5.10) into 14 input and output additions, 3 multiplications, and one polynomial product modulo ($z^2 + 1$), while the 7-point DFT reduces into 30 input and output additions, 3 multiplications, and the polynomial products modulo ($z^2 + z + 1$) and modulo ($z^2 - z + 1$). Therefore, nesting these two DFTs to compute a 35-point DFT requires 9 multiplications, 3 polynomial products modulo ($z_1^2 + 1$), 3 polynomial products modulo ($z_2^2 + z_2 + 1$) and modulo ($z_2^2 - z_2 + 1$), and the polynomial products modulo ($z_1^2 + 1$), ($z_2^2 + z_2 + 1$) and modulo ($z_1^2 + 1$), ($z_2^2 - z_2 + 1$). This defines an algorithm with 54 complex multiplications and 305 additions, as opposed to 54 multiplications and 333 additions for conventional nesting.

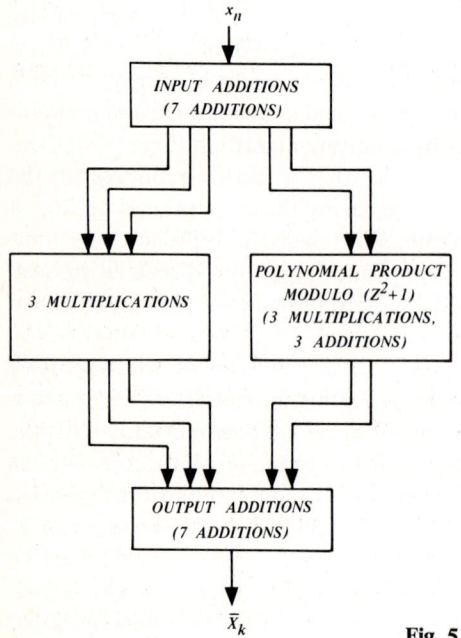

Fig. 5.10. 5-point DFT algorithm

In Table 5.4, we give the number of real additions for various DFTs computed by split nesting. The nesting and split nesting techniques require the same number of multiplications, and it can be seen, by comparing Tables 5.3 and 5.4,

5.4 The Winograd Fourier Transform Algorithm (WFTA)   141

Table 5.4. Number of real additions for one-dimensional DFTs computed by the Winograd Fourier transform algorithm and split nesting

DFT size N	Number of real additions A	Additions per point
240	4848	20.20
420	10680	25.43
504	13580	26.94
840	23460	27.93
1008	30364	30.12
2520	86764	34.43

that, for large DFTs, the split nesting method eliminates 10 to 15% of the additions required by the conventional nesting approach. Additional reduction is obtained when the split nesting method is used to compute large multidimensional DFTs. The implementation of the split nesting technique can be greatly simplified by storing composite split nested DFT algorithms, thus avoiding the complex data manipulations required by split nesting. With this approach, a 504-point DFT, for instance, can be computed by conventionally nesting an 8-point DFT with a 63-point DFT algorithm that has been optimized by split nesting.

5.4.4 Multidimensional DFTs

Until now, we have considered the use of the Winograd nesting method only for the computation of DFTs of size $N_1 N_2 \ldots N_d$ or $N_1 \times N_2 \times \ldots \times N_d$, where the various factors N_i are mutually prime in pairs. Note, however, that the condition $(N_i, N_u) = 1$ for $i \neq u$ is necessary only to convert one-dimensional DFTs into multidimensional DFTs by Good's algorithm. Thus, the Winograd nesting algorithm can also be employed to compute any multidimensional DFT of size $N_1 \times N_2 \times \ldots \times N_d$ where the factors N_i are not necessarily mutually prime in pairs. If each dimension N_i is composite, with $N_i = N_{i,1} N_{i,2} \ldots N_{i,e_i}$ and $(N_{i,u}, N_{i,v}) = 1$ for $u \neq v$, the index change in Good's algorithm maps the d-dimensional DFT into a multidimensional DFT of dimension $e_1 e_2 \ldots e_d$.

In order to illustrate the impact of this approach, we give in Table 5.5, the number of real operations for various multidimensional DFTs computed by the Winograd algorithm. It can be seen that the Winograd method is particularly effective for this application, since a DFT of size 1008×1008 is calculated with only 6.25 real multiplications per point, or about 2 complex multiplications per point. Moreover, for large multidimensional DFTs, the split nesting technique gives significant additional reduction in number of additions. For a DFT of size 1008×1008, for instance, split nesting reduces the number of real additions

Table 5.5. Number of nontrivial real operations for multidimensional DFTs computed by the Winograd Fourier transform algorithm

DFT size	Number of real multiplications	Number of real additions	Multiplications per point	Additions per point
24 × 24	1080	12096	1.87	21.00
30 × 30	2584	24264	2.87	26.96
40 × 40	4536	44736	2.83	27.96
48 × 48	5704	63720	2.48	27.66
72 × 72	15416	180032	2.97	34.73
120 × 120	41400	517824	2.87	35.96
240 × 240	209824	2683584	3.64	46.59
504 × 504	1254456	17742656	4.94	69.85
1008 × 1008	6350920	93076776	6.25	91.61
120 × 120 × 120	5971536	97203456	3.46	56.25
240 × 240 × 240	68023424	1086647616	4.92	78.61

from 93076776 (conventional nesting) to 64808280 or a saving of about 30% in number of additions.

5.4.5 Programming and Quantization Noise Issues

The Winograd Fourier transform algorithm requires substantially fewer multiplications than the FFT method. This reduction is achieved without any significant increase in number of additions and, in some favorable circumstances, the number of additions for the Winograd technique may also be fewer than for the FFT method. This result is quite remarkable in a theoretical sense because the FFT had long been thought to be the optimum computation technique for DFTs. It is a major achievement in computational complexity theory to have shown that a method radically different from the FFT could be computationally more efficient.

A key issue in the application of the Winograd method concerns its ability to be translated into computer programs that would be more effective than conventional FFT programs. Clearly, the number of arithmetic operations is in no way the only measure of computational complexity and, at the time of this writing (1979), there have been conflicting reports on the relative efficiencies of the WFTA and FFT algorithms, with actual WFTA computer execution times reported to be about ± 30 percent longer than those for the FFT for DFTs of about 500 to 1000 points [5.17, 19]. We shall not attempt to resolve these differences here, but we shall instead try to compare FFT programming with WFTA programming qualitatively. A practical WFTA program can be found in [5.9].

We can first note that, since the WFTA is most effective in reducing the number of multiplications, WFTA programs can be expected to be relatively more efficient when run on machines in which the execution times for multiplication

5.4 The Winograd Fourier Transform Algorithm (WFTA)

are significantly longer than for addition. Another important factor concerns the relative size of FFT and WFTA programs. When the FFT programs are built around a single radix FFT algorithm (usually radix-2 or radix-4 FFT algorithm), the computation proceeds by repetitive use of a subroutine implementing the FFT butterfly operation. Thus, the FFT programs can be very compact and essentially independent of the DFT size, provided that the DFT size is a power of the radix. By contrast, the WFTA uses different computation kernels for each DFT size and each of these is an explicit description of a particular small DFT algorithm, as opposed to the recursive, algorithmic structure used in the FFT. Thus, WFTA programs usually require more instructions than FFT programs for DFTs of comparable size and they must incorporate a subroutine which selects the proper computation kernels as a function of the DFT size. This feature prompts one to organize the program structure in two steps; a generation step and an execution step. The program can then be designed in such a way that most bookkeeping operations such as data routing and kernel selection, as well as precomputation of the multipliers, are done within the generation step and therefore do not significantly impact the execution time.

The WFTA program is divided into five main parts: input data reordering, input additions, multiplications, output additions, and output data reordering. The input and output data reordering requires a number of modular multiplications and additions which can be eliminated by precomputing reordering vectors during the generation step. These stored reordering vectors may, then, be used to rearrange the input and output data during the execution step. The input additions, except for the innermost factor, correspond to a set of additions that is executed for each factor N_i, and operates on N_i input arrays to produce M_i output arrays. Since M_i is generally larger than N_i, the calculations cannot generally be done "in-place". Thus, the generated result of each stage cannot be stored over the input data sequence to the stage. However, it is always possible to assign N_i input storage locations from the M_i output storage locations and, since M_i is not much larger than N_i, this results in an algorithm that is not significantly less efficient than an in-place algorithm, as far as memory utilization is concerned. The calculations corresponding to the innermost factor N_d are executed on scalar data and include all the multiplications required by the algorithm to compute the N-point DFT. If M is the total number of multiplications corresponding to the N-point DFT and M_d is the number of multiplications corresponding to the N_d-point small DFT algorithm, the calculations for the innermost factor N_d reduce to M/M_d DFTs of N_d points. The M_d coefficients here are those of the N_d-point DFT, multiplied by the coefficients of the other small DFT algorithms. In order to avoid recalculating this set of M_d coefficients for each of the M/M_d DFTs of N_d points, one is generally led to precompute, at generation time, a vector of M coefficients divided into M/M_d sets of M_d coefficients. Since these coefficients are simply real or imaginary, a total of M real memory locations are required, or significantly less than for an FFT algorithm in which the coefficients are precomputed.

From this, we can conclude that, although the WFTA is not an in-place algorithm, the total memory requirement for storing data and coefficients can be about the same as that of the FFT algorithm. The program sizes will generally be larger for the WFTA than for the FFT, but remain reasonably small, because the number of instructions grows approximately as the sum of the number of additions corresponding to each small algorithm. Thus, if $N = N_1 N_2 \ldots N_t \ldots N_d$ and if A_i is the number of additions corresponding to a N_i-point DFT, $\sum_i A_i$ is a rough measure of program size. $\sum_i A_i$ grows very slowly with N, as can be verified by noting that $\sum_i A_i = 25$ for $N = 30$ and $\sum_i A_i = 154$ for $N = 1008$.

Thus WFTA program size and memory requirements can remain reasonably small, even for large DFTs, provided that the programs are properly designed to work on array organized data. Hence, the WFTA seems to be particularly well suited for systems, such as APL, which have been designed to process array data efficiently.

Another important issue concerns the computational noise of the Winograd algorithms, and only scant information is currently available on this topic. The preliminary results given in [5.18] tend to indicate that proper scaling at each stage is more difficult than for the FFT because of the fact that all moduli are different and not powers of 2. In the case of fixed point data, this significantly impacts the signal-to-noise ratio of the WFTA, and thus, the WFTA generally requires about one or two more bits for representing the data to give an error similar to the FFT.

5.5 Short DFT Algorithms

This section lists the short DFT algorithms that are most frequently used with the prime factor method or the WFTA. These algorithms compute short N-point one-dimensional DFTs of a complex input sequence x_n

$$\bar{X}_k = \sum_{n=0}^{N-1} x_n W^{nk}, \quad k = 0, \ldots, N-1$$

$$W = e^{-j2\pi/N}, \quad j = \sqrt{-1} \quad (5.93)$$

for $N = 2, 3, 4, 5, 7, 8, 9, 16$.

These algorithms are derived from Rader's algorithm and arranged as follows: input data $x_0, x_1, \ldots, x_{N-1}$ and output data $\bar{X}_0, \bar{X}_1, \ldots, \bar{X}_{N-1}$ in natural order. $m_0, m_1, \ldots, m_{N-1}$ are the results of the M multiplications corresponding to length N.

$t_1, t_2, \ldots$ and $s_1, s_2, \ldots$ are temporary storage for input data and output data, respectively.

5.5 Short DFT Algorithms

The operations are executed in the order t_i, m_i, s_i, $\bar{X}_i$, with indices in natural order. For DFTs of lengths 5, 7, 9, 16, the operations can also be executed using the form shown in Sects. 5.6.4, 5, 7, 8 which embeds the various polynomial products.

The figures between parentheses indicate trivial multiplications by ± 1, $\pm j$

At the end of each of algorithm description for $N = 3, 5, 7, 9$, we give the number of operations for the corresponding algorithm in which the number of non-trivial multiplications is minimized and the output is scaled by a constant factor.

5.5.1 2-Point DFT

2 multiplications (2), 2 additions

$m_0 = 1 \cdot (x_0 + x_1)$ $\qquad m_1 = 1 \cdot (x_0 - x_1)$
$\bar{X}_0 = m_0$
$\bar{X}_1 = m_1$

5.5.2 3-Point DFT

$u = 2\pi/3$, $\quad$ 3 multiplications (1), 6 additions
$t_1 = x_1 + x_2$
$m_0 = 1 \cdot (x_0 + t_1)$ $\qquad m_1 = (\cos u - 1) \cdot t_1$
$m_2 = j \sin u \cdot (x_2 - x_1)$
$s_1 = m_0 + m_1$
$\bar{X}_0 = m_0$
$\bar{X}_1 = s_1 + m_2$
$\bar{X}_2 = s_1 - m_2$

Corresponding algorithm with scaled output:
3 multiplications (2), 8 additions, scaling factor: 2

5.5.3 4-Point DFT

4 multiplications (4), 8 additions

$t_1 = x_0 + x_2$ $\qquad\qquad t_2 = x_1 + x_3$
$m_0 = 1 \cdot (t_1 + t_2)$ $\qquad m_1 = 1 \cdot (t_1 - t_2)$
$m_2 = 1 \cdot (x_0 - x_2)$ $\qquad m_3 = j (x_3 - x_1)$
$\bar{X}_0 = m_0$
$\bar{X}_1 = m_2 + m_3$

$\bar{X}_2 = m_1$
$\bar{X}_3 = m_2 - m_3$

5.5.4 5-Point DFT

$u = 2\pi/5,$ 6 multiplications (1), 17 additions

$t_1 = x_1 + x_4$ $\quad t_2 = x_2 + x_3$ $\quad t_3 = x_1 - x_4$ $\quad t_4 = x_3 - x_2$
$t_5 = t_1 + t_2$

$m_0 = 1 \cdot (x_0 + t_5)$
$m_1 = [(\cos u + \cos 2u)/2 - 1]t_5$
$m_2 = [(\cos u - \cos 2u)/2](t_1 - t_2)$
Polynomial product modulo $(z^2 + 1)$

$m_3 = -j(\sin u)(t_3 + t_4)$
$m_4 = -j(\sin u + \sin 2u) \cdot t_4$
$m_5 = j(\sin u - \sin 2u)t_3$

$s_3 = m_3 - m_4$
$s_5 = m_3 + m_5$

$s_1 = m_0 + m_1$ $\quad s_2 = s_1 + m_2$
$s_4 = s_1 - m_2$

$\bar{X}_0 = m_0$ $\qquad\qquad\qquad \bar{X}_3 = s_4 - s_5$
$\bar{X}_1 = s_2 + s_3$ $\qquad\qquad\quad\, \bar{X}_4 = s_2 - s_3$
$\bar{X}_2 = s_4 + s_5$

Corresponding algorithm with scaled output:
6 multiplications (2), 21 additions, scaling factor: 4

5.5.5 7-Point DFT

$u = 2\pi/7,$ 9 multiplications (1), 36 additions

$t_1 = x_1 + x_6$ $\qquad t_2 = x_2 + x_5$ $\qquad t_3 = x_3 + x_4$
$t_4 = t_1 + t_2 + t_3$ $\quad\; t_5 = x_1 - x_6$ $\qquad t_6 = x_2 - x_5$
$t_7 = x_4 - x_3$ $\qquad\; t_8 = t_1 - t_3$ $\qquad\; t_9 = t_3 - t_2$
$t_{10} = t_5 + t_6 + t_7$ $\quad t_{11} = t_7 - t_5$ $\qquad t_{12} = t_6 - t_7$

$m_0 = 1 \cdot (x_0 + t_4)$

$m_1 = [(\cos u + \cos 2u + \cos 3u)/3 - 1] t_4$

$t_{13} = -t_8 - t_9$
$m_2 = [(2\cos u - \cos 2u - \cos 3u)/3] t_8$
$m_3 = [(\cos u - 2\cos 2u + \cos 3u)/3] t_9$
$m_4 = [(\cos u + \cos 2u - 2\cos 3u)/3] t_{13}$

$s_0 = -m_2 - m_3$ \qquad Polynomial product
$s_1 = -m_2 - m_4$ \qquad modulo $(z^2 + z + 1)$

$m_5 = -j[(\sin u + \sin 2u - \sin 3u)/3]t_{10}$

$t_{14} = -t_{11} - t_{12}$
$m_6 = j[(2\sin u - \sin 2u + \sin 3u)/3]t_{11}$
$m_7 = j[\sin u - 2\sin 2u - \sin 3u)/3]t_{12}$
$m_8 = j[(\sin u + \sin 2u + 2\sin 3u)/3]t_{14}$

$s_2 = -m_6 - m_7$ \qquad Polynomial product
$s_3 = m_6 + m_8$ \qquad modulo $(z^2 - z + 1)$

$s_4 = m_0 + m_1$ \qquad $s_5 = s_4 - s_0$ \qquad $s_6 = s_4 + s_1$
$s_7 = s_4 + s_0 - s_1$ \qquad $s_8 = m_5 - s_2$ \qquad $s_9 = m_5 - s_3$
$s_{10} = m_5 + s_2 + s_3$

$\bar{X}_0 = m_0$ \qquad $\bar{X}_1 = s_5 + s_8$ \qquad $\bar{X}_2 = s_6 + s_9$ \qquad $\bar{X}_3 = s_7 - s_{10}$
$\bar{X}_4 = s_7 + s_{10}$ \qquad $\bar{X}_5 = s_6 - s_9$ \qquad $\bar{X}_6 = s_5 - s_8$

Corresponding algorithm with scaled output:
9 multiplications (2), 43 additions, scaling factor: 6

5.5.6 8-Point DFT

$u = 2\pi/8$, \qquad 8 multiplications (6), 26 additions

$t_1 = x_0 + x_4$ \qquad $t_2 = x_2 + x_6$ \qquad $t_3 = x_1 + x_5$
$t_4 = x_1 - x_5$ \qquad $t_5 = x_3 + x_7$ \qquad $t_6 = x_3 - x_7$
$t_7 = t_1 + t_2$ \qquad $t_8 = t_3 + t_5$

$m_0 = 1 \cdot (t_7 + t_8)$ \qquad $m_1 = 1 \cdot (t_7 - t_8)$
$m_2 = 1 \cdot (t_1 - t_2)$ \qquad $m_3 = 1 \cdot (x_0 - x_4)$
$m_4 = \cos u \cdot (t_4 - t_6)$ \qquad $m_5 = j(t_5 - t_3)$

148   5. Linear Filtering Computation of Discrete Fourier Transforms

$m_6 = j(x_6 - x_2)$ $\qquad\qquad\qquad m_7 = -j \sin u \cdot (t_4 + t_6)$

$s_1 = m_3 + m_4$ $\qquad\qquad s_2 = m_3 - m_4 \qquad\qquad s_3 = m_6 + m_7$

$s_4 = m_6 - m_7$

$\bar{X}_0 = m_0 \qquad\qquad\qquad \bar{X}_1 = s_1 + s_3 \qquad\qquad \bar{X}_2 = m_2 + m_5$

$\bar{X}_3 = s_2 - s_4 \qquad\qquad \bar{X}_4 = m_1 \qquad\qquad\qquad \bar{X}_5 = s_2 + s_4$

$\bar{X}_6 = m_2 - m_5 \qquad\qquad \bar{X}_7 = s_1 - s_3$

5.5.7 9-Point DFT

$u = 2\pi/9,$ 11 multiplications (1), 44 additions

$t_1 = x_1 + x_8 \qquad\qquad t_2 = x_2 + x_7 \qquad\qquad t_3 = x_3 + x_6$

$t_4 = x_4 + x_5 \qquad\qquad t_5 = t_1 + t_2 + t_4 \qquad\qquad t_6 = x_1 - x_8$

$t_7 = x_7 - x_2 \qquad\qquad t_8 = x_3 - x_6 \qquad\qquad t_9 = x_4 - x_5$

$t_{10} = t_6 + t_7 + t_9 \qquad t_{11} = t_1 - t_2 \qquad\qquad t_{12} = t_2 - t_4$

$t_{13} = t_7 - t_6 \qquad\qquad t_{14} = t_7 - t_9$

$m_0 = 1 \cdot (x_0 + t_3 + t_5)$

$m_1 = (3/2) t_3$

$m_2 = -t_5/2$

$t_{15} = -t_{12} - t_{11}$

$m_3 = [(2 \cos u - \cos 2u - \cos 4u)/3] t_{11}$

$m_4 = [(\cos u + \cos 2u - 2 \cos 4u)/3] t_{12}$

$m_5 = [(\cos u - 2 \cos 2u + \cos 4u)/3] t_{15}$

$s_0 = -m_3 - m_4 \qquad\qquad$ Polynomial product

$s_1 = m_5 - m_4 \qquad\qquad$ modulo $(z^2 + z + 1)$

$m_6 = -j \sin 3u \cdot t_{10}$

$m_7 = -j \sin 3u \cdot t_8$

$t_{16} = -t_{13} + t_{14}$

$m_8 = j \sin u \cdot t_{13}$

$m_9 = j \sin 4u \cdot t_{14}$

$m_{10} = j \sin 2u \cdot t_{16}$

$s_2 = -m_8 - m_9 \qquad\qquad$ Polynomial product

$s_3 = m_9 - m_{10} \qquad\qquad$ modulo $(z^2 - z + 1)$

5.5 Short DFT Algorithms   149

$s_4 = m_0 + m_2 + m_2$ $\qquad s_5 = s_4 - m_1$
$s_6 = s_4 + m_2$ $\qquad s_7 = s_5 - s_0$
$s_8 = s_1 + s_5$ $\qquad s_9 = s_0 - s_1 + s_5$
$s_{10} = m_7 - s_2$ $\qquad s_{11} = m_7 - s_3$
$s_{12} = m_7 + s_2 + s_3$

$\bar{X}_0 = m_0$ $\qquad \bar{X}_1 = s_7 + s_{10}$ $\qquad \bar{X}_2 = s_8 - s_{11}$
$\bar{X}_3 = s_6 + m_6$ $\qquad \bar{X}_4 = s_9 + s_{12}$ $\qquad \bar{X}_5 = s_9 - s_{12}$
$\bar{X}_6 = s_6 - m_6$ $\qquad \bar{X}_7 = s_8 + s_{11}$ $\qquad \bar{X}_8 = s_7 - s_{10}$

Corresponding algorith with scaled output:
11 multiplications (3), 45 additions, scaling factor: 2

5.5.8 16-Point DFT

$u = 2\pi/16,$ 18 multiplications (8), 74 additions

$t_1 = x_0 + x_8$ $\qquad t_2 = x_4 + x_{12}$ $\qquad t_3 = x_2 + x_{10}$
$t_4 = x_2 - x_{10}$ $\qquad t_5 = x_6 + x_{14}$ $\qquad t_6 = x_6 - x_{14}$
$t_7 = x_1 + x_9$ $\qquad t_8 = x_1 - x_9$ $\qquad t_9 = x_3 + x_{11}$
$t_{10} = x_3 - x_{11}$ $\qquad t_{11} = x_5 + x_{13}$ $\qquad t_{12} = x_5 - x_{13}$
$t_{13} = x_7 + x_{15}$ $\qquad t_{14} = x_7 - x_{15}$ $\qquad t_{15} = t_1 + t_2$
$t_{16} = t_3 + t_5$ $\qquad t_{17} = t_{15} + t_{16}$ $\qquad t_{18} = t_7 + t_{11}$
$t_{19} = t_7 - t_{11}$ $\qquad t_{20} = t_9 + t_{13}$ $\qquad t_{21} = t_9 - t_{13}$
$t_{22} = t_{18} + t_{20}$ $\qquad t_{23} = t_8 + t_{14}$ $\qquad t_{24} = t_8 - t_{14}$
$t_{25} = t_{10} + t_{12}$ $\qquad t_{26} = t_{12} - t_{10}$

$m_0 = 1 \cdot (t_{17} + t_{22})$ $\qquad m_1 = 1 \cdot (t_{17} - t_{22})$
$m_2 = 1 \cdot (t_{15} - t_{16})$ $\qquad m_3 = 1 \cdot (t_1 - t_2)$
$m_4 = 1 \cdot (x_0 - x_8)$ $\qquad m_5 = \cos 2u \cdot (t_{19} - t_{21})$
$m_6 = \cos 2u \cdot (t_4 - t_6)$

$m_7 = \cos 3u \cdot (t_{24} + t_{26})$
$m_8 = (\cos u + \cos 3u) \cdot t_{24}$
$m_9 = (\cos 3u - \cos u) \cdot t_{26}$ Polynomial product
$s_7 = m_8 - m_7$ $\qquad s_8 = m_9 - m_7$ modulo $(z^2 + 1)$

$m_{10} = j \cdot (t_{20} - t_{18})$ $\qquad m_{11} = j \cdot (t_5 - t_3)$
$m_{12} = j \cdot (x_{12} - x_4)$ $\qquad m_{13} = -j \sin 2u \cdot (t_{19} + t_{21})$

$$m_{14} = -j \sin 2u \cdot (t_4 + t_6)$$

$$m_{15} = -j \sin 3u \cdot (t_{23} + t_{25})$$
$$m_{16} = j (\sin 3u - \sin u) \cdot t_{23}$$
$$m_{17} = -j (\sin u + \sin 3u) \cdot t_{25}$$

$S_{15} = m_{15} + m_{16}$ $\quad\quad$ $S_{16} = m_{15} - m_{17}$ $\quad\quad$ Polynomial product modulo$(z^2 + 1)$

$S_1 = m_3 + m_5$	$S_2 = m_3 - m_5$	$S_3 = m_{11} + m_{13}$
$S_4 = m_{13} - m_{11}$	$S_5 = m_4 + m_6$	$S_6 = m_4 - m_6$
$S_9 = S_5 + S_7$	$S_{10} = S_5 - S_7$	$S_{11} = S_6 + S_8$
$S_{12} = S_6 - S_8$		
$S_{13} = m_{12} + m_{14}$	$S_{14} = m_{12} - m_{14}$	$S_{17} = S_{13} + S_{15}$
$S_{18} = S_{13} - S_{15}$	$S_{19} = S_{14} + S_{16}$	$S_{20} = S_{14} - S_{16}$
$\bar{X}_0 = m_0$	$\bar{X}_1 = S_9 + S_{17}$	$\bar{X}_2 = S_1 + S_3$
$\bar{X}_3 = S_{12} - S_{20}$	$\bar{X}_4 = m_2 + m_{10}$	$\bar{X}_5 = S_{11} + S_{19}$
$\bar{X}_6 = S_2 + S_4$	$\bar{X}_7 = S_{10} - S_{18}$	$\bar{X}_8 = m_1$
$\bar{X}_9 = S_{10} + S_{18}$	$\bar{X}_{10} = S_2 - S_4$	$\bar{X}_{11} = S_{11} - S_{19}$
$\bar{X}_{12} = m_2 - m_{10}$	$\bar{X}_{13} = S_{12} + S_{20}$	$\bar{X}_{14} = S_1 - S_3$
$\bar{X}_{15} = S_9 - S_{17}$		

6. Polynomial Transforms

The main objective of this chapter is to develop fast multidimensional filtering algorithms. These algorithms are based on the use of polynomial transforms which can be viewed as discrete Fourier transforms defined in rings of polynomials. Polynomial transforms can be computed without multiplications using ordinary arithmetic, and produce an efficient mapping of multidimensional convolutions into one-dimensional convolutions and polynomial products.

In this chapter, we first introduce polynomial transforms for the calculation of simple convolutions of size $p \times p$, with p prime. We then extend the definition of polynomial transforms to other cases, and establish that these transforms, which are defined in rings of polynomials, will indeed support convolution. As a final item, we also discuss the use of polynomial transforms for the convolution of complex data sequences and multidimensional data structures of dimension d, with $d > 2$.

6.1 Introduction to Polynomial Transforms

We consider a two-dimensional circular convolution of size $N \times N$, with

$$y_{u,l} = \sum_{m=0}^{N-1} \sum_{n=0}^{N-1} h_{n,m} \, x_{u-n, l-m} \qquad u, l = 0, ..., N-1. \tag{6.1}$$

In order to simplify this expression, we resort to a representation in polynomial algebra. This is realized by noting that (6.1) can be viewed as the one-dimensional polynomial convolution

$$Y_l(z) \equiv \sum_{m=0}^{N-1} H_m(z) X_{l-m}(z) \quad \text{modulo } (z^N - 1) \tag{6.2}$$

$$H_m(z) = \sum_{n=0}^{N-1} h_{n,m} \, z^n, \qquad m = 0, ..., N-1 \tag{6.3}$$

$$X_r(z) = \sum_{s=0}^{N-1} x_{s,r} \, z^s, \qquad r = 0, ..., N-1, \tag{6.4}$$

where $y_{u,l}$ is obtained from the N polynomials

$$Y_l(z) = \sum_{u=0}^{N-1} y_{u,l} \, z^u, \qquad l = 0, ..., N-1 \tag{6.5}$$

by taking the coefficients of z^u in $Y_l(z)$. In order to introduce the concept of polynomial transforms in a simple way, we shall assume in this section that N is an odd prime, with $N = p$. In this case, as shown in Sect. 2.2.4, $z^p - 1$ is the product of two cyclotomic polynomials

$$z^p - 1 = (z - 1)P(z) \tag{6.6}$$

$$P(z) = z^{p-1} + z^{p-2} + \ldots + 1. \tag{6.7}$$

Since $Y_l(z)$ is defined modulo $(z^p - 1)$, it can be computed by reducing $H_m(z)$ and $X_r(z)$ modulo $(z - 1)$ and $P(z)$, computing the polynomial convolutions $Y_{1,l}(z) \equiv Y_l(z)$ modulo $P(z)$, and $Y_{2,l} \equiv Y_l(z)$ modulo $(z - 1)$ on the reduced polynomials and reconstructing $Y_l(z)$ by the Chinese remainder theorem (Sect. 2.2.3) with

$$Y_l(z) \equiv S_1(z)Y_{1,l}(z) + S_2(z)Y_{2,l} \quad \text{modulo } (z^p - 1) \tag{6.8}$$

$$\begin{cases} S_1(z) \equiv 1 & S_2(z) \equiv 0 \quad \text{modulo } P(z) \\ S_1(z) \equiv 0 & S_2(z) \equiv 1 \quad \text{modulo } (z - 1) \end{cases} \tag{6.9}$$

with

$$S_1(z) = [p - P(z)]/p \tag{6.10}$$

and

$$S_2(z) = P(z)/p. \tag{6.11}$$

Computing $Y_l(z)$ is therefore replaced by the simpler problem of computing $Y_{1,l}(z)$ and $Y_{2,l}$. $Y_{2,l}$ can be obtained very simply because it is defined modulo $(z - 1)$. Thus, $Y_{2,l}$ is the convolution product of the scalars $H_{2,m}$ and $X_{2,r}$ obtained by substituting 1 for z in $H_m(z)$ and $X_r(z)$

$$Y_{2,l} = \sum_{m=0}^{p-1} H_{2,m} X_{2,l-m}, \quad l = 0, \ldots, p - 1 \tag{6.12}$$

$$H_{2,m} = \sum_{n=0}^{p-1} h_{n,m} \tag{6.13}$$

$$X_{2,r} = \sum_{s=0}^{p-1} x_{s,r}. \tag{6.14}$$

We now turn to the computation of $Y_{1,l}(z)$. In order to simplify this computation, we introduce a transform $\bar{H}_k(z)$ which has the same structure as the DFT, but with the usual complex exponential operator replaced by one defined as an exponential on the variable z and with all operations defined modulo $P(z)$. This

6.1 Introduction to Polynomial Transforms

transform, which we call a *polynomial transform* [6.1, 2] is defined by the expression

$$\bar{X}_k(z) \equiv \sum_{r=0}^{p-1} X_{1,r}(z) z^{rk} \quad \text{modulo } P(z) \qquad k = 0, \ldots, p-1 \tag{6.15}$$

$$X_{1,r}(z) \equiv X_r(z) \quad \text{modulo } P(z). \tag{6.16}$$

We define similarly an inverse transform by

$$X_{1,r}(z) \equiv \frac{1}{p} \sum_{k=0}^{p-1} \bar{X}_k(z) z^{-rk} \quad \text{modulo } P(z),$$

$$z^{-rk} \equiv z^{(p-1)rk}, \qquad r = 0, \ldots, p-1. \tag{6.17}$$

We shall now establish that the polynomial transforms support circular convolution, and that (6.17) is the inverse of (6.15). This can be demonstrated by calculating the transforms $\bar{H}_k(z)$ and $\bar{X}_k(z)$ of $H_{1,m}(z)$ and $X_{1,r}(z)$ via (6.15), multiplying $\bar{H}_k(z)$ by $\bar{X}_k(z)$ modulo $P(z)$, and computing the inverse transform $Q_l(z)$ of $\bar{H}_k(z) \bar{X}_k(z)$. This can be denoted as

$$Q_l(z) \equiv \sum_{m=0}^{p-1} \sum_{r=0}^{p-1} H_{1,m}(z) X_{1,r}(z) \frac{1}{p} \sum_{k=0}^{p-1} z^{qk} \quad \text{modulo } P(z) \tag{6.18}$$

with $q = m + r - l$. Let $S = \sum_{k=0}^{p-1} z^{qk}$. Since $z^p \equiv 1$, the exponents of z are defined modulo p. For $q \equiv 0$ modulo p, $S = p$. For $q \not\equiv 0$ modulo p, the set of exponents qk defined modulo p is a simple permutation of the integers $0, 1, \ldots, p-1$. Thus, $S \equiv \sum_{k=0}^{p-1} z^k \equiv P(z) \equiv 0$ modulo $P(z)$. This means that the only nonzero case corresponds to $q \equiv 0$ or $r \equiv l - m$ modulo p and that $Q_l(z)$ reduces to the circular convolution

$$Q_l(z) \equiv Y_{1,l}(z) \equiv \sum_{m=0}^{p-1} H_{1,m}(z) X_{1,l-m}(z) \quad \text{modulo } P(z). \tag{6.19}$$

The demonstration that the polynomial transform (6.15) and the inverse polynomial transform (6.17) form a transform pair follows immediately by setting $H_{1,m}(z) = 1$ in (6.19).

Using the foregoing method, $Y_{1,l}(z)$ is computed with three polynomial transforms and p polynomial multiplications $\bar{H}_k(z)\bar{X}_k(z)$ defined modulo $P(z)$. In many digital filtering applications, the input sequence $h_{n,m}$ is fixed and its transform $\bar{H}_k(z)$ can be precomputed. In this case, only two polynomial transforms are required, and the Chinese remainder reconstruction can also be simplified by noting, with (2.87), that

$$S_1(z) \equiv (z-1)/[(z-1) \text{ modulo } P(z)] \equiv (z-1) T_1(z) \tag{6.20}$$

$$T_1(z) = [-z^{p-2} - 2z^{p-3} \ldots - (p-3)z^2 - (p-2)z + 1 - p]/p. \quad (6.21)$$

Since $T_1(z)$ is defined modulo $P(z)$, premultiplication by $T_1(z)$ can be accomplished prior to Chinese remainder reconstruction and merged with the precomputation of the transform of the fixed sequence. Similarly, the multiplication by $1/p$ required in (6.11) for the part of the Chinese remainder reconstruction related to $Y_{2,l}$ can be combined with the computation of the scalar convolution $Y_{2,l}$ in (6.12) so that the Chinese remainder reconstruction reduces to

$$Y_l(z) \equiv (z-1)Y_{1,l}(z) + (z^{p-1} + z^{p-2} + \ldots + 1)Y_{2,l} \bmod (z^p - 1). \quad (6.22)$$

Under these conditions, a convolution of size $p \times p$ is computed as shown in Fig. 6.1. With this procedure, the reduction modulo $(z-1)$ and the Chinese

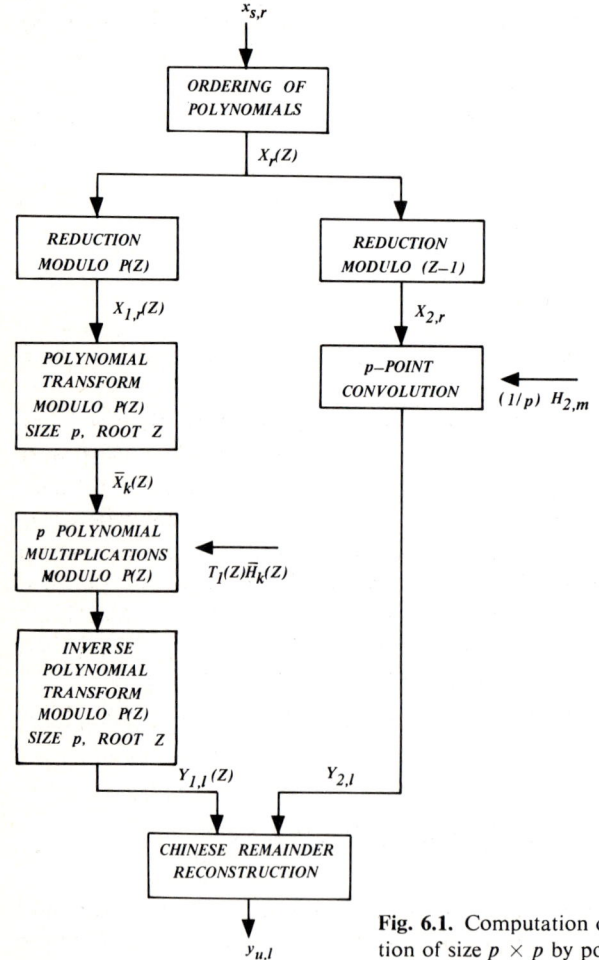

Fig. 6.1. Computation of a two-dimensional convolution of size $p \times p$ by polynomial transforms. p prime

reconstruction are calculated, respectively, by (6.14) and (6.22) without the use of multiplications. The reductions modulo $P(z)$ also require no multiplications because $z^{p-1} \equiv -z^{p-2} - z^{p-3} \ldots - 1$ modulo $P(z)$, which gives for $X_{1,r}(z)$

$$X_{1,r}(z) = \sum_{s=0}^{p-2} (x_{s,r} - x_{s,p-1})z^s = \sum_{s=0}^{p-2} x_{1,s,r} z^s. \tag{6.23}$$

The polynomial transform $\bar{X}_k(z)$ defined by (6.15) requires only multiplications by z^{rk} and additions. Polynomial addition is executed by adding separately the coefficients corresponding to each power of z. For multiplications by powers of z, we note that, since $P(z)$ is a factor of $z^p - 1$,

$$X_{1,r}(z)z^{rk} \equiv [X_{1,r}(z)z^{rk} \text{ modulo } (z^p - 1)] \text{ modulo } P(z). \tag{6.24}$$

Moreover, the congruence relation $z^p \equiv 1$ implies that rk is defined modulo p. Thus, setting $q \equiv rk$ modulo p and computing $X_{1,r}(z)z^{rk}$ by (6.23) and (6.24) yields

$$X_{1,r}(z) = \sum_{s=0}^{p-1} x_{1,s,r} z^s, \qquad x_{1,p-1,r} = 0 \tag{6.25}$$

$$X_{1,r}(z)z^q \text{ modulo } (z^p - 1) = \sum_{s=0}^{p-q-1} x_{1,s,r} z^{s+q} + \sum_{s=p-q}^{p-1} x_{1,s,r} z^{s+q}$$

$$\equiv \sum_{s=0}^{p-1} x_{1,\langle s-q \rangle,r} z^s \tag{6.26}$$

$$X_{1,r}(z)z^q \text{ modulo } P(z) = \sum_{s=0}^{p-2} (x_{1,\langle s-q \rangle,r} - x_{1,\langle p-q-1 \rangle,r})z^s, \quad x_{1,p-1,r} = 0, \tag{6.27}$$

where the symbols $\langle \rangle$ define $s - q$ and $p - q - 1$ modulo p. Thus, the polynomial transforms are evaluated with additions and simple rotations of p-word polynomials, and the only multiplications required to compute the two-dimensional convolution $y_{u,l}$ correspond to the calculation of one convolution of length p and to the evaluation of the p one-dimensional polynomial products $T_1(z)\bar{H}_k(z)\bar{X}_k(z)$ defined modulo $P(z)$. This means that, if the polynomial products and the convolution are evaluated with the minimum multiplication algorithms defined by theorems 2.21 and 2.22, the convolution of size $p \times p$, with p prime, is computed with only $2p^2 - p - 2$ multiplications. It can indeed be shown that this is the theoretical minimum number of multiplications for a convolution of size $p \times p$, with p prime [6.3].

6.2 General Definition of Polynomial Transforms

In order to motivate the development of polynomial transforms, we have, until now, restricted our discussion to polynomial transforms of size $p \times p$, with p

prime. A much more general definition of polynomial transforms can be obtained by considering a polynomial convolution of length N defined modulo a polynomial $P(z)$, where $P(z)$ is no longer a factor of $z^p - 1$, with p prime. Then, it follows from the demonstration of the circular convolution property given in Sect. 6.1 that a length-N circular convolution can be computed by polynomial transforms of length N and root $G(z)$ provided the three following conditions are met:

— $G^N(z) \equiv 1$ modulo $P(z)$ (6.28)

— N and $G(z)$ have inverses modulo $P(z)$ (6.29)

$$-S \equiv \sum_{k=0}^{N-1} [G(z)]^{qk} \text{ modulo } P(z) \equiv \begin{cases} 0 & \text{for } q \not\equiv 0 \text{ modulo } N \\ N & \text{for } q \equiv 0 \text{ modulo } N. \end{cases} \quad (6.30)$$

With these three conditions, a polynomial convolution $Y_l(z)$ is computed by polynomial transforms with

$$Y_l(z) \equiv \sum_{m=0}^{N-1} H_m(z) X_{l-m}(z) \text{ modulo } P(z) \quad (6.31)$$

$$H_m(z) = \sum_{n=0}^{b-1} h_{n,m} z^n \quad (6.32)$$

$$X_r(z) = \sum_{s=0}^{b-1} x_{s,r} z^s \quad (6.33)$$

$$\bar{H}_k(z) \equiv \sum_{m=0}^{N-1} H_m(z) [G(z)]^{mk} \text{ modulo } P(z), \quad (6.34)$$

where b is the degree of $P(z)$. $Y_l(z)$ is obtained by evaluating the inverse transform of $\bar{H}_k(z) \bar{X}_k(z)$ by

$$Y_l(z) \equiv \frac{1}{N} \sum_{k=0}^{N-1} \bar{H}_k(z) \bar{X}_k(z) [G(z)]^{-lk} \text{ modulo } P(z). \quad (6.35)$$

The polynomial transforms have the same structure as DFTs, but with complex exponential roots of unity replaced by polynomials $G(z)$ and with all operations defined modulo $P(z)$. Therefore, these transforms have the same general properties as the DFTs (Sect. 4.1.1), and in particular, they have the linearity property, with

$$\{H_n(z)\} + \{X_n(z)\} \xrightleftharpoons{\text{PT}} \{\bar{H}_k(z)\} + \{\bar{X}_k(z)\} \quad (6.36)$$

$$\{\lambda H_n(z)\} \xrightleftharpoons{\text{PT}} \{\lambda \bar{H}_k(z)\}. \quad (6.37)$$

6.2 General Definition of Polynomial Transforms

The principal application of polynomial transforms concerns the computation of two-dimensional circular convolutions. In this case $P(z)$ must be a factor of $z^N - 1$ and it is desirable that $G(z)$ be as simple as possible in order to avoid multiplications in the calculation of the polynomial transforms. Since the factors of $z^N - 1$ are the cyclotomic polynomials (Sect. 2.2.4) when the coefficients are defined in the field of rationals, the various polynomial transforms suitable for the computation of two-dimensional convolutions can be derived by exploiting the properties of cyclotomic polynomials. We shall now show that when a two-dimensional convolution has common factors in both dimensions, it is always possible to define polynomial transforms which have very simple roots $G(z)$ and which can be computed without multiplications.

6.2.1 Polynomial Transforms with Roots in a Field of Polynomials

A circular convolution of size $N \times N$ can be represented as a polynomial convolution of length N where all polynomials are defined modulo $(z^N - 1)$, as shown in (6.2–4). For coefficients in the field of rationals, $z^N - 1$ is the product of d cyclotomic polynomials $P_{e_i}(z)$, where d is the number of divisors e_i of N, including 1 and N, with

$$z^N - 1 = \prod_{i=1}^{d} P_{e_i}(z). \tag{6.38}$$

The degree of each cyclotomic polynomial $P_{e_i}(z)$ is $\phi(e_i)$, where $\phi(e_i)$ is Euler's totient function (Sect. 2.1.3). Since the various polynomials $P_{e_i}(z)$ are irreducible, the polynomial convolution defined modulo $(z^N - 1)$ can be computed separately modulo each polynomial $P_{e_i}(z)$, with reconstruction of the final result by the Chinese remainder theorem. We show first that there is always a polynomial transform of dimension N and root z which supports circular convolution when defined modulo $P_{e_d}(z)$, the largest cyclotomic polynomial factor of $z^N - 1$.

This can be seen by noting that, since $z^N \equiv 1$ modulo $(z^N - 1)$ and $P_{e_d}(z)$ is a factor of $z^N - 1$, $z^N \equiv 1$ modulo $P_{e_d}(z)$. Thus, condition (6.28) is satisfied. Conditions (6.29) are also satisfied because N always has an inverse in ordinary arithmetic (coefficients in the field of rationals) and because $z^{-1} \equiv z^{N-1}$ modulo $P_{e_d}(z)$. Consider now the conditions (6.30). Since $z^N \equiv 1$ modulo $P_{e_d}(z)$, we have $S \equiv \sum_{k=0}^{N-1} z^{qk} \equiv N$ for $q \equiv 0$ modulo N. For $q \not\equiv 0$ modulo N, we have

$$(z^q - 1)S \equiv z^{qN} - 1 \equiv 0 \text{ modulo } P_{e_d}(z). \tag{6.39}$$

The complex roots of $z^q - 1$ are powers of $e^{-j2\pi/q}$, while the complex roots of $P_{e_d}(z)$ are powers of $e^{-j2\pi/N}$. Thus, for $(q, N) = 1$ these complex roots are different and $z^q - 1$ is relatively prime to $P_{e_d}(z)$, which implies by (6.39) that $S \equiv 0$. For $(q, N) \neq 1$, q can always, without loss of generality, be considered as a factor of N. Then, $\phi(N) > \phi(q)$, and the largest polynomial factors of $z^N - 1$

and of $z^q - 1$ are, respectively, the polynomials $P_{e_d}(z)$ and $Q(z)$ of degree $\phi(N)$ and $\phi(q)$. These polynomials are necessarily different because their degrees $\phi(N)$ and $\phi(q)$ are different. Moreover, $Q(z)$ cannot be a factor of $P_{e_d}(z)$ because $P_{e_d}(z)$ is irreducible. Thus, $z^q - 1 \not\equiv 0$ modulo $P_{e_d}(z)$ and $S \equiv 0$ modulo $P_{e_d}(z)$, which completes the proof that conditions (6.30) are satisfied.

Consequently, the convolution $y_{u,l}$ of dimension $N \times N$ is computed by ordering the input sequence into N polynomials of N terms which are reduced modulo $P_{e_d}(z)$ and modulo $P^1(z) = \prod_{i=1}^{d-1} P_{e_i}(z)$. The output samples $y_{u,l}$ are derived by Chinese remainder reconstruction of a polynomial convolution $Y_{1,l}(z)$ modulo $P_{e_d}(z)$ and a polynomial convolution $Y_{2,l}(z)$ modulo $P^1(z)$. $Y_{1,l}(z)$ is computed by polynomial transforms of root z and of dimension N. $Y_{2,l}(z)$ is a polynomial convolution of length N on polynomials of $N - \phi(N)$ terms defined modulo $P^1(z)$. At this point, two methods may be used to compute $Y_{2,l}(z)$. One can either reduce $Y_{2,l}(z)$ modulo the various factors of $P^1(z)$ and define the corresponding polynomial transforms, or one can consider $Y_{2,l}(z)$ as a two-dimensional polynomial product modulo $P^1(z)$, $z^N - 1$.

In order to illustrate this last approach, we consider a two-dimensional convolution of size $p^c \times p^c$, where p is an odd prime. In this case, $N = p^c$ and $\phi(p^c)$ is given by

$$\phi(p^c) = p^{c-1}(p - 1). \tag{6.40}$$

Then,

$$P_{e_d}(z) = z^{p^{c-1}(p-1)} + z^{p^{c-1}(p-2)} + \ldots + 1 \tag{6.41}$$

and

$$z^{p^c} - 1 = (z^{p^{c-1}} - 1)P_{e_d}(z). \tag{6.42}$$

Hence $Y_{1,l}$ is computed by polynomial transforms of length p^c and root z defined modulo $P_{e_d}(z)$, while $Y_{2,l}$ is a convolution of size $p^c \times p^{c-1}$. This convolution can be viewed as a polynomial convolution of length p^{c-1} on polynomials of p^c terms which is in turn evaluated as a polynomial convolution of length p^{c-1} defined modulo $P_{e_d}(z)$ and a convolution of length $p^{c-1} \times p^{c-1}$. The polynomial convolution defined modulo $P_{e_d}(z)$ can be calculated by polynomial transforms of length p^{c-1} with root z^p and the convolution of size $p^{c-1} \times p^{c-1}$ can be evaluated by repeating the same process. Thus, the two-dimensional convolution of size $p^c \times p^c$ can be completely mapped into a set of one-dimensional polynomial products and convolutions by a c-stage algorithm using polynomial transforms. We illustrate this method in Fig. 6.2 by giving the first stage corresponding to a convolution of size $p^2 \times p^2$. In this case, the second stage represents a convolution of size $p \times p$ which is calculated by the method shown in Fig. 6.1.

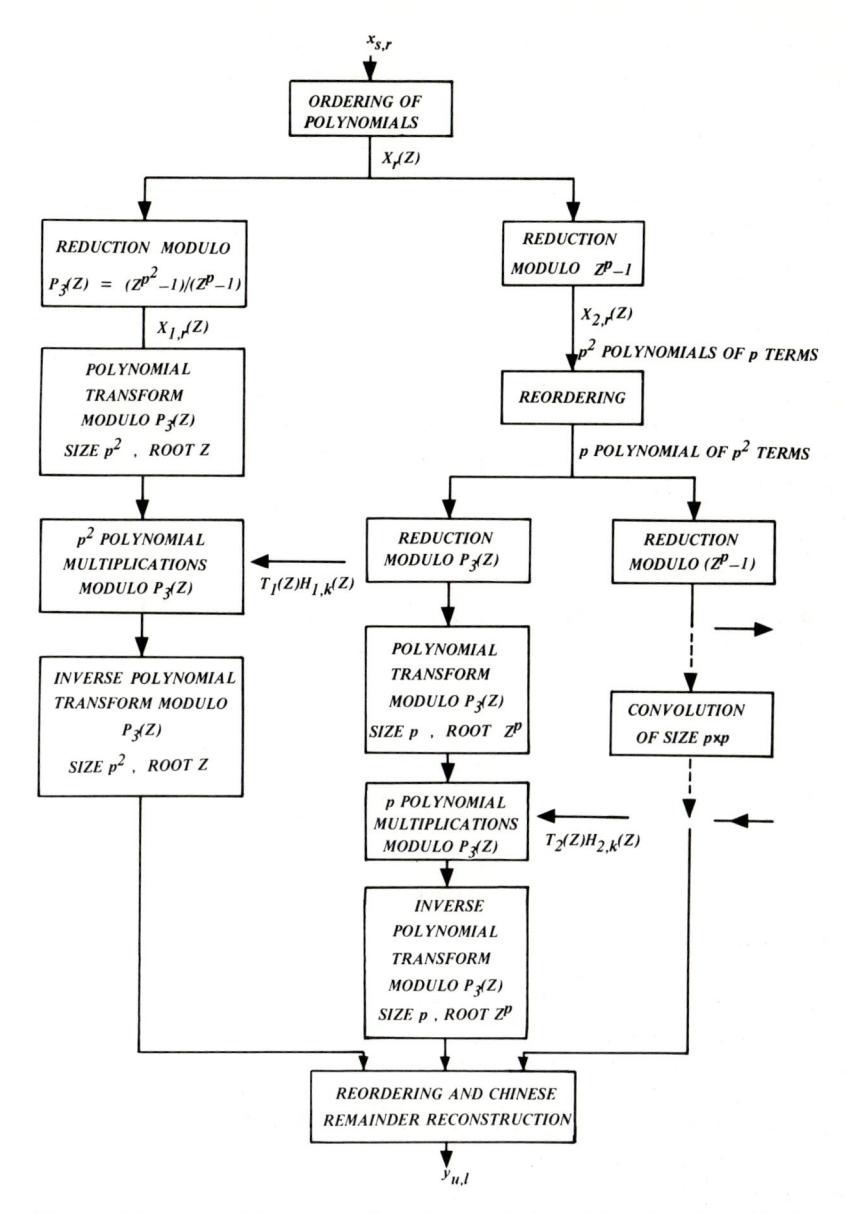

6.2 General Definition of Polynomial Transforms

Fig. 6.2. First stage of the computation of a convolution of size $p^2 \times p^2$. p odd prime

As a second example, Fig. 6.3 gives the first stage of the calculation of a convolution of size $2^t \times 2^t$ by polynomial transforms. In this case, the computation is performed in $t - 1$ stages with polynomial transforms defined modulo $P_{t+1}(z) = z^{2^{t-1}} + 1$, $P_t(z) = z^{2^{t-2}} + 1$, ..., $P_3(z) = z^2 + 1$. These polynomial

160   6. Polynomial Transforms

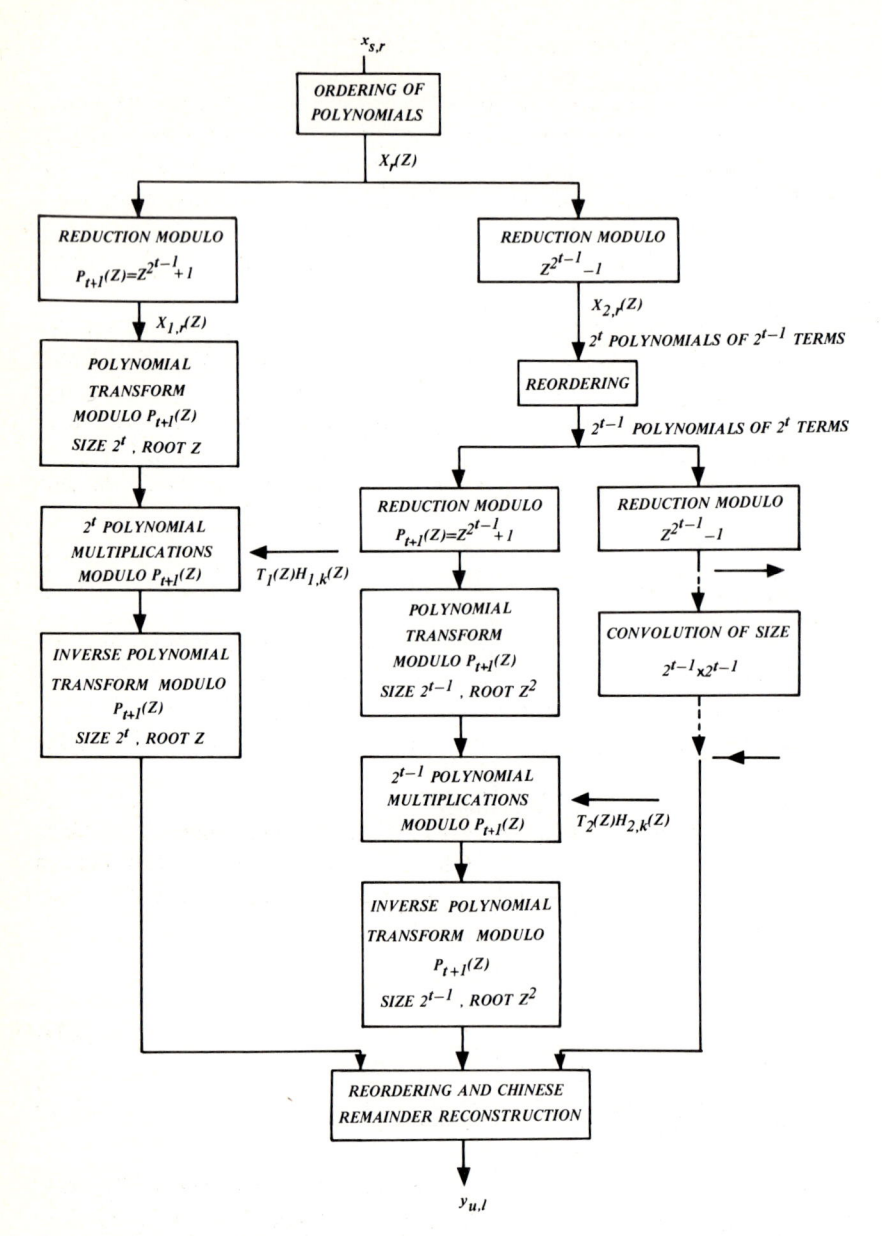

Fig. 6.3. First stage of the computation of a convolution of size $2^t \times 2^t$ by polynomial transforms

transforms are particularly interesting because, due to their power of two sizes, they can be compudted with a reduced number of additions by a radix-2 FFT-type algorithm.

6.2.2 Polynomial Transforms with Composite Roots

Previously, we have restricted our discussion to polynomial transforms having roots $G(z)$ defined in a field or in a ring of polynomials. Additional degrees of freedom are possible by taking advantage of the properties of the field of coefficients. If a length-N_1 polynomial transform supports circular convolution when defined modulo $P(z)$ with root $G(z)$, it is possible to use roots of unity of order N_2 in the field of coefficients for the definition of transforms of length $N_1 N_2$ which also support circular convolution.

Assuming for instance that the coefficients are defined in the field of complex numbers, we can always define DFTs of length N_2 and roots $W = e^{-j2\pi/N_2}$ that support circular convolution. In this case, if $(N_1, N_2) = 1$, the polynomial transform of root $WG(z)$ defined modulo $P(z)$ supports a circular convolution of length N, for $N = N_1 N_2$.

This can be seen by verifying that the three conditions (6.28–30) are met. We note first that, since $W^{N_2} = 1$ and $G(z)^{N_1} \equiv 1$ modulo $P(z)$,

$$[WG(z)]^N = (W^{N_2})^{N_1}[G(z)^{N_1}]^{N_2} \equiv 1 \text{ modulo } P(z). \tag{6.43}$$

Condition (6.29) is also obviously satisfied, since N_1, N_2 and W, $G(z)$ have inverses. In order to meet condition (6.30), we consider S, with

$$S \equiv \sum_{k=0}^{N-1} [WG(z)]^{qk} \text{ modulo } P(z). \tag{6.44}$$

Since $(WG(z))^N \equiv 1$ modulo $P(z)$, the exponents qk are defined modulo N. Thus, $S \equiv N$ for $q \equiv 0$ modulo N. For $q \not\equiv 0$ modulo N, we can always map S into a two-dimensional sum, because N_1 and N_2 are mutually prime. This can be done with

$$k \equiv N_1 k_2 + N_2 k_1 \text{ modulo } N$$
$$k_1 = 0, \ldots, N_1 - 1 \qquad k_2 = 0, \ldots, N_2 - 1 \tag{6.45}$$

$$S \equiv \sum_{k_2=0}^{N_2-1} W^{qN_1 k_2} \sum_{k_1=0}^{N_1-1} G(z)^{qN_2 k_1} \text{ modulo } P(z). \tag{6.46}$$

The existence of the two transforms of lengths N_1 and N_2 with roots $G(Z)$ and W implies that $S \equiv 0$ for $k_1 \not\equiv 0$ modulo N_1 and $k_2 \not\equiv 0$ modulo N_2, and therefore that $S \equiv 0$ for $k \not\equiv 0$ modulo N, which verifies that (6.30) is satisfied.

When N_1 is odd, the condition $(N_1, N_2) = 1$ implies that it is always possible to increase the length of the polynomial transforms to $N_1 N_2$, with $N_2 = 2^t$. The new transforms will usually require some multiplications since the roots $WG(z)$ are no longer simple. We note, however, that this method is particularly useful to compute convolutions of sizes $2N_1 \times 2N_1$ and $4N_1 \times 4N_1$, because in these cases, we have $W = -1$ or $W = -j$ and the polynomial transforms may still be computed without multiplications.

A modified form of the method can be devised by replacing DFTs with number theoretic transforms (Chap. 8) which amounts to defining the polynomial coefficients modulo integers. In this case, it is possible to compute large two-dimensional convolutions with a very small number of multiplications provided that multiplications by powers of two and arithmetic modulo an integer can be implemented easily.

Table 6.1 summarizes the properties of a number of polynomial transforms that can be computed without multiplications and which correspond to two-dimensional convolutions such that both dimensions have a common factor.

Table 6. 1. Polynomial transforms for the compution of two-dimensional convolutions. p, p_1, p_2 odd primes

Transform ring $P(z)$	Transform		Size of convolutions $N \times N$	No. of additions for reductions, polynomial transforms, and Chinese remainder reconstruction	Polynomial products and convolutions
	Length	Root			
$(z^p - 1)/(z - 1)$	p	z	$p \times p$	$2(p^3 + p^2 - 5p + 4)$	p products $P(z)$ 1 convolution p
$(z^p - 1)/(z - 1)$	$2p$	$-z$	$2p \times p$	$4(p^3 + 2p^2 - 6p + 4)$	$2p$ products $P(z)$ 1 convolution $2p$
$(z^{2p} - 1)/(z^2 - 1)$	$2p$	$-z^{p+1}$	$2p \times 2p$	$8(p^3 + 2p^2 - 6p + 4)$	$2p$ products $P(z)$ 1 convolution $2 \times 2p$
$(z^{p^2} - 1)/(z^p - 1)$	p	z^p	$p \times p^2$	$2(p^4 + 2p^3 - 4p^2 - p + 4)$	p products $P(z)$ p products $(z^p - 1)/(z - 1)$ 1 convolution p
$(z^{p^2} - 1)/(z^p - 1)$	p^2	z	$p^2 \times p^2$	$2(2p^5 + p^4 - 5p^3 + p^2 + 6)$	$p(p+1)$ products $P(z)$ p products $(z^p - 1)/(z - 1)$ 1 convolution p
$(z^{2p^2} - 1)/(z^{2p} - 1)$	$2p^2$	$-z^{p^2+1}$	$2p^2 \times 2p^2$	$8(2p^5 + 2p^4 - 6p^3 - p^2 + 5p + 2)$	$2p(p+1)$ products $P(z)$ $2p$ products $(z^{2p} - 1)/(z^2 - 1)$ 1 convolution $2 \times 2p$
$(z^{p_1 p_2} - 1)/(z^{p_2} - 1)$	p_1	z^{p_2}	$p_1 \times p_1 p_2$	$2p_2(p_1^3 + p_1^2 - 5p_1 + 4)$	p_1 products $P(z)$ 1 convolution $p_1 p_2$
$z^{2^{t-1}} + 1$	2^t	z	$2^t \times 2^t$	$2^{2t-1}(3t + 5)$	$3 \cdot 2^{t-1}$ products $P(z)$ 1 convolution $2^{t-1} \times 2^{t-1}$

6.3 Computation of Polynomial Transforms and Reductions

The evaluation of a two-dimensional convolution by polynomial transforms involves the calculation of reductions, Chinese remainder reconstructions, and polynomial transforms. We specify here the number of additions required for these operations in the cases corresponding to Table 6.1.

When $N = p$, p prime, the input sequences must be reduced modulo $(z - 1)$ and modulo $P(z)$, with $P(z) = z^{p-1} + z^{p-2} + \ldots + 1$. Each of these operations is done with $p(p - 1)$ additions, by summing all terms of the input polynomials for the reduction modulo $(z - 1)$, as shown in (6.14) and by subtracting the last word of each input polynomial to all other words, for the reduction modulo $P(z)$, as shown in (6.23). When one of the input sequences, $h_{n,m}$, is fixed, the Chinese remainder reconstruction is defined by (6.22). Since $Y_{1,l}(z)$ is a polynomial of $p - 1$ terms, it can be defined as

$$Y_{1,l}(z) = \sum_{u=0}^{p-2} y_{1,u,l} z^u. \tag{6.47}$$

Thus, (6.22) becomes

$$Y_l(z) = Y_{2,l} - y_{1,0,l}$$
$$+ \sum_{u=1}^{p-2} (y_{1,u-1,l} - y_{1,u,l} + Y_{2,l})z^u + (y_{1,p-2,l} + Y_{2,l})z^{p-1}$$
$$l = 0, \ldots, p - 1, \tag{6.48}$$

which shows that the Chinese remainder reconstruction requires $2p(p - 1)$ additions.

For $N = p^2$, p prime, or for $N = 2^t$, the reductions and Chinese remainder operations are performed similarly and we give in Table 6.2, column 3, the corresponding numbers of additions.

For $N = p$, p prime, the polynomial transform $\bar{X}_k(z)$ defined by (6.15) is computed by

$$\bar{X}_k(z) \equiv X_{1,0}(z) + \sum_{r=1}^{p-1} X_{1,r}(z) z^{rk} \text{ modulo } P(z)$$
$$k = 0, \ldots, p - 2 \tag{6.49}$$

$$\bar{X}_{p-1}(z) \equiv X_{1,0}(z) + \sum_{r=1}^{p-1} X_{1,r}(z) z^{r(p-1)} \text{ modulo } P(z), \tag{6.50}$$

where the p polynomials $X_{1,r}(z)$ have $p - 1$ terms. The computation of $\bar{X}_k(z)$ proceeds by first evaluating $R_k(z)$, with

$$R_k(z) \equiv \sum_{r=1}^{p-1} X_{1,r}(z) z^{rk} \text{ modulo } P(z). \tag{6.51}$$

164   6. Polynomial Transforms

Table 6.2. Number of additions for the computation of reductions, Chinese remainder operations, and polynomial transforms. p odd prime

Transform size	Ring	Number of additions for reductions and Chinese remainder operations	Number of additions for polynomial transforms
p	$(z^p - 1)/(z - 1)$	$4p(p - 1)$	$p^3 - p^2 - 3p + 4$
$2p$	$(z^p - 1)/(z - 1)$	$8p(p - 1)$	$2(p^3 - 4p + 4)$
$2p$	$(z^{2p} - 1)/(z^2 - 1)$	$16p(p - 1)$	$4(p^3 - 4p + 4)$
p	$(z^{p^2} - 1)/(z^p - 1)$	$4p^2(p - 1)$	$p(p^3 - p^2 - 3p + 4)$
p^2	$(z^{p^2} - 1)/(z^p - 1)$	$4p^3(p - 1)$	$2p^5 - 2p^4 - 5p^3 + 5p^2 + p + 2$
$2p$	$(z^{2p^2} - 1)/(z^{2p} - 1)$	$16p^2(p - 1)$	$4p(p^3 - 4p + 4)$
$2p^2$	$(z^{2p^2} - 1)/(z^{2p} - 1)$	$16p^3(p - 1)$	$4(2p^5 - p^4 - 6p^3 + 5p^2 + p + 2)$
2^t	$z^{2^{t-1}} + 1$	2^{2t+1}	$t2^{2t-1}$

Since $R_0(z) = \sum_{r=1}^{p-1} X_{1,r}(z)$, $R_0(z)$ is computed with $(p - 2)(p - 1)$ additions. For $k \neq 0$, $R_k(z)$ is the sum of $p - 1$ polynomials, each polynomial being multiplied modulo $P(z)$ by a power of z. $R_k(z)$ is first computed modulo $(z^p - 1)$. Since $z^p \equiv 1$, each polynomial $X_{1,r}(z)$ becomes, after multiplication by z^{rk}, a polynomial of p words where one of the words is zero. Thus, the computation of each $R_k(z)$ modulo $(z^p - 1)$ requires $p^2 - 3p + 1$ additions. Since the reduction of $R_k(z)$ modulo $P(z)$ is performed with $p - 1$ additions, each $R_k(z)$ for $k \neq 0$ and $k \neq p - 1$ is calculated with $p^2 - 2p$ additions. In order to evaluate (6.50), we note that, for $r \neq 0$,

$$\sum_{k=0}^{p-2} z^{rk} \equiv -z^{r(p-1)} \text{ modulo } P(z). \tag{6.52}$$

This implies that, for $r \neq 0$

$$R_{p-1}(z) \equiv -\sum_{k=0}^{p-2}\sum_{r=1}^{p-1} X_{1,r}(z) z^{rk} \text{ modulo } P(z) \tag{6.53}$$

$$\equiv -\sum_{k=0}^{p-2} R_k(z) \text{ modulo } P(z), \tag{6.54}$$

which shows that $R_{p-1}(z)$ is calculated with $(p - 1)(p - 2)$ additions. Finally, $\bar{X}_k(z)$ is obtained with $p(p - 1)$ additions by adding $X_{1,0}(z)$ to the various polynomials $R_k(z)$. Thus, a total of $p^3 - p^2 - 3p + 4$ additions is required to compute a polynomial transform of length p, with p prime.

Polynomial transforms of length N, with N composite, are computed by using an FFT-type algorithm. For $N = 2^t$, the polynomial transform $\bar{X}_k(z)$ of dimension N is defined modulo $P_{t+1}(z)$ with

$$\bar{X}_k(z) \equiv \sum_{r=0}^{N-1} X_{1,r}(z) z^{rk} \text{ modulo } P_{t+1}(z)$$
$$k = 0, \ldots, N-1 \quad (6.55)$$

and

$$P_{t+1}(z) = z^{2^{t-1}} + 1. \quad (6.56)$$

The first stage of radix-2, decimation in time FFT-type algorithm is given by

$$\bar{X}_k(z) \equiv \sum_{r=0}^{N/2-1} X_{1,2r}(z) z^{2rk} + z^k \sum_{r=0}^{N/2-1} X_{1,2r+1}(z) z^{2rk} \text{ modulo } P_{t+1}(z)$$
$$k = 0, \ldots, N/2 - 1 \quad (6.57)$$

$$\bar{X}_{k+N/2}(z) \equiv \sum_{r=0}^{N/2-1} X_{1,2r}(z) z^{2rk} - z^k \sum_{r=0}^{N/2-1} X_{1,2r+1}(z) z^{2rk} \text{ modulo } P_{t+1}(z). \quad (6.58)$$

Thus, the polynomial transform of length 2^t defined modulo $(z^{2^{t-1}} + 1)$ is computed in t stages with a total of $(N^2/2) \log_2 N$ additions.

We summarize in Table 6.2, column 4, the number of additions for various polynomial transforms. Table 6.1, column 5, also gives the total number of additions for reductions, polynomial transforms, and Chinese remainder reconstruction corresponding to various two-dimensional convolutions evaluated by polynomial transforms.

6.4 Two-Dimensional Filtering Using Polynomial Transforms

We have seen in the preceding sections that polynomial transforms map efficiently two-dimensional circular convolutions into one-dimensional polynomial products and convolutions. When the polynomial transforms are properly selected, this mapping is achieved without multiplications and requires only a limited number of additions. Thus, when a two-dimensional convolution is evaluated by polynomial transforms, the processing load is strongly dependent upon the efficiency of the algorithms used for the calculation of polynomial products and one-dimensional convolutions.

One approach that can be employed for evaluating the one-dimensional convolutions involves the use of one-dimensional transforms that support circular convolution, such as DFTs and NTTs. These transforms can also be used to compute polynomial products modulo cyclotomic polynomials $P_{e_t}(z)$ by noticing that, since $P_{e_t}(z)$, defined by (6.38), is a factor of $z^N - 1$, all computations can be carried out modulo $(z^N - 1)$, with a final reduction modulo $P_{e_t}(z)$. With this method, the calculation of a polynomial product modulo $P_{e_t}(z)$ is replaced by that of a polynomial product modulo $(z^N - 1)$, which is a convolution of length N. Hence, the two-dimensional convolution is completely mapped by the polynomial transform method into a set of one-dimensional convolutions that can be evaluated by DFTs and NTTs.

This approach is illustrated in Fig. 6.4 for a convolution of size $p \times p$, with p prime. In this case, the two-dimensional convolution is mapped into $p + 1$ convolutions of length p instead of one convolution of length p plus p polynomial products modulo $(z^p - 1)/(z - 1)$ as with the method in Fig. 6.1.

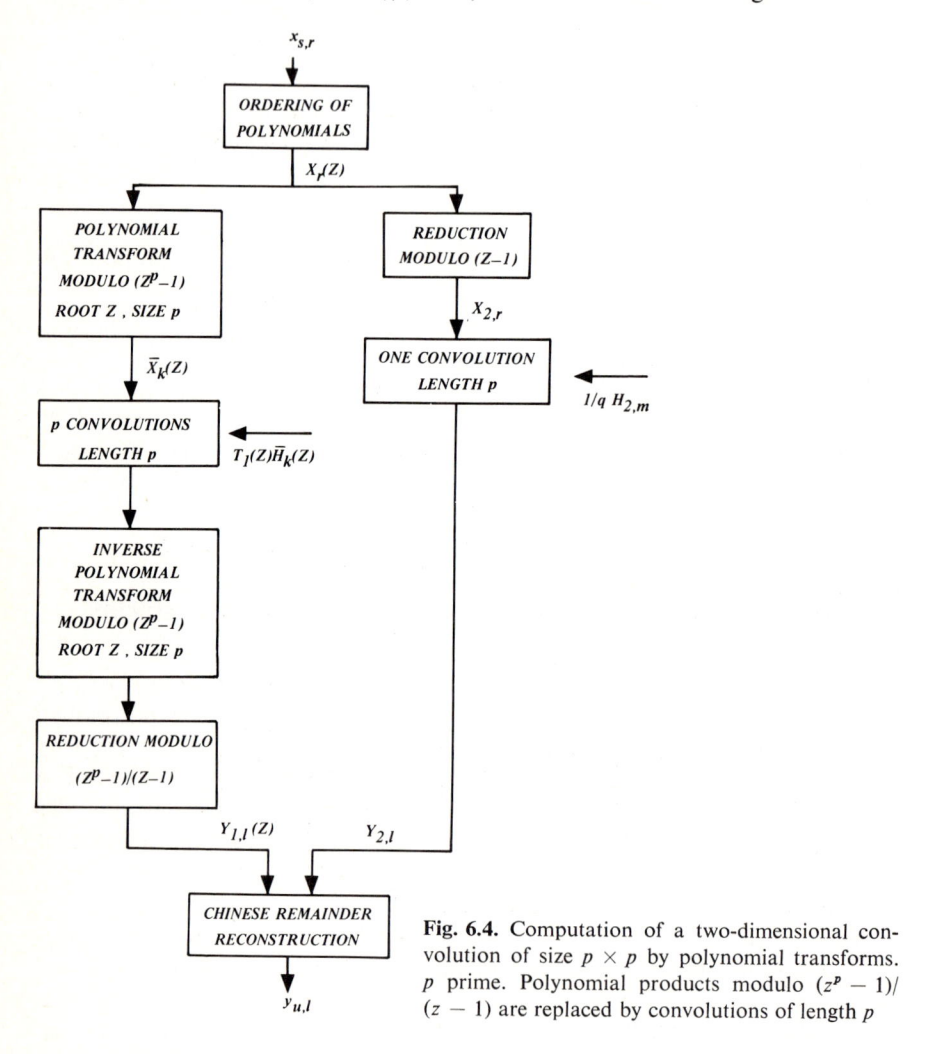

Fig. 6.4. Computation of a two-dimensional convolution of size $p \times p$ by polynomial transforms. p prime. Polynomial products modulo $(z^p - 1)/(z - 1)$ are replaced by convolutions of length p

6.4.1 Two-Dimensional Convolutions Evaluated by Polynomial Transforms and Polynomial Product Algorithms

When two-dimensional convolutions are mapped into one-dimensional convolutions which are evaluated by DFTs and NTTs, the problems associated with

6.4 Two Dimensional Filtering Using Polynomial Transforms 167

the use of these transforms such as roundoff errors for DFTs or modular arithmetic for NTTs are then limited to only a part of the total computation process. We have seen, however, in Chap. 3, that the methods based on interpolation and on the Chinese remainder theorem yield more efficient algorithms than the DFTs or NTTs for some convolutions and polynomial products. It is therefore often advantageous to consider the use of such algorithms in combination with polynomial transforms.

With this method, each convolution or polynomial product algorithm used in a given application must be specifically programmed, and it is desirable to use only a limited number of different algorithms in order to restrict total program size. This can be done by computing the one-dimensional convolutions

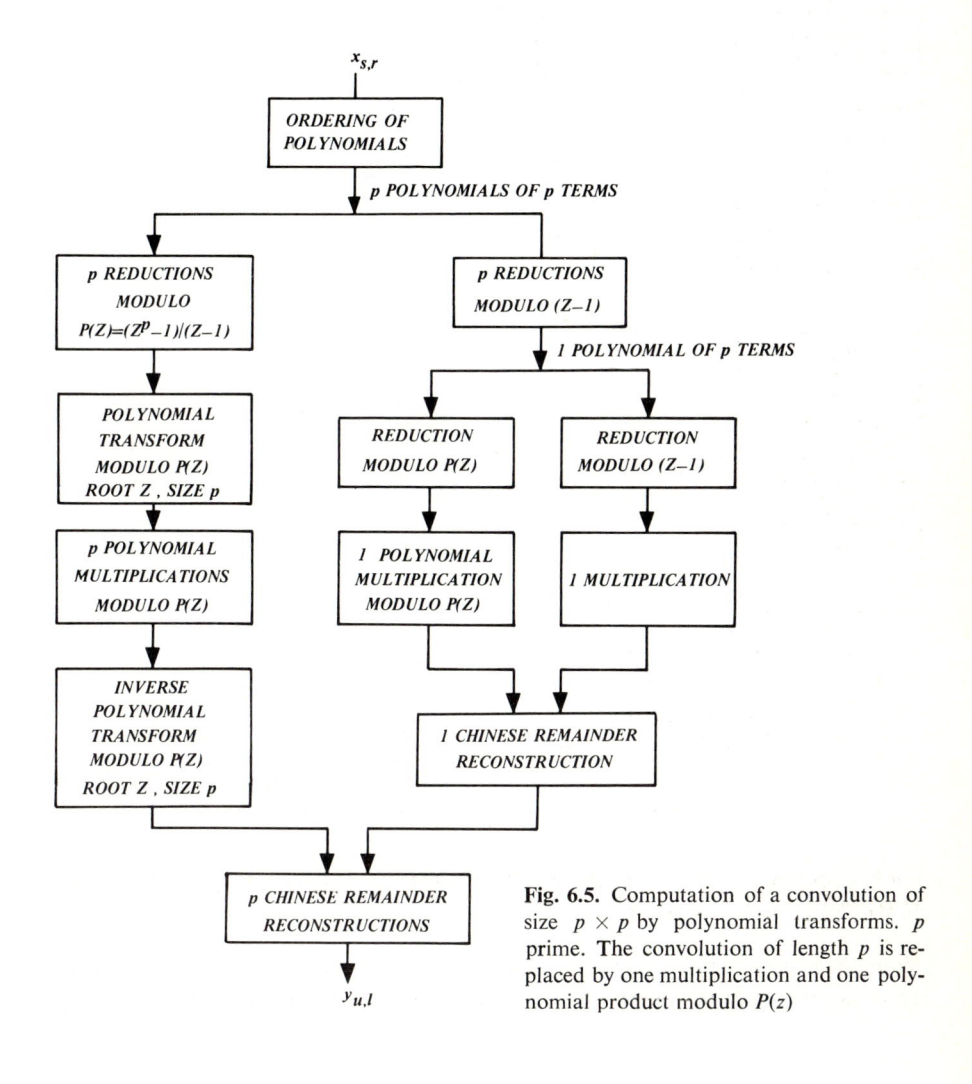

Fig. 6.5. Computation of a convolution of size $p \times p$ by polynomial transforms. p prime. The convolution of length p is replaced by one multiplication and one polynomial product modulo $P(z)$

168   6. Polynomial Transforms

as polynomial products by using the Chinese remainder theorem. For a two-dimensional convolution of dimension $p \times p$, with p prime, the two-dimensional convolution is mapped by polynomial transforms into one convolution of length p plus p polynomial products modulo $P(z)$, with $P(z) = (z^p - 1)/(z - 1)$. However, the same computation can also be done with only one polynomial product algorithm modulo $P(z)$ if the circular convolution of length p is calculated as one multiplication and one polynomial product algorithm modulo $P(z)$ by using the Chinese remainder theorem. Thus, the convolution of size $p \times p$ can be computed as shown in Fig. 6.5 with $p + 1$ polynomial products modulo $P(z)$ and one multiplication instead of p polynomial products modulo $P(z)$ and one convolution of length p.

Table 6.3 gives the number of arithmetic operations for various two-dimensional convolutions computed by polynomial transforms using the convolution and polynomial product algorithms of Sect. 3.7 for which the number of operation is given in Tables 3.1 and 3.2. These data presume that one of the input sequences $h_{n,m}$ is fixed and the operations on this sequence are done only once.

Table 6.3. Number of operations for two-dimensional convolutions computed by polynomial transforms and polynomial product algorithms

Convolution size	Number of multiplications	Number of additions	Multiplications per point	Additions per point
3 × 3 (9)	13	70	1.44	7.78
4 × 4 (16)	22	122	1.37	7.62
5 × 5 (25)	55	369	2.20	14.76
6 × 6 (36)	52	424	1.44	11.78
7 × 7 (49)	121	1163	2.47	23.73
8 × 8 (64)	130	750	2.03	11.72
9 × 9 (81)	193	1382	2.38	17.06
10 × 10 (100)	220	1876	2.20	18.76
14 × 14 (196)	484	5436	2.47	27.73
16 × 16 (256)	634	4774	2.48	18.65
18 × 18 (324)	772	6576	2.38	20.30
24 × 24 (576)	1402	12954	2.43	22.49
27 × 27 (729)	2893	21266	3.97	29.17
32 × 32 (1024)	3658	24854	3.57	24.27
64 × 64 (4096)	17770	142902	4.34	34.89
128 × 128 (16384)	78250	720502	4.78	43.98

6.4.2 Example of a Two-Dimensional Convolution Computed by Polynomial Transforms

In order to illustrate the computation of a two-dimensional convolution by polynomial transforms, we consider a simple convolution of size 3×3. In this

6.4 Two Dimensional Filtering Using Polynomial Transforms

case, we have

$$p = 3 \qquad P(z) = z^2 + z + 1 \qquad z^3 \equiv 1$$

The input sequences are defined by

$$h_{n,0} = \{2, 3, 1\} \qquad x_{s,0} = \{2, 1, 2\}$$
$$h_{n,1} = \{4, 2, 0\} \qquad x_{s,1} = \{1, 3, 0\}$$
$$h_{n,2} = \{3, 1, 4\} \qquad x_{s,2} = \{2, 1, 5\}.$$

These sequences become, after reduction modulo $P(z)$

$$H_{1,n}(z): \{1, 2\} \qquad\qquad X_{1,r}(z): \{0, -1\}$$
$$\{4, 2\} \qquad\qquad\qquad\qquad \{1, 3\}$$
$$\{-1, -3\} \qquad\qquad\qquad\quad \{-3, -4\}$$

The polynomial transforms of $H_{1,n}(z)$ and $X_{1,r}(z)$ are given by

$$\bar{H}_k(z): \{4, 1\} \qquad\qquad \bar{X}_k(z): \{-2, -2\}$$
$$\{-3, 5\} \qquad\qquad\qquad\quad \{-4, 0\}$$
$$\{2, 0\} \qquad\qquad\qquad\qquad \{6, -1\}$$

We note that $T_1(z)$, given by (6.21) is

$$T_1(z) = -(z + 2)/3.$$

Thus, the polynomial multiplications are defined by

$$-(z + 2)\bar{H}_k(z)\bar{X}_k(z)/9 \text{ modulo } P(z): \{4,14\}/9$$
$$\{-44,8\}/9$$
$$\{-26, -10\}/9$$

Computing the inverse polynomial transform of this result yields

$$Y_{1,l}(z) \text{ with } \qquad Y_{1,l}(z): \{-66, 12\}/9$$
$$\{66, 42\}/9$$
$$\{12, -12\}/9.$$

The reductions modulo $(z - 1)$ are given by

$$H_{2,m}: \{6, 6, 8\} \qquad\qquad X_{2,r}: \{5, 4, 8\},$$

which yields the length-3 convolution $Y_{2,l}$

$Y_{2,l}/3$: $\{110, 118, 112\}/3$.

The two-dimensional convolution output $y_{u,l}$ is then generated by the Chinese remainder reconstruction of $Y_{1,l}(z)$ and $Y_{2,l}$ using (6.22),

$y_{u,l}$: $\{44, 28, 38, 32, 42, 44, 36, 40, 36\}$.

6.4.3 Nesting Algorithms

A systematic application of the techniques discussed in the preceding sections allows one to compute large two-dimensional convolutions by use of composite polynomial transforms. In particular, the use of polynomial transforms of length N, with $N = 2^t$, is especially attractive because these transforms can be computed with a reduced number of additions by using a radix-2 FFT-type algorithm.

A nesting algorithm may be devised as an alternative to this approach to construct large two-dimensional convolutions from a limited set of short two-dimensional convolutions (Sect. 3.3.1) [6.4]. Using this method, a two-dimensional convolution $y_{u,l}$ of size $N_1 N_2 \times N_1 N_2$, with N_1 and N_2 mutually prime, can be converted into a four-dimensional convolution of size $(N_1 \times N_1) \times (N_2 \times N_2)$ by a simple index mapping. $y_{u,l}$ is defined as

$$y_{u,l} = \sum_{m=0}^{N_1 N_2 - 1} \sum_{n=0}^{N_1 N_2 - 1} h_{n,m} x_{u-n, l-m}. \tag{6.59}$$

Since N_1 and N_2 are mutually prime, the indices l, m, n, and u can be mapped into two sets of indices l_1, m_1, n_1, u_1 and l_2, m_2, n_2, u_2 by use of an approach based on permutations (Sect. 2.1.2) to obtain

$$\begin{aligned} l &\equiv N_1 l_2 + N_2 l_1 \quad &&\text{modulo } N_1 N_2 \\ m &\equiv N_1 m_2 + N_2 m_1 \quad &&\text{modulo } N_1 N_2 \\ n &\equiv N_1 n_2 + N_2 n_1 \quad &&\text{modulo } N_1 N_2 \quad , \quad l_1, m_1, n_1, u_1 = 0, \ldots, N_1 - 1 \\ u &\equiv N_1 u_2 + N_2 u_1 \quad &&\text{modulo } N_1 N_2 \quad , \quad l_2, m_2, n_2, u_2 = 0, \ldots, N_2 - 1 \end{aligned} \tag{6.60}$$

and

$$y_{N_1 u_2 + N_2 u_1, N_1 l_2 + N_2 l_1} = \sum_{m_1=0}^{N_1-1} \sum_{n_1=0}^{N_1-1} \sum_{m_2=0}^{N_2-1} \sum_{n_2=0}^{N_2-1} h_{N_1 n_2 + N_2 n_1, N_1 m_2 + N_2 m_1} \cdot$$
$$x_{N_1(u_2 - n_2) + N_2(u_1 - n_1), N_1(l_2 - m_2) + N_2(l_1 - m_1)}. \tag{6.61}$$

This four-dimensional convolution can be viewed as a two-dimensional convolution of size $N_1 \times N_1$ where all the scalars are replaced by the two-dimensional polynomials of size $N_2 \times N_2$, $H_{n_1, m_1}(z_1, z_2)$, and $X_{r_1, s_1}(z_1, z_2)$ with

$$H_{n_1,m_1}(z_1, z_2) = \sum_{n_2=0}^{N_2-1} \sum_{m_2=0}^{N_2-1} h_{N_1 n_2+N_2 n_1, N_1 m_2+N_2 m_1} z_1^{m_2} z_2^{n_2} \tag{6.62}$$

$$X_{r_1,s_1}(z_1,z_2) = \sum_{r_2=0}^{N_2-1} \sum_{s_2=0}^{N_2-1} x_{N_1 r_2+N_2 r_1, N_1 s_2+N_2 s_1} z_1^{r_2} z_2^{s_2}$$

$$r_1, s_1 = 0, \ldots, N_1 - 1 \tag{6.63}$$

and $y_{u,l}$ defined by

$$Y_{u_1,l_1}(z_1, z_2) = \sum_{u_2=0}^{N_2-1} \sum_{l_2=0}^{N_2-1} y_{N_1 u_2+N_2 u_1, N_1 l_2+N_2 l_1} z_1^{u_2} z_2^{l_2} \tag{6.64}$$

$$Y_{u_1,l_1}(z_1, z_2) \equiv \sum_{m_1=0}^{N_1-1} \sum_{n_1=0}^{N_1-1} H_{n_1,m_1}(z_1, z_2) X_{u_1-n_1, l_1-m_1}(z_1, z_2)$$

$$\text{modulo } (z_1^{N_2} - 1), (z_2^{N_2} - 1). \tag{6.65}$$

Each polynomial multiplication modulo $(z_1^{N_2} - 1)$, $(z_2^{N_2} - 1)$ corresponds to a convolution of size $N_2 \times N_2$ which is computed with M_2 scalar multiplications and A_2 scalar additions. In the convolution of size $N_1 \times N_1$, all scalars are replaced by polynomials of size $N_2 \times N_2$. Thus, if M_1 and A_1 are the numbers of multiplications and additions for a scalar convolution of size $N_1 \times N_1$, the number of multiplications M and additions A required to evaluate the two-dimensional convolution of size $N_1 N_2 \times N_1 N_2$ becomes

$$M = M_1 M_2 \tag{6.66}$$

$$A = N_2^2 A_1 + M_1 A_2. \tag{6.67}$$

This computation process may be extended recursively to more than two factors provided that all these factors are mutually prime. In practice, the small convolutions of size $N_1 \times N_1$ and $N_2 \times N_2$ are computed by polynomial transforms, and large two-dimensional convolutions can be obtained from a small set of polynomial transform algorithms. A convolution of size 15×15 can, for instance, be computed from convolutions of sizes 3×3 and 5×5. Since these convolutions are calculated by polynomial transforms with 13 multiplications, 70 additions and 55 multiplications, 369 additions, respectively (Table 6.3), nesting the two algorithms yields a total of 715 multiplications and 6547 additions for the convolution of size 15×15.

Table 6.4 itemizes arithmetic operations count for two-dimensional convolutions computed by polynomial transforms and nesting. It can be seen that these algorithms require fewer multiplications and more additions per point than for the approach using composite polynomial transforms and corresponding to Table 6.3. The number of additions here can be further reduced by replacing the

Table 6.4. Number of operations for two-dimensional convolutions computed by polynomial transforms and nesting

Convolution size	Number of multiplications	Number of additions	Multiplications per point	Additions per point
12 × 12 (144)	286	2638	1.99	18.32
20 × 20 (400)	946	14030	2.36	35.07
30 × 30 (900)	2236	35404	2.48	39.34
36 × 36 (1296)	4246	40286	3.28	31.08
40 × 40 (1600)	4558	80802	2.85	50.50
60 × 60 (3600)	12298	192490	3.42	53.47
72 × 72 (5184)	20458	232514	3.95	44.85
80 × 80 (6400)	27262	345826	4.26	54.03
120 × 120 (14400)	59254	1046278	4.11	72.66

Table 6.5. Number of operations for two-dimensional convolutions computed by polynomial transforms and split nesting

Convolution size	Number of multiplications	Number of additions	Multiplications per point	Additions per point
12 × 12 (144)	286	2290	1.99	15.90
20 × 20 (400)	946	10826	2.36	27.06
30 × 30 (900)	2236	28348	2.48	31.50
36 × 36 (1296)	4246	34010	3.28	26.24
40 × 40 (1600)	4558	69534	2.85	43.46
60 × 60 (3600)	12298	129106	3.42	35.86
72 × 72 (5184)	20458	192470	3.95	37.13
80 × 80 (6400)	26254	308494	4.10	48.20
120 × 120 (14400)	59254	686398	4.11	47.67

conventional nesting by a split nesting technique (Sect. 3.3.2). In this case, the number of arithmetic operations becomes as shown in Table 6.5. It can be seen that the number of additions per point in this table is comparable to that obtained with large composite polynomial transforms.

6.4.4 Comparison with Conventional Convolution Algorithms

Polynomial transforms are particularly suitable for the evaluation of real convolutions because they then require only real arithmetic as opposed to complex arithmetic with a DFT approach. Furthermore, when the polynomial products are evaluated by polynomial product algorithms, the polynomial transform approach does not require the use of trigonometric functions. Thus, the computation of two-dimensional convolutions by polynomial transforms can be compared to the nesting techniques [6.4] described in Sect. 3.3.1 which have similar characteristics. It can be seen, by comparing Table 3.4 with Tables 6.3 and 6.5, that the polynomial transform method always requires fewer arithmetic operations than the nesting method used alone, and provides increased efficiency with increasing convolution size. For large convolutions of sizes greater than 100 × 100, the use of polynomial transforms drastically decreases the number of arithmetic operations. In the case of a convolution of 120 × 120, for example, the polynomial transform approach requires about 5 times fewer multiplications and 2.5 times fewer additions than the simple nesting method.

When a convolution is calculated via FFT methods, the computation requires the use of trigonometric functions and complex arithmetic. Thus, a comparison with the polynomial transform method is somewhat difficult, especially when issues such as roundoff error and the relative cost of ancillary operations are considered. A simple comparative evaluation can be made between the two methods by assuming that two real convolutions are evaluated simultaneously with the Rader-Brenner FFT algorithm (Sect. 4.6) and the row-column method. In this case, the number of arithmetic operations corresponding to convolutions with one fixed sequence and precomputed trigonometric functions is listed in Table 4.7. Under these conditions, which are rather favorable to the FFT approach, the number of additions is slightly larger than that of the polynomial transform method while the number of multiplications is about twice that of the polynomial transform approach. Conventional radix-4 FFT algorithms or the Winograd Fourier transform method would also require a significantly larger number of arithmetic operations than the polynomial transform method.

6.5 Polynomial Transforms Defined in Modified Rings

Of all possible polynomial transforms, the most interesting are those defined modulo $(z^N + 1)$, with $N = 2^t$, because these transforms are computed without multiplications and with a reduced number of additions by using a radix-2 FFT-type algorithm. We have seen, in Sect. 6.2.1, that large two-dimensional convolutions are computed with these transforms by using a succession of stages, where each stage is implemented with a set of four polynomial transforms. This approach is very efficient from the standpoint of the number of arithmetic operations, but has the disadvantage of requiring a number of reductions and Chinese remainder reconstructions. In the following, we shall present an interesting variation [6.5] in which a simplification of the original structure is obtained at the expense of increasing the number of operations.

In order to introduce this method, we first establish that a one-dimensional convolution y_l of length N, with $N = 2^t$, can be viewed as a polynomial product modulo $(z^N + 1)$, provided that the input and output sequences are multiplied by powers of W, where W is a root of unity of order $2N$. We consider the circular convolution y_l defined by

6. Polynomial Transforms

$$y_l = \sum_{n=0}^{N-1} h_n x_{l-n}, \qquad l = 0, \ldots, N-1, \tag{6.68}$$

where $l - n$ is defined modulo N. The input sequences are multiplied by W^n and W^m, with $W = e^{-j\pi/N}$, and organized as two polynomials

$$H(z) = \sum_{n=0}^{N-1} h_n W^n z^n \tag{6.69}$$

$$X(z) = \sum_{m=0}^{N-1} x_m W^m z^m. \tag{6.70}$$

We now define a polynomial $A(z)$ of length N

$$A(z) \equiv H(z)X(z) \quad \text{modulo } (z^N + 1) \tag{6.71}$$

$$A(z) = \sum_{l=0}^{N-1} a_l z^l, \tag{6.72}$$

where each coefficient a_l of $A(z)$ corresponds to the products $h_n x_m$ such that $n + m = l$ or $n + m = l + N$. Since $z^N \equiv -1$, we have

$$a_l = \left(\sum_{n=0}^{l} h_n x_{l-n} - W^N \sum_{n=l+1}^{N-1} h_n x_{N+l-n} \right) W^l, \tag{6.73}$$

where $l - n$ is not taken modulo N. Hence, with $W^N = -1$,

$$y_l = a_l W^{-l}, \tag{6.74}$$

which shows that y_l is obtained by simple multiplications of $A(z)$ by W^{-l}. Note that this method is quite general and converts a convolution into a polynomial product modulo $(z^N + 1)$, sometimes called *skew circular convolution*, provided that the input and output sequences are multiplied by powers of any root of order $2N$. Such roots need not be $e^{-j\pi/N}$ and can, for instance, be roots of unity defined in rings of numbers modulo an integer.

We now apply this method to the computation of two-dimensional convolutions. We begin again with a two-dimensional circular convolution $y_{u,l}$ of size $N \times N$, where $N = 2^t$,

$$y_{u,l} = \sum_{m=0}^{N-1} \sum_{n=0}^{N-1} h_{n,m} x_{u-n, l-m} \qquad u, l = 0, \ldots, N-1. \tag{6.75}$$

In polynomial notation, this convolution becomes

$$A_l(z) \equiv \sum_{m=0}^{N-1} H_m(z) X_{l-m}(z) \quad \text{modulo } (z^N + 1) \tag{6.76}$$

$$H_m(z) = \sum_{n=0}^{N-1} h_{n,m} W^n z^n, \qquad W = e^{-j\pi/N} \tag{6.77}$$

$$X_r(z) = \sum_{s=0}^{N-1} x_{s,r} W^s z^s, \qquad m, r = 0, \ldots, N-1 \tag{6.78}$$

$$A_l(z) = \sum_{u=0}^{N-1} a_{u,l} z^u, \qquad l = 0, \ldots, N-1 \tag{6.79}$$

$$y_{u,l} = a_{u,l} W^{-u}. \tag{6.80}$$

The most important part of the calculations corresponds to the evaluation of the polynomial convolution $A_l(z)$ defined modulo $(z^N + 1)$ corresponding to (6.76). We note, however, that we can always define a polynomial transform of length

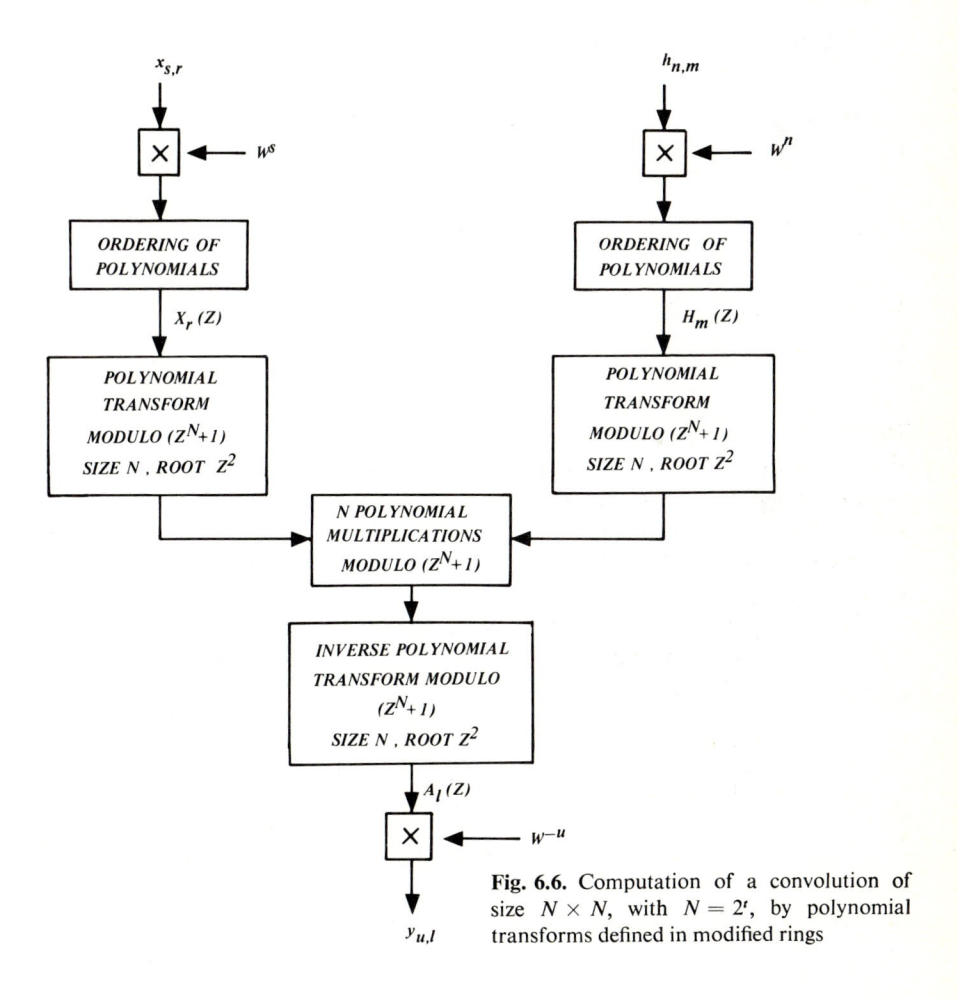

Fig. 6.6. Computation of a convolution of size $N \times N$, with $N = 2^t$, by polynomial transforms defined in modified rings

N modulo $(z^N + 1)$. Hence, the two-dimensional convolution $y_{u,l}$ can be computed with only three polynomial transforms of length N and roots z^2, as shown in Fig. 6.6. When one of the input sequences, $h_{n,m}$, is fixed, the corresponding polynomial transform needs be computed only once and the evaluation of $y_{u,l}$

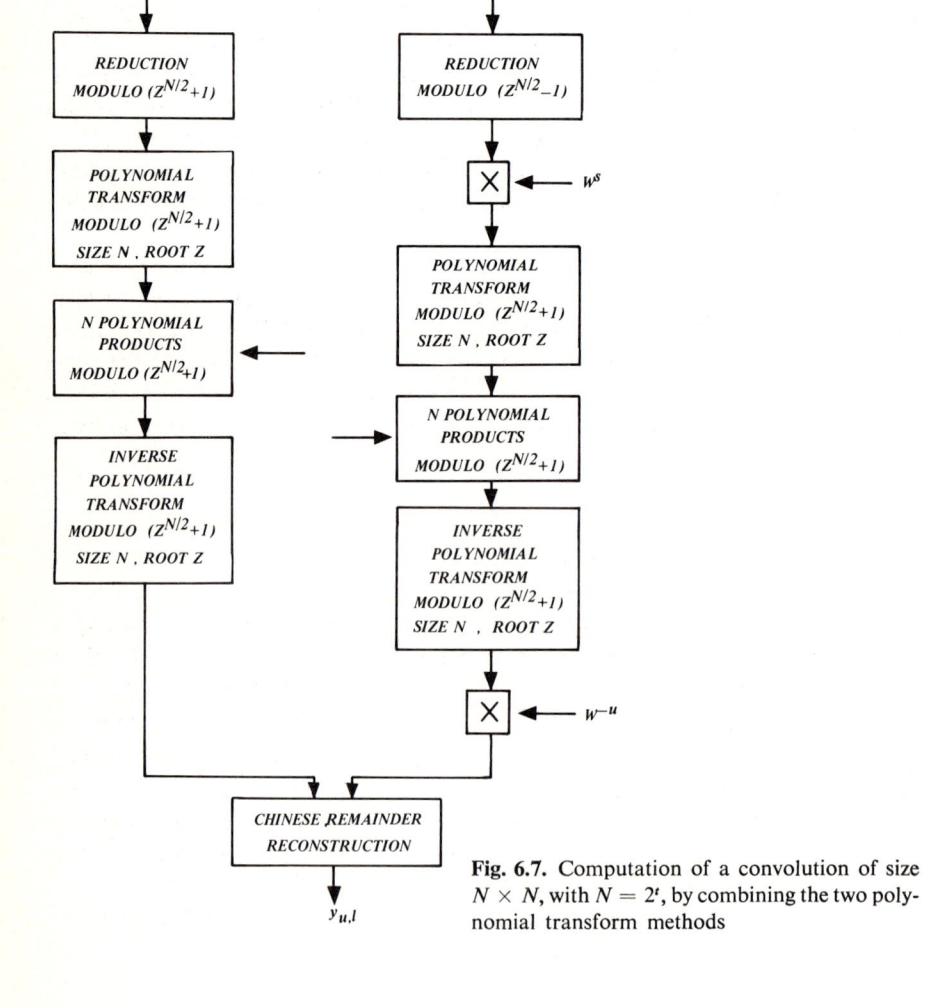

Fig. 6.7. Computation of a convolution of size $N \times N$, with $N = 2^t$, by combining the two polynomial transform methods

reduces to the computation of 2 polynomial transforms of length N, N polynomial multiplications modulo $(z^N + 1)$, and $2N$ scalar multiplications by W^s and W^{-u}.

Hence the overall structure of the algorithm is very simple, and all reductions and Chinese remainder operations have been eliminated at the expense of the multiplications by W^s and W^{-u}, by using polynomial multiplications modulo $(z^N + 1)$ instead of modulo $(z^{N/2} + 1)$, $(z^{N/4} + 1)$ .... As with the method corresponding to Fig. 6.3, the polynomial transforms are computed here with a reduced number of additions by using a FFT-type radix-2 algorithm.

The approach based on modified rings of polynomials can be improved by combining it with the method described in Sect. 6.2.1. This may be done by computing the convolution $y_{u,t}$ as a polynomial convolution of N terms defined modulo $(z^{N/2} + 1)$ plus a polynomial convolution of N terms defined modulo $(z^{N/2} - 1)$, with the latter polynomial convolution converted into a polynomial convolution modulo $(z^{N/2} + 1)$ by multiplications by W^s and W^{-u}, with $W = e^{-j2\pi/N}$. In this case, the algorithmic structure, as shown in Fig. 6.7, uses 4 length-N polynomial transforms defined modulo $(z^{N/2} + 1)$ and only N multiplications by W^s and W^{-u} instead of $2N$ multiplications as with the preceding method. Moreover, this algorithm replaces the computation of N polynomial multiplications modulo $(z^N + 1)$ by that of $2N$ polynomial multiplications modulo $(z^{N/2} + 1)$, which is obviously simpler.

A further reduction in number of operations could also be obtained by using additional stages with reductions modulo $(z^{N/4} + 1)$, $(z^{N/8} + 1)$, ... and complete decomposition would yield the original scheme of Fig. 6.3. Thus, the translation of polynomial products defined modulo $(z^N - 1)$ into polynomial products defined modulo $(z^N + 1)$ by roots of unity defined in the field of coefficients provides considerable flexibility in trading structural complexity for computational complexity.

6.6 Complex Convolutions

For complex convolutions, the polynomial transform approach can be implemented with two real multiplications per complex multiplication by taking advantage of the fact that $j = \sqrt{-1}$ is real in certain fields. In particular, for rings of polynomials defined modulo $(z^N + 1)$, with N even, $z^N \equiv -1$ and $j = \sqrt{-1} \equiv z^{N/2}$. Thus, the method described in Sect. 3.3.3 can be used to replace a complex two-dimensional convolution with two real two-dimensional convolutions. Consequently, a complex two-dimensional convolution can be computed by polynomial transforms with about twice the computation load of a real convolution and the relative efficiency of the polynomial transform approach compared with the FFT method is about the same as for real convolutions.

6.7 Multidimensional Polynomial Transforms

Multidimensional polynomial transforms can be defined in a way similar to one-dimensional polynomial transforms. In order to support, for instance, the computation of a three-dimensional convolution y_{u,l_1,l_2} of size $p \times p \times p$, with p prime, we redefine (6.2–5) as

$$H_{m_1,m_2}(z) = \sum_{n=0}^{p-1} h_{n,m_1,m_2} z^n, \qquad m_1, m_2 = 0, \ldots, p-1 \tag{6.81}$$

$$X_{r_1,r_2}(z) = \sum_{s=0}^{p-1} x_{s,r_1,r_2} z^s, \qquad r_1, r_2 = 0, \ldots, p-1 \tag{6.82}$$

$$Y_{l_1,l_2}(z) \equiv \sum_{m_1=0}^{p-1} \sum_{m_2=0}^{p-1} H_{m_1,m_2}(z) X_{l_1-m_1,l_2-m_2}(z) \text{ modulo } (z^p - 1)$$

$$l_1, l_2 = 0, \ldots, p-1 \tag{6.83}$$

$$Y_{l_1,l_2}(z) = \sum_{u=0}^{p-1} y_{u,l_1,l_2} z^u \tag{6.84}$$

$$y_{u,l_1,l_2} = \sum_{m_1=0}^{p-1} \sum_{m_2=0}^{p-1} \sum_{n=0}^{p-1} h_{n,m_1,m_2} x_{u-n,l_1-m_1,l_2-m_2} \qquad u = 0, \ldots, p-1. \tag{6.85}$$

The two-dimensional polynomial transform is defined modulo $P(z)$, with $P(z) = (z^p - 1)/(z - 1)$, by the expression

$$\bar{H}_{k_1,k_2}(z) \equiv \sum_{m_1=0}^{p-1} \sum_{m_2=0}^{p-1} H_{1,m_1,m_2}(z) z^{m_1 k_1 + m_2 k_2} \text{ modulo } P(z)$$

$$k_1, k_2 = 0, \ldots, p-1, \tag{6.86}$$

where

$$H_{1,m_1,m_2}(z) \equiv H_{m_1,m_2}(z) \text{ modulo } P(z), \tag{6.87}$$

with a similar definition for the inverse transform. The two-dimensional polynomial transform in (6.86) supports circular convolution because z is a root of order N in the field of polynomials modulo $P(z)$. Hence, it may be used to compute the polynomial convolution $Y_{1,l_1,l_2}(z)$ with

$$Y_{1,l_1,l_2}(z) \equiv Y_{l_1,l_2}(z) \text{ modulo } P(z). \tag{6.88}$$

$Y_{l_1,l_2}(z)$ can be obtained from $Y_{1,l_1,l_2}(z)$ by a Chinese remainder reconstruction with

$$Y_{2,l_1,l_2} \equiv Y_{l_1,l_2}(z) \text{ modulo } (z - 1) \tag{6.89}$$

$$Y_{2,l_1,l_2} = \sum_{m_1=0}^{p-1} \sum_{m_2=0}^{p-1} H_{2,m_1,m_2} X_{2,l_1-m_1,l_2-m_2} \tag{6.90}$$

$$H_{2,m_1,m_2} = \sum_{n=0}^{p-1} h_{n,m_1,m_2} \tag{6.91}$$

$$X_{2,r_1,r_2} = \sum_{s=0}^{p-1} x_{s,r_1,r_2} \tag{6.92}$$

and

$$Y_{l_1,l_2}(z) \equiv Y_{1,l_1,l_2}(z) S_1(z) + Y_{2,l_1,l_2} S_2(z) \text{ modulo } P(z) \tag{6.93}$$

with

$$\begin{cases} S_1(z) \equiv 1, & S_2(z) \equiv 0 \quad \text{modulo } P(z) \\ S_1(z) \equiv 0, & S_2(z) \equiv 1 \quad \text{modulo } (z-1). \end{cases} \tag{6.94}$$

Thus, the convolution of size $p \times p \times p$ is mapped by polynomial transforms into p^2 polynomial products modulo $P(z)$ and one convolution of size $p \times p$. A diagram of the computation process is shown in Fig. 6.8. It should be noted that the convolution of size $p \times p$ can, in turn, be computed by polynomial transforms of length p as indicated in Fig. 6.1.

The two-dimensional polynomial transforms can be calculated as $2p$ polynomial transforms of length p by the row-column method, with $2p(p^3 - p^2 - 3p + 4)$ additions. Under these conditions, the convolution of size $p \times p \times p$ is evaluated with $4p^4 + 2p^3 - 14p^2 + 6p + 8$ additions, $p(p+1)$ polynomial products modulo $(z^p - 1)/(z - 1)$, and one convolution of length p. Using the same technique, a convolution of size $p \times p \times p \times p$ would be computed with $6p^5 + 2p^4 - 20p^3 + 10p^2 + 6p + 8$ additions, $p^3 + p^2 + p$ polynomial multiplications modulo $(z^p - 1)/(z - 1)$, and one convolution of length p. We give in Table 6.6 the number of operations for some multidimensional convolutions computed by polynomial transforms, which will be used for the calculation of DFTs by the method described in Sect. 7.2.

Table 6.6. Number of operations for short multidimensional convolutions computed by polynomial transforms

Convolution size	Total number of additions	Total number of multiplications
$3 \times 3 \times 3$	40	325
$3 \times 3 \times 3 \times 3$	121	1324
$6 \times 6 \times 6$	320	3896
$6 \times 6 \times 6 \times 6$	1936	31552

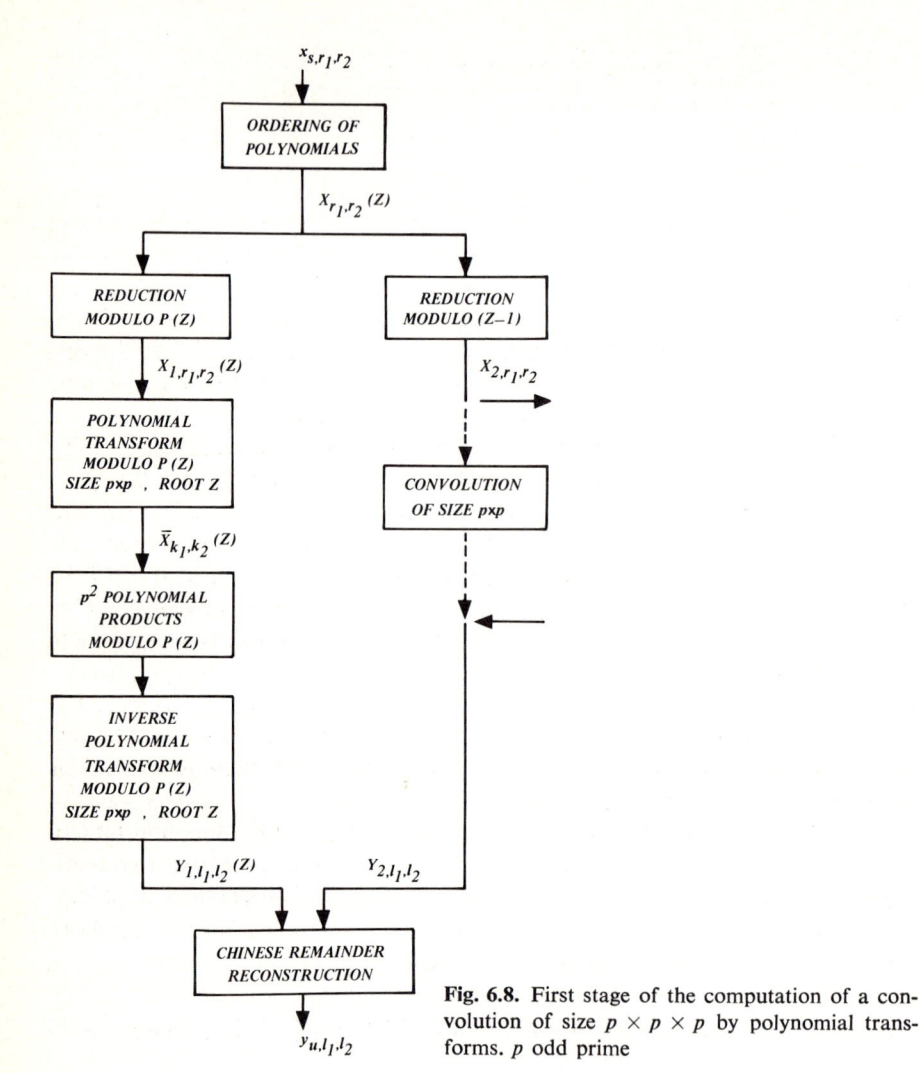

180   6. Polynomial Transforms

Fig. 6.8. First stage of the computation of a convolution of size $p \times p \times p$ by polynomial transforms. p odd prime

A similar approach can be employed to develop multidimensional polynomial transforms from one-dimensional polynomial transforms of length N, with N composite.

7. Computation of Discrete Fourier Transforms by Polynomial Transforms

As indicated in the previous chapter, polynomial transforms can be used to efficiently map multidimensional convolutions into one-dimensional convolutions and polynomial products. In this chapter, we shall see that polynomial transforms can also be used to map multidimensional DFTs into one-dimensional DFTs. This mapping is very efficient because it is accomplished using ordinary arithmetic without multiplications, and because it can be implemented by FFT-type algorithms when the dimensions are composite.

This method, which is significantly simpler than the conventional multidimensional DFT approaches such as the row-column method or the Winograd algorithm' applies only to DFTs having common factors in several dimensions. For one-dimensional DFTs or for multidimensional DFTs having no common factors in several dimensions, we show that polynomial transforms can still be used to reduce the amount of computation by converting the DFTs into multidimensional correlations and by evaluating these multidimensional correlations with polynomial transforms.

In practice, both polynomial transform methods significantly decrease the number of arithmetic operations and can be combined to define procedures with optimum efficiency for large multidimensional DFTs that have common factors in several dimensions.

It may be shown that the DFT algorithms to be discussed in this chapter and the convolution algorithms presented in the preceding chapter are closely related. We shall clarify the relationship between the two sets of polynomial transform algorithms in Appendix A.

7.1 Computation of Multidimensional DFTs by Polynomial Transforms

We consider a two-dimensional DFT, $\bar{X}_{k_1,k_2}$, of size $N \times N$

$$\bar{X}_{k_1,k_2} = \sum_{n_1=0}^{N-1} \sum_{n_2=0}^{N-1} x_{n_1,n_2} W^{n_1 k_1} W^{n_2 k_2},$$
$$W = e^{-j2\pi/N}, \quad k_1, k_2 = 0, \ldots, N-1, \quad j = \sqrt{-1}. \tag{7.1}$$

The conventional row-column method (Sect. 4.4) computes $\bar{X}_{k_1,k_2}$ as N one-dimensional DFTs along dimension k_2 and N one-dimensional DFTs along dimension k_1. Hence this method maps $\bar{X}_{k_1,k_2}$ into $2N$ DFTs of length N and, if M_1 is the number of complex multiplications required to compute the length-N DFT, the total number M of multiplications corresponding to $\bar{X}_{k_1,k_2}$ is $M = 2NM_1$. The DFT $\bar{X}_{k_1,k_2}$ can also be evaluated by use of the Winograd nesting algorithm (Chap. 5) as a DFT of size N in which each scalar is replaced by a vector of N terms and each multiplication is replaced by a DFT of length N. In this case, $M = M_1^2$, and the performance of the nesting method is better than the row-column approach when $M_1 < 2N$.

We shall now show that the number of multiplications is significantly decreased when $\bar{X}_{k_1, k_2}$ is mapped into a set of one-dimensional DFTs by a polynomial transform method.

7.1.1 The Reduced DFT Algorithm

In order to compute $\bar{X}_{k_1, k_2}$ by polynomial transforms [7.1, 2], we shall represent this DFT in polynomial algebra by replacing (7.1) with the following set of three equations:

$$\bar{X}_{k_1}(z) \equiv \sum_{n_1=0}^{N-1} X_{n_1}(z) W^{n_1 k_1} \text{ modulo } (z^N - 1) \tag{7.2}$$

$$X_{n_1}(z) = \sum_{n_2=0}^{N-1} x_{n_1, n_2} z^{n_2} \tag{7.3}$$

$$\bar{X}_{k_1, k_2} \equiv \bar{X}_{k_1}(z) \text{ modulo } (z - W^{k_2}) \qquad k_1, k_2 = 0, \ldots, N-1. \tag{7.4}$$

It can easily be verified that (7.2–4) are equivalent to (7.1) by noting that the definition of (7.4) modulo $(z - W^{k_2})$ is equivalent to substituting W^{k_2} for z in (7.2) and (7.3). It should also be noted that although the definition of $\bar{X}_{k_1}(z)$ modulo $(z^N - 1)$ is superfluous at this stage, it is valid, since $z^N \equiv W^{Nk_2} = 1$.

In order to simplify the presentation, we assume now that N is an odd prime, with $N = p$. Thus, $z^p - 1$ is the product of two cyclotomic polynomials

$$z^p - 1 = (z - 1) P(z) \tag{7.5}$$

$$P(z) = z^{p-1} + z^{p-2} + \ldots + 1. \tag{7.6}$$

For $k_2 \equiv 0$, we have $z \equiv 1$, and $\bar{X}_{k_1, 0}$ is a simple DFT of length N obtained by reducing $X_{n_1}(z)$ modulo $(z - 1)$, with

$$\bar{X}_{k_1, 0} = \sum_{n_1=0}^{p-1} \left(\sum_{n_2=0}^{p-1} x_{n_1, n_2} \right) W^{n_1 k_1}. \tag{7.7}$$

For $k_2 \not\equiv 0$ modulo p, W^{k_2} is always a root of $P(z)$, since

$$P(z) = \prod_{k_2=1}^{p-1} (z - W^{k_2}) \tag{7.8}$$

and $\bar{X}_{k_1, k_2}$ may be obtained by substituting W^{k_2} for z in (7.2). Since $z - W^{k_2}$ is a factor of $P(z)$ and $P(z)$ is a factor of $z^p - 1$, (7.4) becomes

7.1 Computation of Multidimensional DFTs by Polynomial Transforms

$$\bar{X}_{k_1,k_2} \equiv \{[\bar{X}_{k_1}(z) \text{ modulo } (z^p - 1)] \text{ modulo } P(z)\} \text{ modulo } (z - W^{k_2}). \quad (7.9)$$

Hence, for $k_2 \neq 0$, (7.2–4) reduce to

$$\bar{X}_{k_1}^1(z) \equiv \sum_{n_1=0}^{p-1} X_{n_1}^1(z) W^{n_1 k_1} \text{ modulo } P(z) \qquad k_1 = 0, \ldots, p - 1 \quad (7.10)$$

$$X_{n_1}^1(z) = \sum_{n_2=0}^{p-2} (x_{n_1,n_2} - x_{n_1,p-1}) z^{n_2} \equiv X_{n_1}(z) \text{ modulo } P(z) \quad (7.11)$$

$$\bar{X}_{k_1,k_2} \equiv \bar{X}_{k_1}^1(z) \text{ modulo } (z - W^{k_2}) \qquad k_2 = 1, \ldots, p - 1. \quad (7.12)$$

Since p is an odd prime and $k_2 \neq 0$, the permutation $k_2 k_1$ modulo p maps all values of k_1, and we obtain, by replacing k_1 with $k_2 k_1$,

$$\bar{X}_{k_2 k_1}^1(z) \equiv \sum_{n_1=0}^{p-1} X_{n_1}^1(z) W^{k_2 n_1 k_1} \text{ modulo } P(z) \quad (7.13)$$

$$\bar{X}_{k_2 k_1, k_2} \equiv \bar{X}_{k_2 k_1}^1(z) \text{ modulo } (z - W^{k_2}). \quad (7.14)$$

$\bar{X}_{k_2 k_1, k_2}$ is obtained by replacing z by W^{k_2} in (7.14). Therefore, we can substitute z for W^{k_2} in (7.13), with

$$\bar{X}_{k_2 k_1}^1(z) \equiv \sum_{n_1=0}^{p-1} X_{n_1}^1(z) z^{n_1 k_1} \text{ modulo } P(z), \quad (7.15)$$

where the right-hand side of the equation is independent of k_2. $\bar{X}_{k_2 k_1}^1(z)$ is recognized as a polynomial transform of length p which is computed without multiplications, with only $p^3 - p^2 - 3p + 4$ additions. Hence, the only multiplications required for evaluating the DFT of size $p \times p$ are those corresponding to (7.14) and to the length-p DFT defined by (7.7).

In order to specify the operations corresponding to (7.14), we note that the p polynomials $\bar{X}_{k_2 k_1}^1(z)$ are defined modulo $P(z)$ and are therefore of degree $p - 2$. Hence these polynomials can be represented as

$$\bar{X}_{k_2 k_1}^1(z) = \sum_{l=0}^{p-2} y_{k_1,l} z^l. \quad (7.16)$$

If we substitute $\bar{X}_{k_2 k_1}^1(z)$ defined by (7.16) into (7.14), we obtain

$$\bar{X}_{k_2 k_1, k_2} = \sum_{l=0}^{p-2} y_{k_1,l} W^{k_2 l}, \qquad k_2 = 1, \ldots, p - 1, \quad (7.17)$$

which represents p DFTs of p terms corresponding to the p values of k_1. This means that the polynomial transform maps without multiplications a DFT of size $p \times p$, with p prime, into $p + 1$ DFTs of p terms, instead of $2p$ DFTs of p

7. Computation of Discrete Fourier Transforms by Polynomial Transforms

terms with the row-column method. Thus, the number M of multiplications becomes

$$M = (p + 1)M_1, \tag{7.18}$$

where M_1 is the number of complex multiplications corresponding to a length-p DFT. This polynomial transform approach is illustrated in Fig. 7.1.

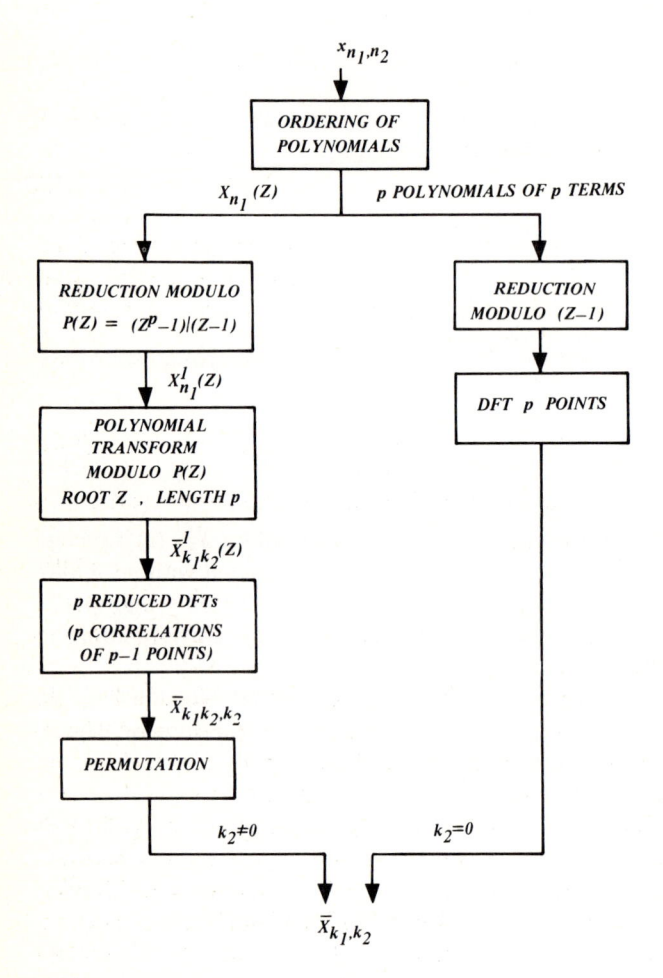

Fig. 7.1. Computation of a DFT of size $p \times p$ by polynomial transforms. p prime. Reduced DFT algorithm

The number of multiplications can be further reduced by noting that, in the p DFTs defined by (7.17), the last input term is equal to zero and the first output

7.1 Computation of Multidimensional DFTs by Polynomial Transforms

term is not computed. The simplification of (7.17) is based upon the fact that, for $k_2 \neq 0$ and p prime,

$$\sum_{l=1}^{p-1} W^{k_2 l} = -1. \tag{7.19}$$

Hence (7.17) can be rewritten as

$$\bar{X}_{k_2 k_1, k_2} = \sum_{l=1}^{p-1} y^1_{k_1, l} W^{k_2 l}, \quad k_2 = 1, \ldots, p-1, \tag{7.20}$$

where

$$y^1_{k_1, l} = y_{k_1, l} - y_{k_1, 0}, \quad y^1_{k_1, p-1} = -y_{k_1, 0}$$
$$l = 1, \ldots, p-2. \tag{7.21}$$

In the DFTs defined by (7.20), the first input and output terms are missing. Thus, these DFTs are usually called *reduced* DFTs. They can be computed as correlations of length $p-1$ by using Rader's algorithm [7.3] (Sect. 5.2). In this case, if g is a primitive root modulo p, l and k_2 are redefined by

$$\begin{cases} l \equiv g^u \mod p \\ k_2 \equiv g^v \mod p \end{cases} \quad u, v = 0, \ldots, p-2. \tag{7.22}$$

Under these conditions, the reduced DFT (7.20) is converted into a correlation

$$\bar{X}_{k_1 g^v, g^v} = \sum_{u=0}^{p-2} y^1_{k_1, g^u} W^{g^{u+v}}. \tag{7.23}$$

The sequence $y^1_{k_1, l}$ can be constructed from the sequence $y_{k_1, l}$ without additions by noting that it is equivalent to a multiplication of $\bar{X}_{k_2 k_1}(z)$ by z^{-1} modulo $(z^p - 1)$, followed by a reduction modulo $P(z)$ and a multiplication by z. In practice, the multiplication by z^{-1} may be combined with the ordering of the input polynomials. The reduction modulo $P(z)$ can, therefore, be executed without additions as part of the computation of the polynomial transform. We have seen in Sect. 5.2 that when a length-p DFT, with p prime, is computed by Rader's algorithm, the calculations reduce to one correlation of $p-1$ terms plus one scalar multiplication. Thus, if the conventional DFT is computed with M_1 complex multiplications, the corresponding reduced DFT defined by (7.23) is calculated with $M_1 - 1$ complex multiplications and the number M of complex multiplications required to evaluate the DFT of size $p \times p$ by polynomial transforms reduces to

$$M = (p+1)M_1 - p. \tag{7.24}$$

This is about half the number of multiplications corresponding to the row-column method and always less (except for $p = M_1$) than the number of multiplications required by the Winograd algorithm. When the DFTs and reduced DFTs of size p are evaluated by Rader's algorithm, all complex multiplications reduce to multiplications by pure real or pure imaginary numbers and can be implemented with only two real multiplications. In this case, the number of real multiplications required to evaluate the DFT of size $p \times p$ by polynomial transforms becomes $2(p + 1)M_1 - 2p$.

7.1.2 General Definition of the Algorithm

In the foregoing, we have restricted our discussion to DFTs of size $p \times p$, with p prime. A similar polynomial transform approach can also be applied to DFTs of size $N \times N$, with N composite, by developing algorithms as in Sect. 6.2.1, with polynomial transforms defined modulo the various cyclotomic polynomials $P_{e_i}(z)$ which are factors of $z^N - 1$ [7.1]. Since the most important form of the general algorithm concerns transforms corresponding to $N = 2^t$, we shall restrict detailed discussion to this case, and simply give a summary of results for other cases of interest.

Assuming N is a power of 2, with $N = 2^t$, we represent once again the DFT $\bar{X}_{k_1,k_2}$ of size $N \times N$ by a set of three polynomial equations

$$\bar{X}_{k_1}(z) \equiv \sum_{n_1=0}^{N-1} X_{n_1}(z) W^{n_1 k_1} \text{ modulo } (z^N - 1) \tag{7.25}$$

$$X_{n_1}(z) = \sum_{n_2=0}^{N-1} x_{n_1,n_2} z^{n_2} \tag{7.26}$$

$$\bar{X}_{k_1,k_2} \equiv \bar{X}_{k_1}(z) \text{ modulo } (z - W^{k_2}). \tag{7.27}$$

Since N is even, $z^N - 1$ is the product of the two polynomials $z^{N/2} - 1$ and $z^{N/2} + 1$. The complex roots of $z^{N/2} + 1$ are W^{k_2}, for k_2 odd, and we have

$$z^{N/2} + 1 = \prod_{k_2 \text{ odd}} (z - W^{k_2}). \tag{7.28}$$

Therefore, for k_2 odd, $z - W^{k_2}$ is a factor of $z^{N/2} + 1$ which, in turn, is itself a factor of $z^N - 1$. Thus, for k_2 odd, (7.25–27) can be reduced modulo $(z^{N/2} + 1)$ to become

$$\bar{X}^1_{k_1}(z) \equiv \sum_{n_1=0}^{N-1} X^1_{n_1}(z) W^{n_1 k_1} \text{ modulo } (z^{N/2} + 1) \tag{7.29}$$

$$X^1_{n_1}(z) = \sum_{n_2=0}^{N/2-1} (x_{n_1,n_2} - x_{n_1,n_2+N/2}) z^{n_2} \equiv X_{n_1}(z) \text{ modulo } (z^{N/2} + 1) \tag{7.30}$$

$$\bar{X}_{k_1,k_2} \equiv \bar{X}^1_{k_1}(z) \text{ modulo } (z - W^{k_2}), \quad k_2 \text{ odd}. \tag{7.31}$$

7.1 Computation of Multidimensional DFTs by Polynomial Transforms

Since k_2 is odd and N is a power of two, the permutation $k_2 k_1$ modulo N maps all values of k_1 and we obtain, by replacing k_1 with $k_2 k_1$,

$$\bar{X}^1_{k_2 k_1}(z) \equiv \sum_{n_1=0}^{N-1} X^1_{n_1}(z) W^{k_2 n_1 k_1} \text{ modulo } (z^{N/2} + 1) \tag{7.32}$$

$$\bar{X}_{k_2 k_1, k_2} \equiv \bar{X}^1_{k_2 k_1}(z) \text{ modulo } (z - W^{k_2}), \quad k_2 \text{ odd.} \tag{7.33}$$

Since (7.33) is defined modulo $(z - W^{k_2})$, we have $z \equiv W^{k_2}$. Hence, we can substitute z for W^{k_2} in (7.32). This gives

$$\bar{X}^1_{k_2 k_1}(z) \equiv \sum_{n_1=0}^{N-1} X^1_{n_1}(z) z^{n_1 k_1} \text{ modulo } (z^{N/2} + 1), \tag{7.34}$$

which indicates that $\bar{X}^1_{k_2 k_1}(z)$ can be computed as a polynomial transform of length N, with $N = 2^t$, and of root z defined modulo $(z^{N/2} + 1)$. We have shown in Sect. 6.2.1 that such a transform may be calculated using a radix-2 FFT-type algorithm with only $(N^2/2) \log_2 N$ additions and without multiplications. The N polynomials $\bar{X}^1_{k_2 k_1}(z)$ are of degree $N/2 - 1$ because they are defined modulo $(z^{N/2} + 1)$. Therefore, we can represent these polynomials as

$$\bar{X}^1_{k_2 k_1}(z) = \sum_{l=0}^{N/2-1} y_{k_1, l} z^l. \tag{7.35}$$

Then, substituting (7.35) into (7.33) yields

$$\bar{X}_{k_2 k_1, k_2} = \sum_{l=0}^{N/2-1} y_{k_1, l} W^{l k_2}, \quad k_2 \text{ odd} \tag{7.36}$$

and, since k_2 is odd

$$\bar{X}_{(2u+1)k_1, 2u+1} = \sum_{l=0}^{N/2-1} y_{k_1, l} W^l W^{2ul} \tag{7.37}$$

with $k_2 = 2u + 1$. Equation (7.37) represents N DFTs of length $N/2$ where the input sequence is multiplied pointwise by $1, W, W^2, \ldots$. These DFTs are identical to the reduced DFT that appears in the first stage of a decimation in frequency radix-2 FFT decomposition and are sometimes called *odd* DFTs [7.4]. Therefore, for k_2 odd, the DFT of size $N \times N$ is computed by N reductions modulo $(z^{N/2} + 1)$, one polynomial transform of length N, and N odd DFTs of length $N/2$, the only multiplications being those corresponding to the odd DFTs.

For k_2 even, with $k_2 = 2u$, the DFT $\bar{X}_{k_1, k_2}$ of size $N \times N$ becomes a simple DFT of size $N \times (N/2)$, with

$$\bar{X}_{k_1, 2u} = \sum_{n_1=0}^{N-1} \sum_{n_2=0}^{N/2-1} (x_{n_1, n_2} + x_{n_1, n_2 + N/2}) W^{n_1 k_1} W^{2u n_2}$$

$$u = 0, \ldots, N/2 - 1. \tag{7.38}$$

By reversing the role of k_1 and $2u$, this DFT can be represented in polynomial notation as

$$\bar{X}_{2u}(z) \equiv \sum_{n_2=0}^{N/2-1} X_{n_2}(z) W^{2un_2} \text{ modulo } (z^N - 1) \qquad (7.39)$$

$$X_{n_2}(z) = \sum_{n_1=0}^{N-1} (x_{n_1,n_2} + x_{n_1,n_2+N/2}) z^{n_1} \qquad (7.40)$$

$$\bar{X}_{k_1, 2u} \equiv \bar{X}_{2u}(z) \text{ modulo } (z - W^{k_1}). \qquad (7.41)$$

We may then use the same polynomial transform method as above to compute $\bar{X}_{k_1, 2u}$. The polynomial $z^N - 1$ factors into the two polynomials $z^{N/2} - 1$ and $z^{N/2} + 1$ and the roots of $z^{N/2} - 1$ correspond to W^{k_1}, k_1 even. Therefore, for k_1 even, with $k_1 = 2v$, $\bar{X}_{k_1, 2u}$ reduces to a simple DFT of size $(N/2) \times (N/2)$

$$\bar{X}_{2v, 2u} = \sum_{n_1=0}^{N/2-1} \sum_{n_2=0}^{N/2-1} (x_{n_1,n_2} + x_{n_1,n_2+N/2} + x_{n_1+N/2,n_2} + x_{n_1+N/2,n_2+N/2}) W^{2vn_1} W^{2un_2}$$

$$v = 0, ..., N/2 - 1. \qquad (7.42)$$

For k_1 odd, the W^{k_1} are the roots of $z^{N/2} + 1$ and (7.39, 40) can be defined modulo $(z^{N/2} + 1)$ instead of modulo $(z^N - 1)$. In this case, $\bar{X}_{k_1, 2u}$ can be computed using a polynomial transform of length $N/2$ in a way similar to that discussed above for $\bar{X}_{k_1, k_2}$, k_2 odd. This is accomplished with

$$\bar{X}^1_{2uk_1}(z) \equiv \sum_{n_2=0}^{N/2-1} X^1_{n_2}(z) z^{2un_2} \text{ modulo } (z^{N/2} + 1)$$

$$u = 0, ..., N/2 - 1 \qquad (7.43)$$

$$X^1_{n_2}(z) \equiv X_{n_2}(z) \text{ modulo } (z^{N/2} + 1)$$

$$= \sum_{n_1=0}^{N/2-1} (x_{n_1,n_2} + x_{n_1,n_2+N/2} - x_{n_1+N/2,n_2} - x_{n_1+N/2,n_2+N/2}) z^{n_1} \qquad (7.44)$$

$$\bar{X}_{k_1, 2uk_1} \equiv \bar{X}^1_{2uk_1} \text{ modulo } (z - W^{k_1}), \quad k_1 \text{ odd}. \qquad (7.45)$$

Following this procedure, the DFT of size $N \times N$ is computed as shown in Fig. 7.2 with reductions modulo $(z^{N/2} - 1)$ and $(z^{N/2} + 1)$, two polynomial transforms, $3N/2$ reduced DFTs of $N/2$ terms, and one DFT of size $(N/2) \times (N/2)$. This last DFT can in turn be computed by the same method, and, by repeating this process, the $(N \times N)$-point DFT is completely evaluated in $(\log_2 N) - 1$ stages by polynomial transforms. With the conventional row-column method, the first stage of a radix-2 FFT algorithm reduces the DFT of size $N \times N$ into $2N$ odd DFTs of $N/2$ terms, N DFTs of length $N/2$, and one DFT of size $N/2 \times N/2$. This corresponds to about twice as many DFTs and therefore to about twice as many multiplications as with the polynomial transform method.

7.1 Computation of Multidimensional DFTs by Polynomial Transforms 189

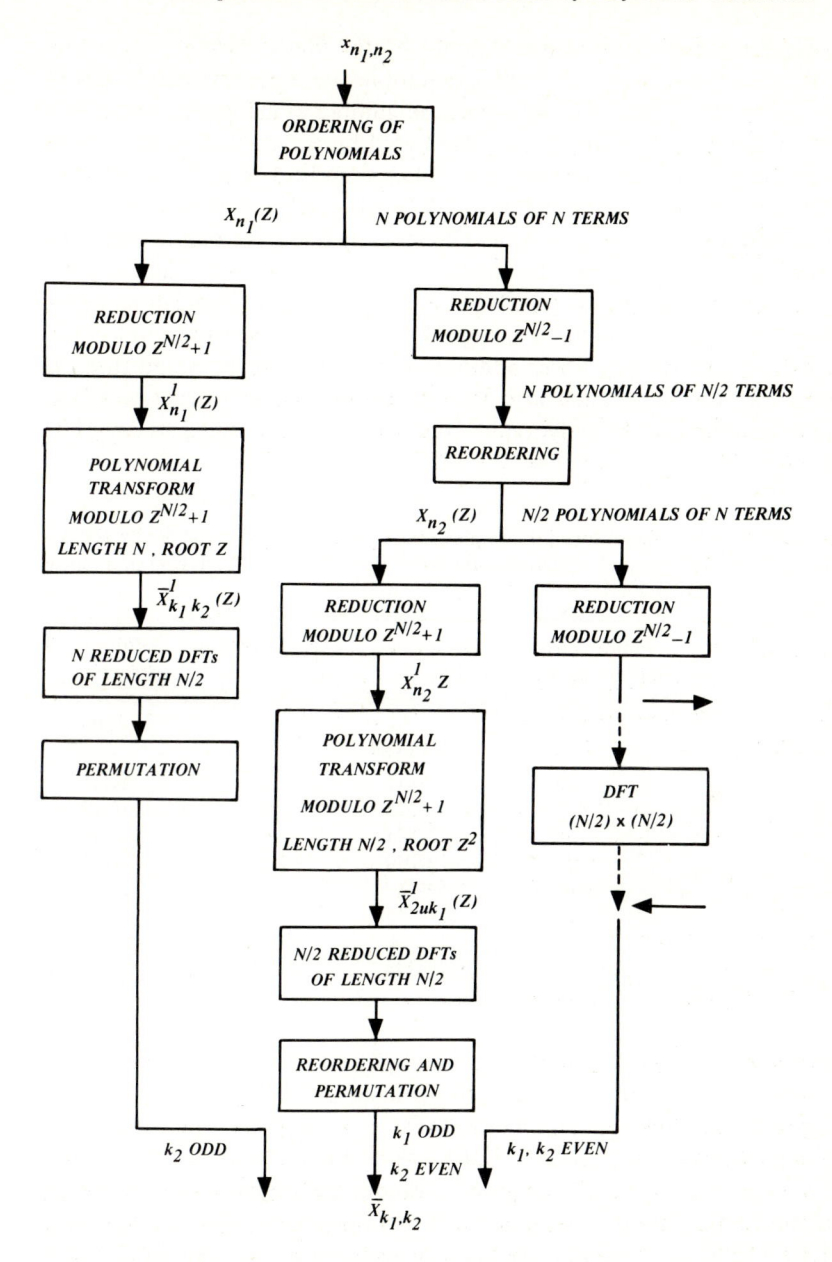

Fig. 7.2. Computation of a DFT of size $N \times N$ by polynomial transforms. $N = 2^r$. Reduced DFT algorithm

When the reduced DFTs are computed by the Rader-Brenner algorithm [7.5], as discussed in Sect. 4.3, all complex multiplications are implemented with only two real multiplications. In this case, the number of arithmetic operations for DFTs of size $N \times N$, with $N = 2^t$, computed by polynomial transforms, is as given in Table 7.1. The entries in this table are derived from the number of operations corresponding to the reduced DFTs given in Table 4.4 and from the number of operations for reductions and Chinese remainder reconstruction given in Table 6.2. It can be verified by comparison with Table 4.5 that the polynomial transform method requires only about half as many multiplications as the conventional row-column method using the same FFT algorithm and is implemented with significantly fewer additions. It should also be noted that the polynomial transform approach with $N = 2^t$ retains the basic structure of the FFT algorithm because the polynomial transforms are computed by an FFT-type partition.

Table 7.1. Number of real operations for complex DFTs of size $N \times N$ computed by polynomial transforms with the reduced Rader-Brenner DFT algorithm. $N = 2^t$. Trivial multiplications by $\pm 1, \pm j$ are not counted

DFT size	Number of multiplications	Number of additions	Multiplications per point	Additions per point
2×2	0	16	0	4.00
4×4	0	128	0	8.00
8×8	48	816	0.75	12.75
16×16	432	4528	1.69	17.69
32×32	2736	24944	2.67	24.36
64×64	15024	125040	3.67	30.53
128×128	76464	599152	4.67	36.57
256×256	371376	2790512	5.67	42.58
512×512	1747632	12735600	6.67	48.58
1024×1024	8039088	57234544	7.67	54.58

The same general approach can also be employed to compute DFTs of size $N \times N$, with $N = p^c$, p an odd prime. If, for instance, $N = p^2$, $z^{p^2} - 1$ factors into the three cyclotomic polynomials $P_1(z) = z - 1$, $P_2(z) = z^{p-1} + z^{p-2} + \ldots + 1$, and $P_3(z) = z^{p(p-1)} + z^{p(p-2)} + \ldots + 1$. In this case, a DFT of size $p^2 \times p^2$ is computed as shown in Fig. 7.3 with one polynomial transform of p^2 terms modulo $P_3(z)$, one polynomial transform of p terms modulo $P_3(z)$, $p^2 + p$ reduced DFTs of length p^2, and one DFT of size $p \times p$. This last DFT can in turn be evaluated by polynomial transforms. In this approach, each of the reduced DFTs of size p^2 is such that only the first $p(p-1)$ input samples are nonzero and that the output samples with indices multiple of p are not computed. These reduced DFTs are equivalent to one correlation of $p(p-1)$ terms plus one reduced DFT of length p.

7.1 Computation of Multidimensional DFTs by Polynomial Transforms 191

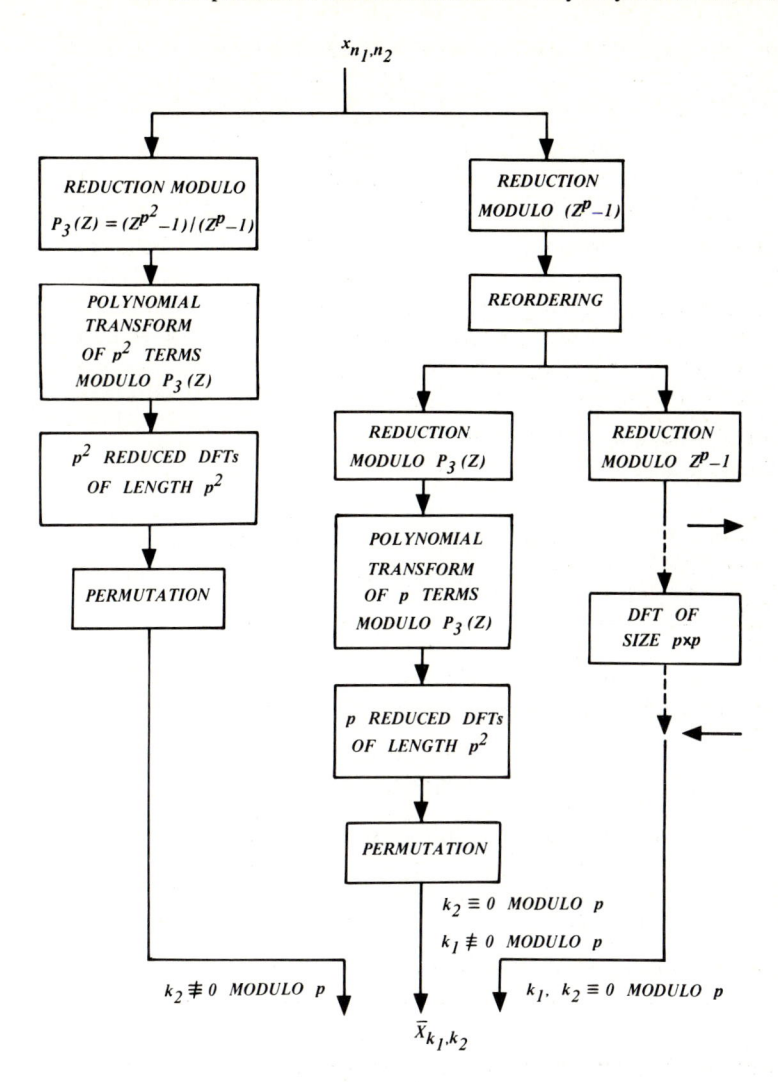

Fig. 7.3. Computation of a DFT of size $p^2 \times p^2$ by polynomial transforms. p odd prime

We summarize in Table 7.2 the main properties of various two-dimensional DFTs computed by the polynomial transform method. In this table, the operations count for execution of polynomial transforms and reductions is derived from Table 6.2. We also list in Table 7.3 the number of arithmetic operations for multidimensional DFTs computed by polynomial transforms. In this table, we have presumed the use of the reduced DFT algorithms of Sect. 7.4 (Table 7.8) for $N = 8, 9, 16$ and of the short convolution algorithms of size $N - 1$ given in Sect. 3.8.1 (Table 3.1) for $N = 5$. The reduced DFT algorithm of 7 points is

7. Computation of Discrete Fourier Transforms by Polynomial Transforms

Table 7.2. Main parameters for DFTs of size $N \times N$ computed by polynomial transforms

N	Polynomial transforms	Number of additions for polynomial transforms and reductions	DFTs and reduced DFTs
$N = p$, p prime	1 polynomial transform of p terms modulo $(z^p - 1)/(z - 1)$	$p^3 + p^2 - 5p + 4$	1 DFT of p terms; p reduced DFTs of p terms (p correlations of $p - 1$ terms)
$N = p^2$, p prime	1 polynomial transform of p^2 terms modulo $(z^{p^2} - 1)/(z^p - 1)$; 1 polynomial transform of p terms modulo $(z^{p^2} - 1)/(z^p - 1)$; 1 polynomial transform of p terms modulo $(z^p - 1)/(z - 1)$	$2p^5 + p^4 - 5p^3 + p^2 + 6$	$p^2 + p$ reduced DFTs of p^2 terms; p reduced DFTs of p terms (p correlations of $p - 1$ terms); 1 DFT of p terms
$N = 2^t$	1 polynomial transform of 2^t terms modulo $(z^{2^{t-1}} + 1)$; 1 polynomial transform of 2^{t-1} terms modulo $(z^{2^{t-1}} + 1)$	$(3t + 5)2^{2(t-1)}$	$3 \cdot 2^{t-1}$ reduced DFTs of dimension 2^t; 1 DFT of size $2^{t-1} \times 2^{t-1}$
$N = p_1 p_2$, p_1, p_2 primes	1 polynomial transform of $p_1 p_2$ terms; p_2 polynomial transforms of p_1 terms; p_1 polynomial transforms of p_2 terms	$p_1^2 p_2^2 (p_1 + p_2 + 2)$ $-5p_1 p_2 (p_1 + p_2)$ $+4(p_1^2 + p_2^2)$	$p_1 p_2 + p_1 + p_2$ reduced DFTs of $p_1 p_2$ terms; p_1 reduced DFTs of p_1 terms; p_2 reduced DFTs of p_2 terms; 1 DFT of $p_1 p_2$ terms

Table 7.3. Number of complex operations for simple two-dimensional DFTs evaluated by polynomial transforms. Trivial multiplications are given between parentheses. Each complex multiplication is implemented with two real multiplications

DFT size	Number of multiplications	Number of additions
2×2	4 (4)	8
3×3	9 (1)	36
4×4	16 (16)	64
5×5	31 (1)	221
7×7	65 (1)	635
8×8	64 (40)	408
9×9	105 (1)	785
16×16	304 (88)	2264

obtained as a 6-point convolution computed with 8 complex multiplications and 34 complex additions by nesting convolutions of 2 and 3 terms. In Table 7.3, the DFTs corresponding to $N = 2, 3, 4$ are computed by simple nesting and the DFT corresponding to $N = 9$ is computed partly by nesting and partly by polynomial transforms.

7.1.3 Multidimensional DFTs

A similar polynomial transform approach can also be developed to compute DFTs of dimension greater than 2. Consider, for instance, a DFT $\bar{X}_{k_1,k_2,k_3}$ of size $N \times N \times N$

$$\bar{X}_{k_1,k_2,k_3} = \sum_{n_1=0}^{N-1} \sum_{n_2=0}^{N-1} \sum_{n_3=0}^{N-1} x_{n_1,n_2,n_3} W^{n_1 k_1} W^{n_2 k_2} W^{n_3 k_3}$$

$$k_1, k_2, k_3 = 0, \ldots, N-1. \tag{7.46}$$

In polynomial notation, this DFT becomes

$$\bar{X}_{k_1,k_2}(z) \equiv \sum_{n_1=0}^{N-1} \sum_{n_2=0}^{N-1} X_{n_1,n_2}(z) W^{n_1 k_1} W^{n_2 k_2} \text{ modulo } (z^N - 1) \tag{7.47}$$

$$X_{n_1,n_2}(z) = \sum_{n_3=0}^{N-1} x_{n_1,n_2,n_3} z^{n_3} \tag{7.48}$$

$$\bar{X}_{k_1,k_2,k_3} \equiv \bar{X}_{k_1,k_2}(z) \text{ modulo } (z - W^{k_3}). \tag{7.49}$$

When N is an odd prime, with $N = p$, this DFT reduces to a DFT of size $p \times p$ for $k_3 = 0$. For $k_3 \neq 0$, $\bar{X}_{k_1,k_2,k_3}$ can be computed by a two-dimensional polynomial transform with

$$\bar{X}_{k_3 k_1, k_3 k_2}(z) \equiv \sum_{n_1=0}^{p-1} \sum_{n_2=0}^{p-1} X^1_{n_1,n_2}(z) z^{n_1 k_1} z^{n_2 k_2} \text{ modulo } P(z) \tag{7.50}$$

$$P(z) = (z^p - 1)/(z - 1) \tag{7.51}$$

$$X^1_{n_1,n_2}(z) \equiv X_{n_1,n_2}(z) \text{ modulo } P(z) \tag{7.52}$$

$$\bar{X}_{k_3 k_1, k_3 k_2, k_3} \equiv \bar{X}_{k_3 k_1, k_3 k_2}(z) \text{ modulo } (z - W^{k_3}). \tag{7.53}$$

Therefore, the DFT of size $p \times p \times p$ is mapped by a two-dimensional polynomial transform into a DFT of size $p \times p$ plus p^2 reduced DFTs of dimension p. For a DFT of dimension d, the same process is applied recursively and the DFT is mapped into one DFT of length p plus $p + p^2 + \ldots + p^{d-1}$ odd DFTs of length p. Thus, if M_1 is the number of complex multiplications for a DFT of length p, the number of complex multiplications M corresponding to a DFT of dimension d with length p in all dimensions becomes

$$M = 1 + (M_1 - 1)(p^d - 1)/(p - 1). \tag{7.54}$$

The same DFT is computed with $dp^{d-1}M_1$ complex multiplications by the row-column method. Therefore, the number of multiplications is approximately reduced by a factor of d when the row-column method is replaced by the polynomial transform approach. Thus, the efficiency of the polynomial transform method, relative to the row-column algorithm, is proportional to d. A similar result is also obtained when the polynomial transform method is compared to a nesting technique, since the number of multiplications for nesting is M_1^d, with $M_1 > p$ for $p \neq 3$. This point is illustrated more clearly by considering the case of a DFT of size $7 \times 7 \times 7$ which is computed with 457 complex multiplications by polynomial transforms, as opposed to 1323 and 729 multiplications when the calculations are done by the row-column method and the nesting algorithm, respectively.

A similar polynomial transform approach applies to any d-dimensional DFT with common factors in several dimensions and we give the number of complex arithmetic operations in Table 7.4 for some complex three-dimensional DFTs computed by polynomial transforms.

Table 7.4. Number of complex operations for simple three-dimensional DFTs evaluted by polynomial transforms. Trivial multiplications are given between parentheses. Each complex multiplication is implemented with two real multiplications

DFT size	Number of multiplications	Number of additions
$2 \times 2 \times 2$	8 (8)	24
$3 \times 3 \times 3$	27 (1)	162
$4 \times 4 \times 4$	64 (64)	384
$5 \times 5 \times 5$	156 (1)	1686
$7 \times 7 \times 7$	457 (1)	6767
$8 \times 8 \times 8$	512 (288)	4832
$9 \times 9 \times 9$	963 (1)	10383
$16 \times 16 \times 16$	4992 (1184)	52960

7.1.4 Nesting and Prime Factor Algorithms

We have seen in Chap. 5 that large DFTs of size $N \times N$ can be computed by nesting small DFTs of size $N_i \times N_i$ or by using a prime factor algorithm, when the various N_i are factors of N which are mutually prime [7.6, 7]. These methods can be used in combination with polynomial transforms as an alternative to using large polynomial transforms.

If we consider the simple case corresponding to $N = N_1 N_2$, with $(N_1, N_2) = 1$, the DFT of size $N_1 N_2 \times N_1 N_2$ can be transformed into a four-dimensional

DFT of size $(N_1 \times N_1) \times (N_2 \times N_2)$ by using Good's mapping algorithm [7.6]. With this approach, the four-dimensional DFT is, in turn, computed using Winograd nesting [7.7] by calculating, by polynomial transforms, a DFT of size $N_1 \times N_1$ in which each scalar is replaced by an array of $N_2 \times N_2$ terms and each multiplication is replaced by a DFT of size $N_2 \times N_2$ computed by polynomial transforms. Thus, if M_1, M_2, M and A_1, A_2, A are, respectively, the number of complex multiplications and additions required to evaluate the DFTs of sizes $N_1 \times N_1$, $N_2 \times N_2$, and $N_1 N_2 \times N_1 N_2$, we have

$$M = M_1 M_2 \tag{7.55}$$

$$A = N_2^2 A_1 + M_1 A_2. \tag{7.56}$$

The four-dimensional DFT of size $(N_1 \times N_1) \times (N_2 \times N_2)$ can also be computed by the row-column method as N_1^2 DFTs of dimension $N_2 \times N_2$ plus N_2^2 DFTs of dimension $N_1 \times N_1$. In this case, we have

$$M = N_1^2 M_2 + N_2^2 M_1 \tag{7.57}$$

$$A = N_1^2 A_2 + N_2^2 A_1. \tag{7.58}$$

Since $M_1 \geqslant N_1^2$ and $M_2 \geqslant N_2^2$, the nesting method generally requires more addition than the row-column method, except when $M_1 = N_1^2$. However, for

Table 7.5. Number of real operations for complex multidimensional DFTs evaluated by polynomial transforms and nesting. Trivial multiplications by ± 1, $\pm j$ are not counted

DFT size	Number of multiplications	Number of additions	Multiplications per point	Additions per point
24×24	1072	11952	1.86	20.75
30×30	2224	26712	2.47	29.68
36×36	3328	35488	2.57	27.38
40×40	3888	48688	2.43	30.43
48×48	5296	59184	2.30	25.69
56×56	8240	121264	2.63	38.67
63×63	13648	204920	3.44	51.63
72×72	13360	166576	2.58	32.13
80×80	18672	247568	2.92	38.68
112×112	39344	607952	3.14	48.47
120×120	35632	553392	2.47	38.43
144×144	63664	844048	3.07	40.70
240×240	169456	2688912	2.94	46.68
504×504	873520	16353584	3.44	64.38
1008×1008	4149424	80267312	4.08	79.00
$120 \times 120 \times 120$	4312512	99966528	2.50	57.85
$240 \times 240 \times 240$	42050240	977859648	3.04	70.74

short DFTs, M_1 and M_2 are not much larger than N_1^2 and N_2^2, and the nesting method requires fewer multiplications than the row-column algorithm, and a number of additions which is about the same. Thus, the nesting algorithm is generally better suited for DFTs of moderate sizes whereas the prime factor technique is best for large DFTs.

With both methods, large DFTs can be evaluated using a small set of short length DFTs computed by polynomial transforms. Moreover, additional computational savings can be obtained by splitting the calculations with the techniques discussed in Sects. 5.3.3 and 5.4.3.

Table 7.5 gives the number of real operations for complex multidimensional DFTs computed by nesting the small multidimensional DFTs evaluated by polynomial transforms for which data are tabulated in Tables 7.3 and 7.4. It can be seen by comparison with Table 7.1 that this method requires only about half the number of multiplications of the large polynomial transform approach with size $N = 2^t$, but uses more additions. We shall see, however, that when this method is combined with split nesting and another polynomial transform method, significant additional reduction in the number of operations is made possible.

7.1.5 DFT Computation Using Polynomial Transforms Defined in Modified Rings of Polynomials

We have seen that multidimensional DFTs with common factors in several dimensions can be efficiently converted by polynomial transforms into one-dimensional DFTs and reduced DFTs. This method is particularly worthwhile for DFTs with dimensions which are powers of two, because the polynomial transforms and the DFTs can then be calculated with a minimum number of operations by a radix-2 FFT-type algorithm. The main disadvantage of this method, however, is that it is implemented with a number of different polynomial transforms.

We shall now show that the implementation can be greatly simplified, at the expense of a slightly larger number of arithmetic operations, by modifying the definition of the rings with a premultiplication of the input data sequence by powers of a root of -1. In order to introduce this method [7.8], we consider a DFT $\bar{X}_{k_1,k_2}$ of dimension $N \times N$, with $N = 2^t$,

$$\bar{X}_{k_1,k_2} = \sum_{n_1=0}^{N-1} \sum_{n_2=0}^{N-1} x_{n_1,n_2} W^{2n_1k_1} W^{2n_2k_2}$$

$$W = e^{-j\pi/N} \qquad k_1, k_2 = 0, \ldots, N-1, \qquad (7.59)$$

where the symbol W represents $e^{-j\pi/N}$ instead of $e^{-j2\pi/N}$ for reasons that will be apparent later.

We first rewrite (7.59) as

$$\bar{X}_{k_1,k_2} = \sum_{n_1=0}^{N-1} \sum_{n_2=0}^{N-1} x_{n_1,n_2} W^{-n_2} W^{2n_1 k_1} W^{(2k_2+1)n_2}, \qquad (7.60)$$

which is equivalent to premultiplying the input data samples by W^{-n_2} and computing a modified two-dimensional DFT which is a regular DFT along dimension k_1 and an odd DFT along dimension k_2. In order to simplify the computation of (7.60), we replace (7.60) by the following equivalent polynomial representation:

$$X_{n_1}(z) = \sum_{n_2=0}^{N-1} x_{n_1,n_2} W^{-n_2} z^{n_2} \qquad (7.61)$$

$$\bar{X}_{k_1}(z) \equiv \sum_{n_1=0}^{N-1} X_{n_1}(z) W^{2n_1 k_1} \text{ modulo } (z^N + 1) \qquad (7.62)$$

$$\bar{X}_{k_1,k_2} \equiv \bar{X}_{k_1}(z) \text{ modulo } (z - W^{2k_2+1}). \qquad (7.63)$$

It can be verified easily that (7.61–63) are a valid representation of (7.60) by substituting $X_{n_1}(z)$, defined by (7.61) into (7.62) and by replacing z by W^{2k_2+1}. We note that the definition of (7.62) modulo $(z^N + 1)$ is not necessary at this stage. However, this definition is valid because $z^N \equiv W^{N(2k_2+1)} = -1$ and because all the roots of $z^N + 1$ are given by $z = W^{2k_2+1}$ for $k_2 = 0, \ldots, N-1$. Since $2k_2 + 1$ is odd and since $N = 2^t$, the permutation $(2k_2 + 1)k_1$ modulo N maps all values of k_1 for $k_1 = 0, \ldots, N-1$. With this permutation, we obtain

$$\bar{X}_{(2k_2+1)k_1}(z) \equiv \sum_{n_1=0}^{N-1} X_{n_1}(z) W^{2(2k_2+1)n_1 k_1} \text{ modulo } (z^N + 1) \qquad (7.64)$$

$$\bar{X}_{(2k_2+1)k_1,k_2} \equiv \bar{X}_{(2k_2+1)k_1}(z) \text{ modulo } (z - W^{2k_2+1}). \qquad (7.65)$$

Equation (7.65) amounts to a simple substitution of W^{2k_2+1} for z. Therefore, we can replace W^{2k_2+1} by z in (7.64). This gives

$$\bar{X}_{(2k_2+1)k_1}(z) \equiv \sum_{n_1=0}^{N-1} X_{n_1}(z) z^{2n_1 k_1} \text{ modulo } (z^N + 1), \qquad (7.66)$$

which is recognized as a polynomial transform of length N defined modulo $(z^N + 1)$. This transform is computed without multiplications and with $2N^2 \log_2 N$ real additions using a radix-2 FFT-type algorithm.

When employing this method, the only multiplications required for the DFT of size $N \times N$ are the N^2 premultiplications by W^{-n_2} and the multiplications required for the evaluation of (7.65). The number of operations corresponding to (7.65) can be quantified by noting that $\bar{X}_{(2k_2+1)k_1}(z)$ represents N polynomials of N terms which can be defined by

$$\bar{X}_{(2k_2+1)k_1}(z) = \sum_{l=0}^{N-1} y_{k_1,l} z^l. \qquad (7.67)$$

Consequently, (7.65) becomes

$$\bar{X}_{(2k_2+1)k_1,k_2} = \sum_{l=0}^{N-1} y_{k_1,l} W^l W^{2lk_2}, \tag{7.68}$$

which represents N odd DFTs of length N.

Following this procedure, a DFT of size $N \times N$, with $N = 2^t$, is computed as shown in Fig. 7.4, with N^2 premultiplications by W^{-n_2}, one polynomial transform of length N, one permutation, and N reduced DFTs of N terms. The reduced DFTs can be calculated using any convenient FFT algorithm. If we assume that these DFTs are evaluated by a simple radix-2 FFT algorithm in which the trivial multiplications by ± 1 and $\pm j$ are counted as general multiplications, a length-N DFT is evaluated with $2N\log_2 N$ real multiplications and $3N\log_2 N$ real additions. In this case, the DFT of size $N \times N$ is computed with M_1 real multiplications and A_1 real additions, where

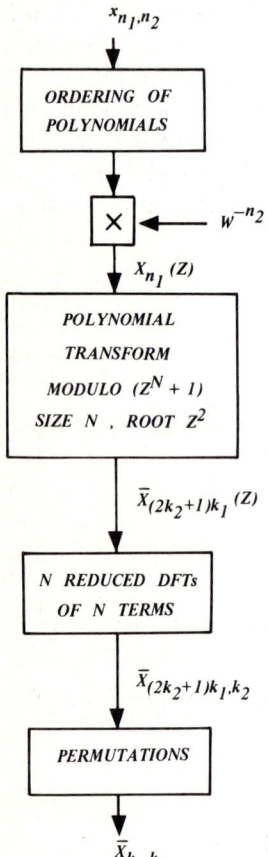

Fig. 7.4. Computation of a DFT of size $N \times N$ by polynomial transforms defined in modified rings of polynomials. $N = 2^t$

7.1 Computation of Multidimensional DFTs by Polynomial Transforms

$$M_1 = 2N^2(4 + \log_2 N) \tag{7.69}$$

$$A_1 = N^2(4 + 5\log_2 N). \tag{7.70}$$

If the DFT of size $N \times N$ is evaluated by the row-column method, the number of multiplications M_2 and additions A_2 become

$$M_2 = 4N^2 \log_2 N \tag{7.71}$$

$$A_2 = 6N^2 \log_2 N, \tag{7.72}$$

which demonstrates that the polynomial transform approach is better than the row-column method for $N > 16$ and reduces the number of multiplications by half for large transforms. It should also be noted that the polynomial transform approach reduces the number of additions by about 15% for large transforms.

Therefore, the foregoing polynomial transform method reduces significantly the number of arithmetic operations while retaining the structural simplicity of the row-column radix-2 FFT algorithm. In practice, the reduced DFTs will usually be calculated via the Rader-Brenner algorithm [7.5] because all complex multiplications are then implemented with only two real multiplications.

The number of arithmetic operations can be further reduced by modifying the ring structure for only part of the procedure. This may be realized by computing one or several stages with the method described in Sect. 7.1.2 and by completing the calculations with the modified ring technique. In the case of a one-stage process, the DFT $\bar{X}_{k_1,k_2}$ of size $N \times N$ is redefined by

$$\bar{X}_{k_1,k_2} = \sum_{n_1=0}^{N-1} \sum_{n_2=0}^{N-1} x_{n_1,n_2} W^{n_1 k_1} W^{n_2 k_2}$$
$$W = e^{-j2\pi/N} \qquad k_1, k_2 = 0, \ldots, N-1, \tag{7.73}$$

where W takes its usual representation. In polynomial notation, $\bar{X}_{k_1,k_2}$ becomes

$$\bar{X}_{k_1}(z) \equiv \sum_{n_1=0}^{N-1} X_{n_1}(z) W^{n_1 k_1} \text{ modulo } (z^N - 1) \tag{7.74}$$

$$X_{n_1}(z) = \sum_{n_2=0}^{N-1} x_{n_1,n_2} z^{n_2} \tag{7.75}$$

$$\bar{X}_{k_1,k_2} \equiv \bar{X}_{k_1}(z) \text{ modulo } (z - W^{k_2}). \tag{7.76}$$

For k_2 odd, $\bar{X}_{k_1,k_2}$ is calculated as in Sect. 7.1.2 by a polynomial transform of N terms defined modulo $(z^{N/2} + 1)$

$$\bar{X}^1_{k_2 k_1}(z) \equiv \sum_{n_1=0}^{N-1} X^1_{n_1}(z) z^{n_1 k_1} \text{ modulo } (z^{N/2} + 1) \tag{7.77}$$

$$X^1_{n_1}(z) \equiv X_{n_1}(z) \text{ modulo } (z^{N/2} + 1) \tag{7.78}$$

$$\bar{X}_{k_2 k_1, k_2} \equiv \bar{X}_{k_2 k_1}^1 \text{ modulo } (z - W^{k_2}), \qquad k_2 \text{ odd}. \tag{7.79}$$

For k_2 even, $\bar{X}_{k_1, k_2}$ reduces to a DFT of size $N \times (N/2)$ which is computed using a ring translation technique

$$X_{n_1}^2(z) = \sum_{n_2=0}^{N/2-1} (x_{n_1, n_2} + x_{n_1, n_2 + N/2}) W^{-n_2} z^{n_2} \tag{7.80}$$

$$\bar{X}_{(k_2+1)k_1}(z) \equiv \sum_{n_1=0}^{N-1} X_{n_1}^2(z) z^{n_1 k_1} \text{ modulo } (z^{N/2} + 1) \tag{7.81}$$

$$\bar{X}_{(k_2+1)k_1, k_2} \equiv \bar{X}_{(k_2+1)k_1}(z) \text{ modulo } (z - W^{k_2+1}), \qquad k_2 \text{ even}, \tag{7.82}$$

which indicates that the DFT $\bar{X}_{k_1, k_2}$ of size $N \times N$ is computed as shown in Fig. 7.5 with only $N^2/2$ premultiplications by W^{-n_2}, plus two polynomial transforms

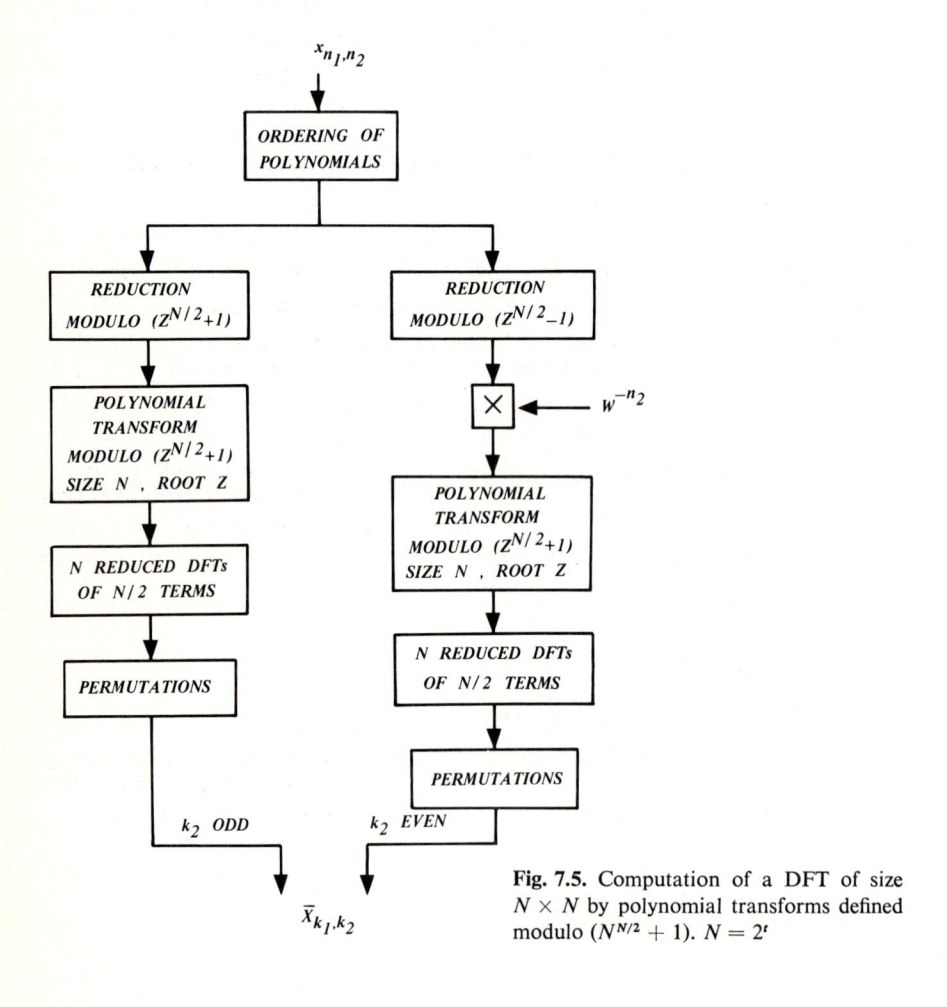

Fig. 7.5. Computation of a DFT of size $N \times N$ by polynomial transforms defined modulo $(N^{N/2} + 1)$. $N = 2^t$

defined modulo ($z^{N/2} + 1$) and $2N$ reduced DFTs of size $N/2$. When the reduced DFTs are computed by a simple radix-2 FFT algorithm, the number of real multiplications M_3 and real additions A_3 become

$$M_3 = 2N^2(2 + \log_2 N) \tag{7.83}$$

$$A_3 = N^2(2 + 5\log_2 N). \tag{7.84}$$

With this scheme, an additional reduction in number of arithmetic operations is obtained at the expense of using two polynomial transforms instead of one. The same method can be used recursively by reducing the DFT of size $N \times (N/2)$ into a DFT of size $(N/2) \times (N/2)$ plus $N/2$ reduced DFTs of N terms and, with additional stages, into DFTs of sizes $(N/4) \times (N/4)$, $(N/8) \times (N/8)$, ..., each additional stage reducing the number of arithmetic operations at the expense of an additional number of different polynomial transforms. When the decomposition is complete, the method becomes identical to that described in Sect. 7.1.2. Thus, there is considerable flexibility in trading structural complexity for computational complexity by selecting the number of stages.

7.2 DFTs Evaluated by Multidimensional Correlations and Polynomial Transforms

We have seen in the preceding sections that multidimensional DFTs can be efficiently partitioned by polynomial transforms into one-dimensional DFTs and reduced DFTs. This method is mainly applicable to DFTs having common factors in two or more dimensions and therefore does not apply readily to one-dimensional DFTs. In this section, we shall present a second way of computing DFTs by polynomial transforms [7.1, 2, 9]. This method is based on the decomposition of a composite DFT into multidimensional correlations via the Winograd [7.7] algorithm and on the computation of these multidimensional correlations by polynomial transforms when they have common factors in several dimensions. This method is applicable in general to multidimensional DFTs and also to some one-dimensional DFTs.

7.2.1 Derivation of the Algorithm

We consider a two-dimensional DFT $\bar{X}_{k_1,k_2}$ of size $N_1 \times N_2$. This DFT may either be a genuine two-dimensional DFT or a one-dimensional DFT of length $N_1 N_2$, with N_1 and N_2 mutually prime, which has been mapped into a two-dimensional DFT structure by using Good's algorithm [7.6] (Sect. 5.3). $\bar{X}_{k_1,k_2}$ is defined by

$$\bar{X}_{k_1,k_2} = \sum_{n_1=0}^{N_1-1} \sum_{n_2=0}^{N_2-1} x_{n_1,n_2} W_1^{n_1 k_1} W_2^{n_2 k_2}$$

$$W_1 = e^{-j2\pi/N_1}, \qquad W_2 = e^{-j2\pi/N_2}, \qquad j = \sqrt{-1}$$
$$k_1 = 0, ..., N_1 - 1 \qquad\qquad k_2 = 0, ..., N_2 - 1. \tag{7.85}$$

In order to simplify the presentation, we shall assume that N_1 and N_2 are prime. For $k_2 = 0$, $\bar{X}_{k_1,k_2}$ becomes a DFT of length N_1

$$\bar{X}_{k_1,0} = \sum_{n_1=0}^{N_1-1} \left(\sum_{n_2=0}^{N_2-1} x_{n_1,n_2} \right) W_1^{n_1 k_1} \qquad k_1 = 0, ..., N_1 - 1. \tag{7.86}$$

For $k_2 \neq 0$, we consider first the case corresponding to $k_1 = 0$. Then, $\bar{X}_{k_1,k_2}$ becomes

$$\bar{X}_{0,k_2} = \sum_{n_2=0}^{N_2-1} \left(\sum_{n_1=0}^{N_1-1} x_{n_1,n_2} \right) W_2^{n_2 k_2} \qquad k_2 = 1, ..., N_2 - 1 \tag{7.87}$$

and, since $1 + W_2 + W_2^2 + ... + W_2^{N-1} = 0$,

$$\bar{X}_{0,k_2} = \sum_{n_2=1}^{N_2-1} \left[\sum_{n_1=0}^{N_1-1} (x_{n_1,n_2} - x_{n_1,0}) \right] W_2^{n_2 k_2}. \tag{7.88}$$

Since N_2 is a prime, and $n_2, k_2 \neq 0$, we can map $\bar{X}_{0,k_2}$ into a correlation of length $N_2 - 1$ by using Rader's algorithm [7.3] with

$$n_2 \equiv g^{u_2} \text{ modulo } N_2$$
$$k_2 \equiv g^{v_2} \text{ modulo } N_2 \qquad u_2, v_2 = 0, ..., N_2 - 2 \tag{7.89}$$

$$\bar{X}_{0,g^{v_2}} = \sum_{u_2=0}^{N_2-2} \left[\sum_{n_1=0}^{N_1-1} (x_{n_1,g^{u_2}} - x_{n_1,0}) \right] W_2^{g^{u_2+v_2}}, \tag{7.90}$$

where g is a primitive root of order $N_2 - 1$ modulo N_2. When $k_1, k_2 \neq 0$, $\bar{X}_{k_1,k_2}$ becomes a two-dimensional correlation of size $(N_1 - 1) \times (N_2 - 1)$

$$\bar{X}_{h^{v_1},g^{v_2}} = \sum_{u_1=0}^{N_1-2} \sum_{u_2=0}^{N_2-2} (x_{h^{u_1},g^{u_2}} - x_{h^{u_1},0} - x_{0,g^{u_2}} + x_{0,0}) W_1^{h^{u_1+v_1}} W_2^{g^{u_2+v_2}}$$
$$n_1 \equiv h^{u_1} \text{ modulo } N_1, \qquad k_1 \equiv h^{v_1} \text{ modulo } N_1. \tag{7.91}$$

Using the Winograd method [7.7], the DFT of size $N_1 \times N_2$ is calculated by nesting the DFTs of lengths N_1 and N_2, which is equivalent to computing the two-dimensional correlation (7.91) via the Agarwal-Cooley nesting algorithm [7.10] (Sect. 3.3.1). We have seen, however, in Chap. 6 that when a two-dimensional convolution or correlation has common factors in several dimensions, the number of arithmetic operations can be significantly reduced by replacing the conventional nesting structure with a polynomial transform method. Thus, one can expect to reduce the computational complexity of a DFT of size $N_1 \times N_2$ if the derived two-dimensional correlation is evaluated by polynomial trans-

7.2 DFTs Evaluated by Multidimensional Correlations   203

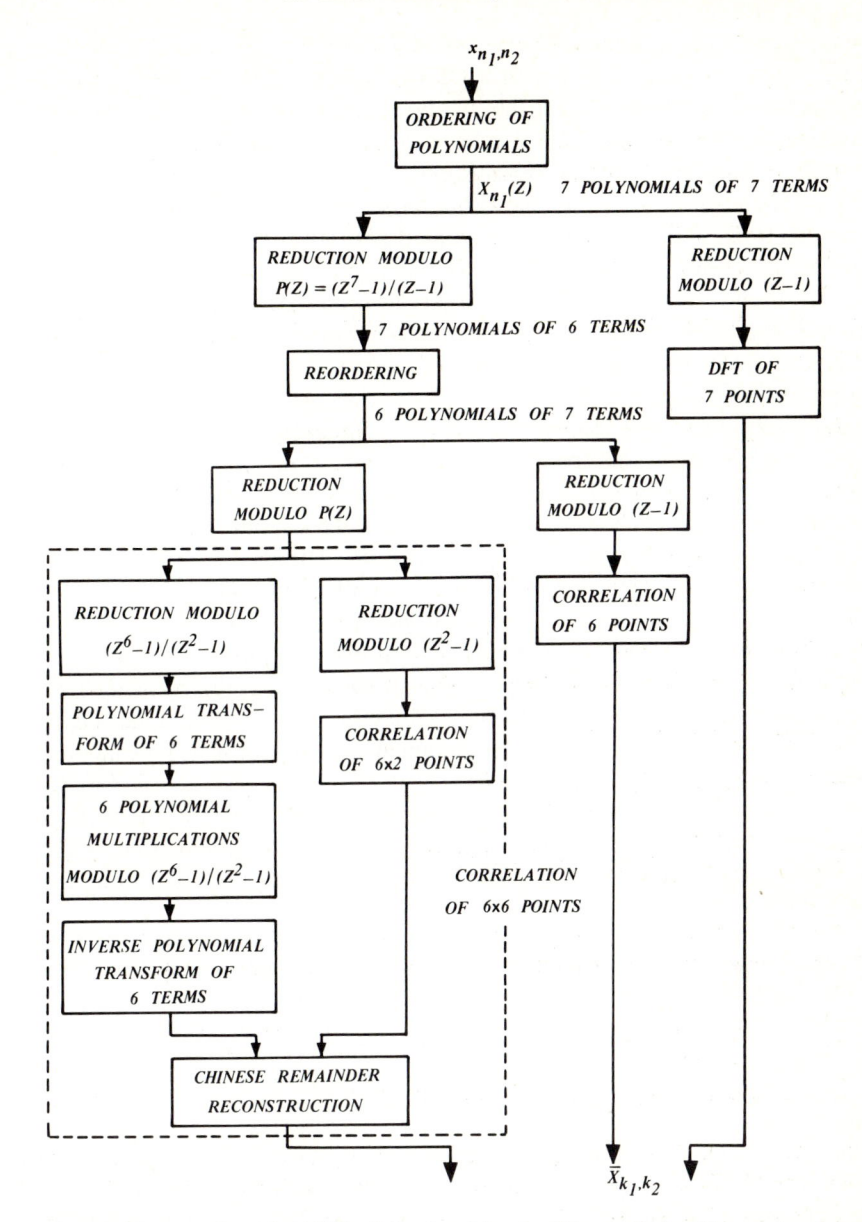

Fig. 7.6. Computation of a DFT of size 7×7 by the Winograd algorithm and polynomial transforms

forms. The same technique can also be applied recursively to accommodate the case of more than two factors or factors that are composite (Sect. 3.3.1).

When $N_1 = N_2$, all factors in both dimensions are common and a polynomial transform mapping of the two-dimensional correlation is always realizable. We illustrate this method in Fig. 7.6 for a DFT of size 7×7. Since the DFT of 7 terms is reduced by Rader's algorithm to one multiplication and one correlation of 6 terms, the DFT of size 7×7 can be mapped into one DFT of 7 terms, one correlation of 6 terms, and one correlation of size 6×6. If the correlations of 6 terms are then calculated by an algorithm requiring 8 complex multiplications, the complete DFT of size 7×7 is evaluated with 81 complex multiplications using the Winograd nesting algorithm. In this approach, the correlation of size 6×6 is computed via nesting, with 64 multiplications. However, if the (6×6)-point correlation is evaluated by polynomial transforms, only 52 complex multiplications are required and the DFT of size 7×7 is evaluated with only 69 multiplications instead of 81.

Thus, the polynomial transform mapping of multidimensional correlations provides an alternate solution to the polynomial transform DFT mapping method discussed in Sect. 7.1. It should be noted, however, that the latter method is always more efficient whenever it is applicable. This result is due to the fact that the polynomial transform mapping of the correlations is based on smaller extension fields. This point can be illustrated by noting that a DFT of size 7×7 can be computed with only 65 multiplications by the method of Sect. 7.1, as opposed to 69 multiplications with the method discussed here. Therefore, the utility of the method based on the polynomial transform mapping of multidimensional correlations is limited to the evaluation of multidimensional DFTs having no common factors in the different dimensions and for which the method of Sect. 7.1 is not applicable.

If we consider, for instance, a two-dimensional DFT of size 7×9 (or a one-dimensional DFT of dimension 63), this DFT cannot be computed by the method of Sect. 7.1 because 7 and 9 have no common factors. Employing the

Table 7.6. Number of real operations for complex DFTs computed by multidimensional correlations and polynomial transforms. Trivial multiplications by $\pm 1, \pm j$ are not counted

DFT size	Number of multiplications	Number of additions	Multiplications per point	Additions per point
63	172	1424	2.73	22.60
80	188	1340	2.35	16.75
504	1380	14668	2.74	29.10
1008	3116	34956	3.09	34.68
5×5	64	452	2.56	18.08
7×7	136	1300	2.78	26.53
9×9	216	1816	2.67	22.42
16×16	496	4752	1.98	18.56
63×63	11680	208904	2.94	52.63
$5 \times 5 \times 5$	346	3490	2.77	27.92
$7 \times 7 \times 7$	1000	14048	3.04	40.96

Winograd algorithm, this DFT is evaluated by nesting a DFT of 9 terms with a DFT of 7 terms. Using Rader's algorithm, the DFT of 7 terms is converted into a process with one multiplication and one correlation of 6 terms, while the DFT of 9 terms is reduced to 5 multiplications and one correlation of 6 terms. Thus, the Winograd algorithm computes the DFT of size 7×9 as 5 DFTs of 7 terms, one correlation of 6 terms, and one correlation of size 6×6, with a total of 198 real multiplications. Alternatively, if the correlation of size 6×6 is computed by polynomial transforms, the total number of multiplications is reduced to 174.

In Table 7.6, we tabulate the number of real operations for complex DFTs computed by the polynomial transform mapping of multidimensional correlations. It can be seen by comparison with Table 5.3 that this method requires significantly fewer arithmetic operations than the conventional Winograd algorithm. In the case of a DFT of 1008 points, for example, the numbers of operations are reduced to 3116 real multiplications and 34956 real additions, as opposed to 3548 multiplications and 34668 additions for the Winograd algorithm.

7.2.2 Combination of the Two Polynomial Transform Methods

For large multidimensional DFTs, the two polynomial transform methods can be combined by converting the multidimensional DFT into a set of one-dimensional DFTs by use of a polynomial transform mapping and, then, by computing these one-dimensional DFTs via a multidimensional correlation polynomial transform mapping. With this technique, a DFT of size 63×63, for instance, is calculated by nesting DFTs of size 7×7 and 9×9 evaluated by the first polynomial transform method. Hence, the DFT of size 7×7 is partitioned into 1 multiplication plus 8 correlations of 6 terms, and the DFT of size 9×9 is mapped into 33 multiplications plus 12 correlations of 6 terms. Thus, the DFT of size 63×63 is computed with 33 multiplications, 276 correlations of 6 terms, and 96 correlations of size 6×6. When the (6×6)-point correlations are computed by polynomial transforms, the DFT of size 63×63 is calculated with only 11344 real multiplications as opposed to 13648 multiplications when the first polynomial transform method is used alone and 19600 multiplications for the conventional Winograd nesting algorithm. It should be noted that combining the two polynomial transform methods also reduces the number of additions.

Table 7.7 lists the number of real operations for complex DFTs computed by combining the two polynomial transform methods with the split nesting technique. It can be seen by comparison with Table 7.1 that the combined polynomial transform method requires about half the number of multiplications of the first polynomial transform method for large transforms. In practice, the number of multiplications required by this method is always very small, as exemplified by a DFT of size 1008×1008 which is calculated with only 3.39 real multiplications per point or about one complex multiplication per point.

It should be noted however that this low computation requirement is ob-

Table 7.7. Number of real operations for complex DFTs computed by combining the two polynomial transform methods. Trivial multiplications by ± 1, $\pm j$ are not counted

DFT size	Number of multiplications	Number of additions	Multiplications per point	Additions per point
80	188	1340	2.35	16.75
240	596	4980	2.48	20.75
504	1380	14668	2.74	29.10
840	2580	24804	3.07	29.53
1008	3116	32244	3.09	31.99
2520	8340	95532	3.31	37.90
5040	17732	208108	3.52	41.29
63 × 63	11344	193480	2.86	48.75
80 × 80	16944	231344	2.65	36.15
120 × 120	35632	553392	2.47	38.43
240 × 240	153904	2542896	2.67	44.15
504 × 504	726064	15621424	2.86	61.50
1008 × 1008	3449024	71455456	3.39	70.33
80 × 80 × 80	1451616	28134656	2.84	54.96
120 × 120 × 120	4312512	103038528	2.50	57.85
240 × 240 × 240	39221088	925433712	2.84	66.94

tained at the expense of a relatively complex structure and, thus, it is expected that most practical applications will use the first polynomial transform method with polynomial transforms of sizes which are powers of two. In this case, the number of operations is still substantially lower than with the conventional methods, but the structure of the algorithm remains simple and comparable in complexity to a conventional FFT algorithm implemented with the row-column method.

7.3 Comparison with the Conventional FFT

The calculation of a DFT by polynomial transforms is based upon the use of roots of unity in fields of polynomials. Conversely, the polynomial transforms can also be viewed as DFTs defined in fields of polynomials. If we consider the simple scheme of Sect. 7.1.5, a DFT of size $N \times N$, with $N = 2^t$, is evaluated with $2N^2$ multiplications by powers of W, plus one polynomial transform of N terms defined modulo $(z^N + 1)$ and N DFTs of N terms. This method makes use of multiplications by W^{-n_2} and W^l to translate rings of polynomials modulo $(z^N - 1)$ into fields of polynomials modulo $(z^N + 1)$ and is equivalent to a row-column FFT method in which the N first DFTs of N terms are replaced by a polynomial transform. Therefore, this approach eliminates the multiplications in the N first DFTs of the row-column method, thereby saving $(N^2/2) \log_2 N$ complex multiplications, or $2N^2 \log_2 N$ real multiplications and $N^2 \log_2 N$ real additions, while retaining the simple structure of the FFT implementation. The method given in Sect. 7.1.2 is essentially a generalization of this technique, which is derived by using a complete decomposition to eliminate the multiplications by W^{-n_2}.

When a large multidimensional DFT is evaluated by combining polynomial transforms and nesting, as in Sects. 7.1.4 and 7.2.1, this method can be considered as a generalization of the Winograd algorithm in which small multidimensional DFTs and correlations having common factors in several dimensions are systematically partitioned into one-dimensional DFTs and correlations by polynomial transform mappings.

In practice, significant computational savings are obtained by computing DFTs by polynomial transforms. This can be seen by comparing the data given in Tables 7.1 and 7.7 with those in Table 4.5 which corresponds to two-dimensional DFTs calculated by the Rader-Brenner FFT algorithm and the row-column method. It can be seen that the number of multiplications is reduced by a factor of about 2 for large DFTs computed by the first polynomial transform method used alone and by a factor of about 4 when the two polynomial transform methods are combined. In both cases the number of additions is comparable to and sometimes smaller than the number corresponding to the FFT approach.

A comparison with the Winograd-Fourier transform algorithm also demonstrates a significant advantage in favor of polynomial transform methods. For example, a DFT of size 1008×1008 is computed by the WFTA altorithm with 6.25 real multiplications and 91.61 additions per point. This contrasts with the first polynomial transform technique which requires 7.67 multiplications and 54.58 additions per point for a DFT of size 1024×1024 and the combination of the two polynomial transform methods which requires 3.39 multiplications and 70.33 additions per point for a DFT of size 1008×1008.

7.4 Odd DFT Algorithms

When a DFT is evaluated by polynomial transforms, it is partitioned into one-dimensional DFTs, reduced DFTs, correlations, and polynomial products. The correlations and polynomial products are formed by the application of the Rader algorithm (Sect. 5.2) and are therefore implemented with two real multiplications per complex multiplication. These correlations and polynomial products can then be computed by the algorithms of Sects. 3.7.1 and 3.7.2 by replacing the real data with complex data and by inverting one of the input sequences. The corresponding number of complex operations for this process are given by Tables 3.1 and 3.2.

The small DFT algorithms are given in Sect. 5.5 for $N = 2, 3, 4, 5, 7, 8, 9,$

16. For $N = 2^t$, with $N > 16$, the one-dimensional DFTs can be computed by the Rader-Brenner algorithm (Sect. 4.3) and the corresponding number of operations is given in Table 4.3.

For the reduced DFT algorithms, we have already seen in Sect. 7.1.1 that, when N is a prime, the reduced DFTs become correlations of $N - 1$ terms. Thus, these reduced DFTs may be computed by the algorithms of Sect. 3.7.1 with the corresponding number of complex operations given in Table 3.1. Large odd DFTs corresponding to $N = 2^t$ can be computed by the Rader-Brenner algorithm as shown in Sect. 4.3 with an operation count given in Table 4.4.

We define in Sects. 7.4.1–4 reduced DFT algorithms for $N = 4, 8, 9, 16$. These algorithms are derived from the short DFT algorithms of Sect. 5.5 and compute $q^{t-1}(q-1)$ output samples of a DFT of length $N = q^t$. The reduced DFT is defined by

Table 7.8. Number of real operations for complex DFTs and reduced DFTs. Trivial multiplications by ± 1, $\pm j$ are given between parentheses

Size N	Number of multiplications	Number of additions	
2	4 (4)	4	
3	6 (2)	12	
4	8 (8)	16	
5	12 (2)	34	
7	18 (2)	72	
8	16 (12)	52	DFTs
9	22 (2)	88	
16	36 (16)	148	
32	104 (36)	424	
64	272 (76)	1104	
128	672 (156)	2720	
256	1600 (316)	6464	
512	3712 (636)	14976	
1024	8448 (1276)	34048	
3	4 (0)	8	
4	4 (4)	4	
5	10 (0)	30	
7	16 (0)	68	
8	8 (4)	20	Reduced
9	16 (0)	56	DFTs
16	20 (4)	64	
32	68 (20)	212	
64	168 (40)	552	
128	400 (80)	1360	
256	928 (160)	3232	
512	2112 (320)	7488	
1024	4736 (640)	17024	

7.4 Odd DFT Algorithms

$$\bar{X}_k = \sum_{n=0}^{q^{t-1}(q-1)-1} x_n W^{nk}, \quad 1 \leq k \leq N-1 \quad k \not\equiv 0 \text{ modulo } q$$

$$W = e^{-j2\pi/N}, \quad j = \sqrt{-1}, \tag{7.92}$$

where the input sequence is labelled x_n, the output sequence is labelled $\bar{X}_k$, and the last q^{t-1} input samples are zero. Input and output additions must be executed in the specified index numerical order. Table 7.8 summarizes the number of real operations for various complex DFTs and reduced DFTs used as building blocks in the polynomial transform algorithms. Trivial multiplications by $\pm 1, \pm j$ are given in parentheses.

7.4.1 Reduced DFT Algorithm. $N = 4$

2 complex multiplications (2), 2 complex additions

$m_0 = 1 \cdot x_0$
$m_1 = -j \cdot x_1$
$\bar{X}_1 = m_0 + m_1$
$\bar{X}_3 = m_0 - m_1.$

7.4.2 Reduced DFT Algorithm. $N = 8$

4 complex multiplications (2), 10 complex additions. $u = \pi/4$

$m_0 = 1 \cdot x_0$ $\qquad m_1 = (x_1 - x_3) \cos u$
$m_2 = -j \cdot x_2$ $\qquad m_3 = -j(x_1 + x_3) \sin u$
$s_1 = m_0 + m_1$ $\qquad s_2 = m_0 - m_1$
$s_3 = m_2 + m_3$ $\qquad s_4 = m_2 - m_3$
$\bar{X}_1 = s_1 + s_3$ $\quad \bar{X}_3 = s_2 - s_4$ $\quad \bar{X}_5 = s_2 + s_4$ $\quad \bar{X}_7 = s_1 - s_3.$

7.4.3 Reduced DFT Algorithm. $N = 9$

8 complex multiplications (0). 28 complex additions. $u = 2\pi/9$

$t_1 = x_4 + x_5 \qquad\qquad t_2 = x_4 - x_5$
$m_0 = (x_0 + x_0 - x_3)/2$
$m_1 = \left(\dfrac{2 \cos u - \cos 2u - \cos 4u}{3}\right)(x_1 - x_2)$
$m_2 = \left(\dfrac{\cos u + \cos 2u - 2 \cos 4u}{3}\right)(x_2 - t_1)$
$m_3 = \left(\dfrac{\cos u - 2 \cos 2u + \cos 4u}{3}\right)(t_1 - x_1)$
$m_4 = -jx_3 \sin 3u$

$$m_5 = -j(x_1 + x_2) \sin u$$
$$m_6 = -j(x_2 + t_2) \sin 4u$$
$$m_7 = j(x_1 - t_2) \sin 2u$$

$$s_2 = m_1 + m_2 + m_0 \qquad\qquad s_3 = -m_2 + m_3 + m_0$$
$$s_4 = -m_1 - m_3 + m_0 \qquad\qquad s_5 = m_4 + m_5 + m_6$$
$$s_6 = -m_6 + m_7 + m_4 \qquad\qquad s_7 = -m_5 - m_7 + m_4$$

$$\bar{X}_1 = s_2 + s_5 \qquad \bar{X}_2 = s_3 - s_6 \qquad \bar{X}_4 = s_4 + s_7$$
$$\bar{X}_5 = s_4 - s_7 \qquad \bar{X}_7 = s_3 + s_6 \qquad \bar{X}_8 = s_2 - s_5.$$

7.4.4 Reduced DFT Algorithm. $N = 16$

10 complex multiplications (2), 32 complex additions $u = \pi/8$

$$t_1 = x_1 + x_7 \qquad t_2 = x_1 - x_7 \qquad t_3 = x_3 + x_5 \qquad t_4 = x_5 - x_3$$

$$m_0 = 1 \cdot x_0 \qquad m_1 = (x_2 - x_6)\cos 2u \qquad m_2 = (t_2 + t_4) \cos 3u$$
$$m_3 = (\cos u + \cos 3u)\, t_2 \qquad\qquad\qquad m_4 = (\cos 3u - \cos u)\, t_4$$
$$m_5 = -j\, x_4 \qquad m_6 = -j(x_2 + x_6) \sin 2u \qquad m_7 = -j(t_1 + t_3) \sin 3u$$
$$m_8 = j(\sin 3u - \sin u)\, t_1 \qquad\qquad\qquad m_9 = -j(\sin u + \sin 3u)\, t_3$$

$$s_1 = m_0 + m_1 \qquad\qquad s_2 = m_0 - m_1 \qquad\qquad s_3 = m_3 - m_2$$
$$s_4 = m_4 - m_2 \qquad\qquad s_5 = s_1 + s_3 \qquad\qquad s_6 = s_1 - s_3$$
$$s_7 = s_2 + s_4 \qquad\qquad s_8 = s_2 - s_4 \qquad\qquad s_9 = m_5 + m_6$$
$$s_{10} = m_5 - m_6 \qquad\qquad s_{11} = m_7 + m_8 \qquad\qquad s_{12} = m_7 - m_9$$
$$s_{13} = s_9 + s_{11} \qquad\qquad s_{14} = s_9 - s_{11} \qquad\qquad s_{15} = s_{10} + s_{12}$$
$$s_{16} = s_{10} - s_{12}$$

$$\bar{X}_1 = s_5 + s_{13} \qquad \bar{X}_3 = s_8 - s_{16} \qquad \bar{X}_5 = s_7 + s_{15}$$
$$\bar{X}_7 = s_6 - s_{14} \qquad \bar{X}_9 = s_6 + s_{14} \qquad \bar{X}_{11} = s_7 - s_{15}$$
$$\bar{X}_{13} = s_8 + s_{16} \qquad \bar{X}_{15} = s_5 - s_{13}.$$

8. Number Theoretic Transforms

Most of the fast convolution techniques discussed so far are essentially algebraic methods which can be implemented with any type of arithmetic. In this chapter, we shall show that the computation of convolutions can be greatly simplified when special arithmetic is used. In this case, it is possible to define number theoretic transforms (NTT) which have a structure similar to the DFT, but with complex exponential roots of unity replaced by integer roots and all operations defined modulo an integer. These transforms have the circular convolution property and can, in some instances, be computed using only additions and multiplications by a power of two. Hence, significant computational savings can be realized if NTTs are executed in computer structures which efficiently implement modular arithmetic.

We begin by presenting a general definition of NTTs and by introducing the two most important NTTs, the Mersenne transform and the Fermat number transform (FNT). Then, we generalize our definition of the NTT to include complex transforms and pseudo transforms. Finally, we conclude the chapter by discussing several implementation issues and establishing a theoretical relationship between NTTs and polynomial transforms.

8.1 Definition of the Number Theoretic Transforms

Let x_m and h_n be two N-point integer sequences. Our objective is to compute the circular convolution y_l of dimension N

$$y_l = \sum_{n=0}^{N-1} h_n x_{l-n}. \tag{8.1}$$

In most practical cases, h_n and x_m are not sequences of integers, but it is always possible, by proper scaling, to reduce these sequences to a set of integers. We shall first assume that all arithmetic operations are performed modulo a prime number q, in the field $GF(q)$. If h_n and x_m are so scaled that $|y_l|$ never exceeds $q/2$, y_l has the same numerical value modulo q that would be obtained in normal arithmetic.

Under these conditions, the calculation of y_l can be simplified by introducing a number theoretic transform [8.1–3] having the same structure as a DFT, but with the complex exponentials replaced by an integer g and with all operations performed modulo q. The direct NTT of h_n is, thus,

$$\bar{H}_k \equiv \sum_{n=0}^{N-1} h_n g^{nk} \text{ modulo } q, \qquad (8.2)$$

with a similar relation for the NTT $\bar{X}_k$ of x_m. Since q is a prime, N has an inverse N^{-1} modulo q, and we define an inverse transform as

$$a_l \equiv N^{-1} \sum_{k=0}^{N-1} \bar{H}_k g^{-lk} \text{ modulo } q, \qquad (8.3)$$

where

$$N N^{-1} \equiv 1 \text{ modulo } q. \qquad (8.4)$$

Note that, since q is a prime, g has also an inverse g^{-1} modulo q. Thus, the notation g^{-lk} is valid.

We would now like to establish the conditions which must be met for the transform (8.2) to support circular convolution. Computing the NTTs $\bar{H}_k$ and $\bar{X}_k$ of h_n and x_m, multiplying $\bar{H}_k$ by $\bar{X}_k$, and evaluating the inverse transform a_l of $\bar{H}_k \bar{X}_k$ yields

$$a_l \equiv N^{-1} \sum_{n=0}^{N-1} \sum_{m=0}^{N-1} h_n x_m \sum_{k=0}^{N-1} g^{(n+m-l)k} \text{ modulo } q. \qquad (8.5)$$

Let

$$S \equiv \sum_{k=0}^{N-1} g^{(n+m-l)k} \text{ modulo } q. \qquad (8.6)$$

If the NTTs support convolution, then (8.5) must reduce to (8.1) and we must have $S \equiv N$ for $t = n + m - l \equiv 0$ modulo N and $S \equiv 0$ for $t \not\equiv 0$ modulo N. The first condition means that the exponents of g must be defined modulo N, and this implies that

$$g^N \equiv 1 \text{ modulo } q. \qquad (8.7)$$

For $t \not\equiv 0$ modulo N, (8.6) becomes

$$(g^t - 1)S \equiv g^{Nt} - 1 \equiv 0 \text{ modulo } q. \qquad (8.8)$$

Thus, $S \equiv 0$ provided $g^t - 1 \not\equiv 0$ modulo q for $t \not\equiv 0$ modulo N. This implies that g must be a root of unity of order N modulo q, that is to say, g must be an integer such that N is the smallest nonzero integer for which $g^N \equiv 1$ modulo q. Hence the following existence theorem:

Theorem 8.1: A NTT of length N and root g supports circular convolution when defined modulo a prime q if and only if g is a root of unity of order N modulo q.

An immediate consequence of this theorem is that the inverse transform defined by (8.3) is indeed the inverse transform. This follows from theorem 8.1 by choosing the sequence x_m such that $x_0 = 1$, $x_m = 0$ for $m \neq 0$.

Theorem 8.1 also allows one to specify the size of NTTs defined modulo a prime q. We know from Sect. 2.1.3 that there are always primitive roots of order $q - 1$ modulo q and that all the roots must be of order N, with $N|(q - 1)$. Furthermore, the number of roots of order N is given by $\phi(N)$, Euler's totient function. This implies the following theorem.

Theorem 8.2: A NTT of length N and defined modulo a prime q exists if and only if $N|(q - 1)$. This NTT supports circular convolution.

Thus, for any prime q, we are able to find the sizes N for which there is an NTT. The NTT is then completely defined, provided that we can find a root of order N. This is done by using the methods given in Sect. 2.1.3.

8.1.1 General Properties of NTTs

Previously, we have restricted our discussion to NTTs defined modulo a prime q. In practice, this definition is unnecessarily restrictive, since NTTs can also be defined in a ring of numbers modulo an integer q, where q is a composite number. In order to specify the existence conditions for such NTTs, we proceed once again as above, by defining this new NTT class by (8.2) and the inverse NTT by (8.3) and (8.4).

Note, however, that in order to define the inverse NTT, we need the inverses N^{-1} and g^{-1} of N and g. Since q is composite, such inverses exist if and only if N and g are mutually prime with q. Now if we try to establish the circular convolution property by (8.5) and (8.6), we must have, as above,

$$g^N \equiv 1 \text{ modulo } q. \tag{8.9}$$

This condition ensures the existence of an inverse g^{-1} of g, since

$$g\, g^{N-1} \equiv 1 \text{ modulo } q. \tag{8.10}$$

The last condition which must be satisfied to support the circular convolution property corresponds to (8.8) and implies not only that g is a root of order N modulo q, but also, since q is composite, that $[(g^t - 1), q] = 1$. Hence, the following existence theorem may be defined.

Theorem 8.3: An NTT of length N and root g, defined modulo a composite integer q, supports circular convolution if and only if the following conditions are met:

$g^N \equiv 1$ modulo q

$NN^{-1} \equiv 1$ modulo q

$[(g^t - 1), q] = 1$ for $t = 1, ..., N - 1$.

Note that the condition $[(g^t - 1), q)] = 1$ is stronger than just stating that g must be a root of order N modulo q. This can be seen, for instance, in the case corresponding to $q = 15$. In this case, 2 is a root of order 4 modulo 15, since the 4 powers of two $2^0, 2^1, 2^2, 2^3$ are all distinct and $2^4 \equiv 1$ modulo 15. However we have $2^2 - 1 = 3$, and 3 is not relatively prime to 15. In practice, the condition $[(g^t - 1), q] = 1$ can be replaced by a more restrictive condition by noting that it corresponds to the need to ensure that

$$S \equiv \sum_{k=0}^{N-1} g^{tk} \equiv 0 \text{ modulo } q, \quad \text{for} \quad t = 1, ..., N - 1. \tag{8.11}$$

The following theorem, due to Erdelsky [8.4], specifies the conditions required for the existence of NTTs which support circular convolution.

Theorem 8.4: An NTT of length N and root g, defined modulo a composite integer q, supports circular convolution if and only if the following conditions are met:

$g^N \equiv 1$ modulo q

$NN^{-1} \equiv 1$ modulo q

$[(g^d - 1), q] = 1$ for every integer d such that N/d is a prime.

(Or equivalently $\sum_{k=0}^{N-1} g^{dk} \equiv 0$ modulo q for every d such that N/d is a prime).

Proof of this theorem can be found in [8.4].

Consider now the simplest composite numbers, which correspond to a power of a prime

$$q = q_1^{r_1}, \quad q_1 \text{ prime.} \tag{8.12}$$

In this case, the condition $NN^{-1} \equiv 1$ implies that N must be relatively prime with q_1. Moreover, g is of necessity relatively prime to q_1, because the condition $g^N \equiv 1$ modulo q implies $g^N \equiv 1$ modulo q_1. Therefore, for each g relatively prime to q, we have, by Euler's theorem (theorem 2.3),

$$g^{\phi(q)} \equiv 1 \text{ modulo } q. \tag{8.13}$$

And, since $\phi(q) = q_1^{r_1-1}(q_1 - 1)$,

$$N | (q_1 - 1). \tag{8.14}$$

We now can demonstrate the following theorem which establishes the existence of an NTT defined modulo $q_1^{r_1}$ and of length $q_1 - 1$.

Theorem 8.5: Given an NTT which supports circular convolution when defined modulo q_1, q_1 prime, with the root g_1 and the length $q_1 - 1$, there is always an NTT of length $q_1 - 1$ when defined modulo $q_1^{r_1}$. This NTT supports circular convolution and its root is $g = g_1^{q_1^{r_1-1}}$.

In order to demonstrate this theorem, we note first that the existence of the NTT defined modulo q_1 implies that $(g_1, q_1) = 1$. Then, Euler's theorem implies that $g^{q_1-1} = g_1^{q_1^{r_1-1}(q_1-1)} \equiv 1$ modulo $q_1^{r_1}$. Moreover, since $q_1 - 1$ has no common factors with q_1, $(q_1 - 1)$ is mutually prime with q_1 and $q_1^{r_1}$ and has therefore an inverse modulo $q_1^{r_1}$. We also note that the existence of the NTT defined modulo q_1 implies that $[(g_1^s - 1), q_1] = 1$ for $s = 1, ..., q_1 - 2$. Thus, $g_1^s - 1$ is not a multiple of q_1, and since, by Fermat's theorem (theorem 2.4) $g_1^{q_1} \equiv g_1$ modulo q_1, we have, by systematically replacing g_1 by $g_1^{q_1}$, $g_1^s - 1 \equiv g_1^{sq_1^{r_1-1}} - 1$ modulo $q_1 \equiv g^s - 1$. This means that $g^s - 1$ has no common factors with q_1 for $s = 1, ..., q_1 - 2$. Hence the three conditions of theorem 8.3 are met and this completes the proof of theorem 8.5.

We can now consider any composite integer q given by its unique prime power factorization.

$$q = q_1^{r_1} q_2^{r_2} ... q_i^{r_i} ... q_e^{r_e}. \tag{8.15}$$

By the Chinese remainder theorem (theorem 2.1), the N-length circular convolution modulo q can be calculated by evaluating separately the N-length convolutions modulo each $q_i^{r_i}$ and performing a Chinese remainder reconstruction to recover the convolution modulo q from the convolutions modulo $q_i^{r_i}$. Therefore, an N-length NTT which supports circular convolution will exist if and only if N-length NTTs exist modulo each factor $q_i^{r_i}$. Theorem 8.5 shows that this is the case if $N | (q_i - 1)$. Thus, N must divide the greatest common divisor (GCD) of the $(q_i - 1)$ and we have the following existence theorem.

Theorem 8.6: A length-N NTT defined modulo q, with $q = q_1^{r_1} ... q_i^{r_i} ... q_e^{r_e}$ supports circular convolution if and only if

$$N | \text{GCD}[(q_1 - 1), (q_2 - 2), ... (q_i - 1), ..., (q_e - 1)]. \tag{8.16}$$

This theorem immediately gives the maximum transform length, which is $\text{GCD}[(p_1 - 1), ..., (p_i - 1), ..., (p_e - 1)]$. It should also be noted that theorem 8.6 gives a simple way of computing the root g of order N modulo q. This is done by first computing the roots $g_{1,i}$ of order N modulo q_i by the methods given in Sect. 2.1.3. The roots $g_{2,i}$ of order N modulo $q_i^{r_i}$ are then obtained by theorem 8.5, with $g_{2,i} = g_{1,i}^{q_1^{r_i-1}}$ and g is derived from the $g_{2,i}$ by a Chinese remainder reconstruction.

The circular convolution property is by far the most important property of the NTTs, because it allows one to replace the direct computation of a convolution by that of three NTTs plus N multiplications in the transform domain. In the general case, NTTs are computed with multiplications by powers of integers g and are, therefore, not significantly simpler than DFTs, except that the complex exponentials in the DFTs are replaced by real integers. Thus, real convolutions are computed via NTTs with real arithmetic, instead of complex arithmetic as with the DFT approach. We shall see, however, in the following sections that when the modulus q is properly selected, the multiplications by powers of g are replaced by simple shifts, thereby simplifying the NTT computations considerably. Another advantage of computing convolutions by NTTs instead of FFTs is that the convolutions are computed exactly, without round-off errors. Thus, the NTT approach is well adapted to high accuracy computations.

Since the NTTs have the same structure as DFTs, they have the same general properties as the DFTs and the reader can refer to Sect. 4.1 for a description of them. We simply note here that the NTT definition implies that the linearity property

$$\{h_n\} + \{x_n\} \xrightleftharpoons{\text{NTT}} \{\bar{H}_k\} + \{X_k\} \tag{8.17}$$

$$\{\lambda h_n\} \xrightleftharpoons{\text{NTT}} \{\lambda \bar{H}_k\}. \tag{8.18}$$

8.2 Mersenne Transforms

In order to simplify the computation of NTTs, we would like the modulus q to be as simple as possible. The most obvious choice is $q = 2^t$. However, in this case, the maximum transform length is 1, which rules out $q = 2^t$. Similarly, when q is even, one of the factors of q is a power of 2 and, by theorem 8.6 the maximum transform length is also 1. Thus, the only cases of interest correspond to q odd. Then, the simplest choice is $q = 2^p - 1$ because arithmetic modulo $(2^p - 1)$ is the well-known one's complement arithmetic which can be implemented easily with binary hardware. If p is composite, with $p = p_1 p_2$ and p_1 a prime, then $2^{p_1} - 1$ is a factor of $2^p - 1$ and the maximum transform length cannot be larger than possible with $2^{p_1} - 1$. Therefore, for $q = 2^p - 1$, the most interesting cases correspond to p prime. The integers $2^p - 1$ with p prime are the Mersenne numbers discussed in Sect. 2.1.5 and the transforms defined modulo Mersenne numbers are called Mersenne transforms [8.5].

8.2.1 Definition of Mersenne Transforms

Theorem 8.6 implies the existence of a length-N NTT which supports circular convolution modulo a Mersenne number $q = 2^p - 1$, p prime, provided that N divides all the $q_i - 1$, where the $q_i^{r_i}$ are the factors of q. Some of the Mersenne numbers are primes. For these numbers, the possible transform lengths are given by $N | (q - 1)$

$$N | (2^p - 2). \tag{8.19}$$

8.2 Mersenne Transforms

We know, by Fermat's theorem (theorem 2.4) that p divides $2^p - 2$. Moreover, 2 is an obvious divisor of $2^p - 2$. Hence we can define NTTs of lengths p and $2p$ modulo prime Mersenne numbers. When q is composite, we know, by theorem 2.14 that every prime factor q_i of a composite Mersenne number is of the form

$$q_i = 2c_i p + 1. \tag{8.20}$$

Thus we have $q_i - 1 = 2c_i p$ and $2p$ divides every $q_i - 1$. This implies, by theorem 8.6, that we can also define NTTs of lengths p and $2p$ modulo composite Mersenne numbers.

In order to complete the definition of the Mersenne transforms, we must now find the roots g of order p and $2p$. For q prime, an obvious root of order p is 2, since the p first powers of 2 corresponding to $1, 2, 2^2, \ldots, 2^{p-1}$ are all distinct and $2^p \equiv 1$ modulo q. For q composite, 2 is also a root of order p modulo q, but we must also insure that the two last conditions of theorem 8.3 are satisfied, that is to say, that p has an inverse p^{-1} and that $\sum_{k=0}^{p-1} 2^{tk} \equiv 0$ modulo q for $t \not\equiv 0$ modulo p.

For the inverse p^{-1} of p, we note that, since $p | (2^p - 2)$,

$$p^{-1} \equiv 2^p - 1 - (2^p - 2)/p \text{ modulo } (2^p - 1). \tag{8.21}$$

For the last condition, we note that, since p is a prime, the set of exponents $t \cdot k$ modulo p in $S = \sum_{k=0}^{p-1} 2^{tk}$ is a simple permutation of k. Thus, for $t \not\equiv 0$ modulo p,

$$\sum_{k=0}^{p-1} 2^{tk} \equiv \sum_{k=0}^{p-1} 2^k \equiv 2^p - 1 \equiv 0 \text{ modulo } (2^p - 1). \tag{8.22}$$

Hence we can define a p-point Mersenne transform having the circular convolution property by

$$\bar{X}_k \equiv \sum_{m=0}^{p-1} x_m 2^{mk} \text{ modulo } (2^p - 1), \quad p \text{ prime} \quad k = 0, \ldots, p-1 \tag{8.23}$$

and the corresponding inverse Mersenne transform

$$x_m = p^{-1} \sum_{k=0}^{p-1} \bar{X}_k 2^{-mk} \text{ modulo } (2^p - 1) \tag{8.24}$$

with

$$2^{-mk} \equiv 2^{(p-1)mk} \text{ modulo } (2^p - 1). \tag{8.25}$$

Thus, a length-p circular convolution is computed as shown in Fig. 8.1 by three Mersenne transforms plus p multiplications in the transform domain. When one of the input sequences, h_n, is fixed, its transform $\bar{H}_k$ can be precalculated and combined with the multiplications by p^{-1} corresponding to the inverse transform.

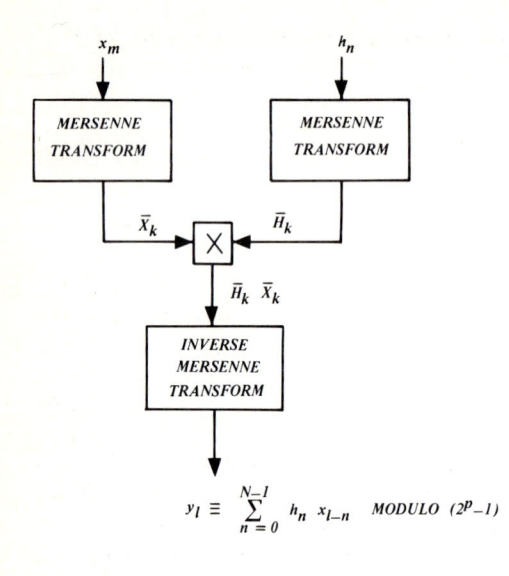

Fig. 8.1. Computation of a length-p circular convolution modulo $(2^p - 1)$ by Mersenne transforms

$$y_l \equiv \sum_{n=0}^{N-1} h_n\, x_{l-n} \quad \text{MODULO } (2^p - 1)$$

In this case, only two Mersenne transforms need to be evaluated. Moreover, since the roots of Mersenne transforms are powers of two, each Mersenne transform is calculated with only $p(p - 1)$ additions and $(p - 1)^2$ shifts and the only general multiplications required to compute the length-p circular convolution are the p multiplications in the transform domain. Hence the use of Mersenne transforms can bring significant computational savings for the evaluation of circular convolutions, even when compared to other efficient methods, such as the FFT approach.

We have seen above that Mersenne transforms can also be defined with length $2p$. It can be seen that -2 is a root of order $2p$ modulo a Mersenne number $q = 2^p - 1$, since the $2p$ first powers of -2 defined by $1, -2, 4, \ldots (-2)^{2p-1}$ are all distinct and $(-2)^{2p} \equiv 1$ modulo q. Thus, for q prime, we can define a Mersenne transform of length $2p$ with root -2. For q composite, since $2p \mid (2^p - 2)$, $2p$ has an inverse $(2p)^{-1}$ defined by

$$(2p)^{-1} \equiv 2^p - 1 - (2^p - 2)/2p \quad \text{modulo } (2^p - 1). \tag{8.26}$$

For q composite, we must also show that $S = \sum_{k=0}^{2p-1} (-2)^{tk} \equiv 0$ modulo q for $t = 1, \ldots, 2p - 1$. We note first that when t is even, $S = \sum_{k=0}^{2p-1} (-2)^{tk} = 2 \sum_{k=0}^{p-1} (-2)^{tk} \equiv 0$ modulo q for $t \not\equiv 0$ modulo p. For t odd, and $t \neq p$, tk modulo $2p$ is a simple permutation and we have

$$\sum_{k=0}^{2p-1} (-2)^{tk} \equiv \sum_{k=0}^{2p-1} (-2)^k = -(2^{2p} - 1)/3 \equiv 0 \text{ modulo } q. \tag{8.27}$$

For $t = p$, $S = 1 - 1 + 1 - 1 \cdots = 0$.

Hence, for any Mersenne number, we can define a Mersenne transform of length $2p$ by

$$\bar{X}_k \equiv \sum_{m=0}^{2p-1} x_m (-2)^{mk} \text{ modulo } (2^p - 1) \tag{8.28}$$

and an inverse transform

$$x_m = (2p)^{-1} \sum_{k=0}^{2p-1} \bar{X}_k (-2)^{-mk} \text{ modulo } (2^p - 1) \tag{8.29}$$

with

$$(-2)^{-mk} \equiv (-2)^{(2p-1)mk} \text{ modulo } (2^p - 1). \tag{8.30}$$

These double-length Mersenne transforms can also be computed without multiplications. When $2^p - 1$ is a prime, it is possible to define Mersenne transforms of dimension larger than $2p$, since the maximum transform length is $2^p - 2$. However, the maximum number of distinct powers of -2 is exactly $2p$ so that the roots g of these larger transforms can no longer be simple powers of two. Thus, these larger transforms require some general multiplications and are therefore less useful than the transforms of lengths p and $2p$. In practice, only the transforms of length p and $2p$ are called Mersenne transforms.

8.2.2 Arithmetic Modulo Mersenne Numbers

Any integer x_m defined modulo a Mersenne number $q = 2^p - 1$ can be represented as a p-bit word, with

$$x_m = \sum_{i=0}^{p-1} x_{m,i} 2^i, \quad x_{m,i} \in (0, 1) \tag{8.31}$$

From the binary representation of x_m given by (8.31), $-x_m$ can be obtained easily by replacing each bit $x_{m,i}$ of x_m with its complement $\bar{x}_{m,i}$. Since $x_{m,i} + \bar{x}_{m,i} = 1$, the integer $\bar{x}_m$ obtained by this complementation is such that

$$x_m + \bar{x}_m = \sum_{i=0}^{p-1} 2^i = 2^p - 1 \equiv 0. \tag{8.32}$$

Hence

$$-x_m = \bar{x}_m.$$

Consider now two integers x_m and h_n, with x_m defined by (8.31) and h_n defined by

$$h_n = \sum_{i=0}^{p-1} h_{n,i} 2^i, \quad h_{n,i} \in (0, 1). \tag{8.33}$$

If we add the two numbers h_n and x_m, we obtain a $(p+1)$-bit integer c_n defined by

$$c_n = h_n + x_m = \sum_{i=0}^{p-1} c_{n,i} 2^i + c_{n,p} 2^p, \qquad c_{n,i} \in (0, 1) \tag{8.34}$$

and, since $2^p \equiv 1$ modulo $(2^p - 1)$,

$$c_n \equiv c_{n,p} + \sum_{i=0}^{p-1} c_{n,i} 2^i. \tag{8.35}$$

Thus, addition modulo a Mersenne number is performed very simply by using a conventional full binary adder of p bits and by folding the most significant carry bit output back into the least significant carry bit input.

The multiplication modulo $(2^p - 1)$ of an integer x_m by a power of two, 2^d, is also done very simply. Assuming that c_n is defined by

$$c_n = x_m 2^d \text{ modulo } (2^p - 1), \tag{8.36}$$

we have

$$c_n = \sum_{i=0}^{p-1-d} x_{m,i} 2^{i+d} + 2^p \sum_{i=p-d}^{p-1} x_{m,i} 2^{i+d-p}, \tag{8.37}$$

and since $2^p \equiv 1$ modulo $(2^p - 1)$,

$$c_n \equiv \sum_{i=0}^{n-1} x_{m,\langle i-d \rangle} 2^i, \tag{8.38}$$

where the index $\langle i - d \rangle$ is taken modulo p. This shows that a multiplication by 2^d amounts to a simple d-bit rotation of a word of p bits.

General multiplications are implemented easily by combining the additions and shifts discussed above. It should also be noted that the binary representation discussed above assigns a double representation to the integer zero, since $x_m \equiv 0$ if the bits $x_{m,i}$ are either all zeros or all ones. Thus, when the final result of a computation modulo $(2^p - 1)$ is converted into normal arithmetic, one must detect the condition corresponding to all bits of x_m being equal to one and set the final result to zero when this condition is realized.

Thus, arithmetic modulo a Mersenne number is not significantly more complex than normal arithmetic when implemented with special purpose hardware and the multiplications by powers of two are considerably more simple than the general multiplications used in other transforms such as the DFT. This provides motivation for replacing FFTs with NTTs in the calculation of convolutions.

8.2.3 Illustrative Example

Given the two length-5 data sequences h_n and x_m, use the 5-point Mersenne transform defined modulo $(2^5 - 1)$ to compute the circular convolution y_l of h_n and x_m. h_n and x_m are defined by

$$h_0 = 1 \quad h_1 = 3 \quad h_2 = 2 \quad h_3 = 2 \quad h_4 = 0 \tag{8.39}$$

$$x_0 = 3 \quad x_1 = 0 \quad x_2 = 2 \quad x_3 = 1 \quad x_4 = 2. \tag{8.40}$$

The 5-point Mersenne transform $\bar{H}_k$ of h_n is defined by

$$\bar{H}_k \equiv \sum_{n=0}^{4} h_n 2^{nk} \text{ modulo } 31 \tag{8.41}$$

with a similar relation for the transform $\bar{X}_k$ of x_m. In matrix notation, (8.41) becomes

$$\begin{bmatrix} \bar{H}_0 \\ \bar{H}_1 \\ \bar{H}_2 \\ \bar{H}_3 \\ \bar{H}_4 \end{bmatrix} = \begin{bmatrix} 1 & 1 & 1 & 1 & 1 \\ 1 & 2 & 4 & 8 & 16 \\ 1 & 4 & 16 & 2 & 8 \\ 1 & 8 & 2 & 16 & 4 \\ 1 & 16 & 8 & 4 & 2 \end{bmatrix} \begin{bmatrix} 1 \\ 3 \\ 2 \\ 2 \\ 0 \end{bmatrix}. \tag{8.42}$$

Thus, the transform sequences $\bar{H}_k$ and $\bar{X}_k$ are given by

$$\bar{H}_0 \equiv 8 \quad \bar{H}_1 \equiv 0 \quad \bar{H}_2 \equiv 18 \quad \bar{H}_3 \equiv 30 \quad \bar{H}_4 \equiv 11 \tag{8.43}$$

$$\bar{X}_0 \equiv 8 \quad \bar{X}_1 \equiv 20 \quad \bar{X}_2 \equiv 22 \quad \bar{X}_3 \equiv 0 \quad \bar{X}_4 \equiv 27. \tag{8.44}$$

Multiplying $\bar{H}_k$ by $\bar{X}_k$ modulo 31 yields

$$\bar{H}_0\bar{X}_0 \equiv 2 \quad \bar{H}_1\bar{X}_1 \equiv 0 \quad \bar{H}_2\bar{X}_2 \equiv 24 \quad \bar{H}_3\bar{X}_3 \equiv 0 \quad \bar{H}_4\bar{X}_4 \equiv 18. \tag{8.45}$$

We must now multiply $\bar{H}_k\bar{X}_k$ by $5^{-1} \equiv 25$ modulo 31. This gives

$$25\bar{H}_0\bar{X}_0 \equiv 19 \quad 25\bar{H}_1\bar{X}_1 \equiv 0 \quad 25\bar{H}_2\bar{X}_2 \equiv 11 \quad 25\bar{H}_3\bar{X}_3 \equiv 0 \quad 25\bar{H}_4\bar{X}_4 \equiv 16. \tag{8.46}$$

Since $2^{-1} \equiv 2^4$ modulo 31, the inverse Mersenne transform is given by

$$y_l \equiv \sum_{k=0}^{4} 25\, \bar{H}_k\bar{X}_k\, 2^{4lk} \text{ modulo } 31, \tag{8.47}$$

or, in matrix notation

$$\begin{bmatrix} y_0 \\ y_1 \\ y_2 \\ y_3 \\ y_4 \end{bmatrix} = \begin{bmatrix} 1 & 1 & 1 & 1 & 1 \\ 1 & 16 & 8 & 4 & 2 \\ 1 & 8 & 2 & 16 & 4 \\ 1 & 4 & 16 & 2 & 8 \\ 1 & 2 & 4 & 8 & 16 \end{bmatrix} \begin{bmatrix} 19 \\ 0 \\ 11 \\ 0 \\ 16 \end{bmatrix}. \tag{8.48}$$

This gives the final result

$$y_0 = 15 \quad y_1 = 15 \quad y_2 = 12 \quad y_3 = 13 \quad y_4 = 9. \tag{8.49}$$

The direct computation in ordinary arithmetic would produce the same result. Note, however, that if we had chosen larger input samples, some of the output samples of the convolution would have been greater than 30. In this case, the Mersenne transform approach would produce erroneous samples because of the reduction modulo 31.

8.3 Fermat Number Transforms

Mersenne transforms have a number of desirable attributes for the computation of convolutions. In particular, these transforms utilize an easily implemented arithmetic and can be computed without multiplications. The principal deficiencies of Mersenne transforms, however, relate to the lack of a fast transform algorithm and to the very rigid relationship between word length and transform length. These limitations stem from the fact that Mersenne transforms support convolution with the simple roots 2 or -2 only for transform lengths p and $2p$, with p prime. Thus, for a word length of p bits, one can, in practice, employ only the two lengths p and $2p$. Since p is a prime, the Mersenne transform of length p cannot be computed with a fast FFT-like computation structure, and only a two-stage fast algorithm can be used for length $2p$. Thus, large Mersenne transforms are computed with a much larger number of additions than FFTs of comparable size.

In order to overcome these problems, one is led to choose another modulus q. The simplest choice, after $q = 2^t$ and $q = 2^p - 1$, is $q = 2^v + 1$, which yields a still relatively simple arithmetic. If v is odd, 3 is a factor of $2^v + 1$ and, by theorem 8.6, the maximum transform length cannot be larger than $3 - 1 = 2$. If v is even, with $v = s2^t$ and s odd, $2^{2^t} + 1$ is a factor of $2^{s2^t} + 1$, and the transform defined modulo($2^{s2^t} + 1$) cannot be larger than the transform defined modulo($2^{2^t} + 1$). This indicates that the best choice corresponds to a modulus $F_t = 2^{2^t} + 1$. The numbers F_t are the Fermat numbers (Sect. 2.1.5) and the NTTs defined modulo F_t are called Fermat number transforms (FNT) [8.5–7].

8.3.1 Definition of Fermat Number Transforms

We have seen in Sect. 2.15 that the first five Fermat numbers, F_0 to F_4, are prime while all other known Fermat numbers are composite. When F_t is a prime, the maximum transform length is $F_t - 1 = 2^{2^t}$, and therefore, all possible transform lengths N correspond to $N | 2^{2^t}$.

When F_t is composite, every prime factor q_i of F_t is of the form

$$q_i = c_i \, 2^{t+2} + 1. \tag{8.50}$$

Hence, theorem 8.6 implies that we can always define an N-length transform modulo a composite Fermat number, provided that

$$N | 2^{t+2}. \tag{8.51}$$

We must now find the roots of these transforms. It is obvious that 2 is root of order 2^{t+1} modulo F_t, since $2^{2^t} \equiv -1$ and since 2^i takes the 2^{t+1} distinct values $1, 2, 2^2, \ldots, -1, -2, \ldots, 2^{2^t-1} - 1$ for $i = 0, 1, 2, \ldots, 2^{t+1} - 1$. This means that when F_t is a prime, we can define an FNT of length $N = 2^{t+1}$ with root 2. For F_t composite, 2 is also a root of order 2^{t+1}, but we must also prove that 2^{t+1} has an inverse and that $2^{2^t} - 1$ is mutually prime with $2^{2^t} + 1$ (theorem 8.4). Since $2^{2^t} \equiv -1$, the inverse of 2^{t+1} modulo F_t is obviously -2^{2^t-t-1}. Moreover, we have $2^{2^t} - 1 = (2^{2^t} + 1) - 2$. This means that any divisor of $2^{2^t} - 1$ and $2^{2^t} + 1$ should also divide 2. Thus, this divisor could only be 2, but this is impossible since $2^{2^t} - 1$ and $2^{2^t} + 1$ are odd.

Under these conditions, we can define a length $- 2^{t+1}$ FNT which supports circular convolution by

$$\bar{X}_k \equiv \sum_{m=0}^{2^{t+1}-1} x_m \, 2^{mk} \text{ modulo } (2^{2^t} + 1) \tag{8.52}$$

and the corresponding inverse FNT

$$x_m \equiv - 2^{2^t-t-1} \sum_{k=0}^{2^{t+1}-1} \bar{X}_k \, 2^{-mk} \text{ modulo } (2^{2^t} + 1) \tag{8.53}$$

with

$$2^{-mk} \equiv -2^{(2^t-1)mk} \text{ modulo}(2^{2^t} + 1). \tag{8.54}$$

If follows immediately that, since 2 is a root of order 2^{t+1} modulo F_t, 2^{2^i} is a root of order 2^{t+1-i}. We can, therefore, always define FNTs of length 2^{t+1-i} with root 2^{2^i}.

We have shown above that it is possible to define FNTs of length 2^{t+2} when F_t is composite. Since the maximum number of distinct powers of ± 2 modulo $(2^{2^t} + 1)$ is equal to 2^{t+1}, the roots of the length $- 2^{t+2}$ FNTs can no longer be simple powers of two. We note that 2 is a root of order 2^{t+1} and therefore that $\sqrt{2}$ is a root of order 2^{t+2}. However, $\sqrt{2}$ has a very simple expression in a ring of Fermat numbers

$$\sqrt{2} \equiv 2^{v/4}(2^{v/2} - 1) \text{ modulo } (2^v + 1), \quad v = 2^t. \tag{8.55}$$

Thus, FNTs have lengths which are powers of two, and can be computed using only additions and multiplications by powers of 2 for sizes up to $N = 2^{t+2}$. Larger FNTs can be defined modulo prime Fermat numbers, but in this case, the roots are no longer simple and the computation of these transforms requires general multiplications. Therefore, most practical applications are restricted to a maximum length equal to 2^{t+2}.

FNTs are superior to Mersenne transforms in several respects. As a first point of difference, it can be noted that FNTs permit much more flexibility in selecting the transform length as a function of the word length than Mersenne transforms. A second advantage of using FNTs relates to the highly composite length of such transforms. This makes it possible to evaluate an FNT with a reduced number of additions by use of a radix-2 FFT-type algorithm. To illustrate this point with a decimation in time algorithm, the length — 2^{t+1} FNT defined by (8.52) can be calculated, in the first stage by

$$\bar{X}_k \equiv \sum_{m=0}^{N/2-1} x_{2m} 2^{2mk} + 2^k \sum_{m=0}^{N/2-1} x_{2m+1} 2^{2mk} \text{ modulo } F_t \tag{8.56}$$

$$\bar{X}_{k+N/2} \equiv \sum_{m=0}^{N/2-1} x_{2m} 2^{2mk} - 2^k \sum_{m=0}^{N/2-1} x_{2m+1} 2^{2mk} \text{ modulo } F_t. \tag{8.57}$$

Thus, the FFT-type computation structure of an FNT is similar to that of a conventional FFT, but with multiplications by complex exponentials replaced by simple shifts. This means that an FNT of length N can be calculated with $N \log_2 N$ real additions and $(N/2) \log_2 N$ shifts.

8.3.2 Arithmetic Modulo Fermat Numbers

Any integer x_m defined modulo a Fermat number $F_t = 2^v + 1$, $v = 2^t$ can be represented as a $(v + 1)$ — bit word

$$x_m = \sum_{i=0}^{v-1} x_{m,i} 2^i + x_{m,v} 2^v, \quad x_{m,i} \in (0, 1)$$

$$x_{m,i} x_{m,v} = 0 \quad \text{for} \quad i \neq v. \tag{8.58}$$

Since $x_m \leqslant 2^v$, $x_{m,v}$ is equal to 1 only if all the $x_{m,i}$ are equal to zero for $i < v$. Negation can be realized by complementing all bits $x_{m,i}$ of x_m, except $x_{m,v}$. Thus, if we treat $x_{m,v}$ separately, we have

$$x_m + \bar{x}_m = \sum_{i=0}^{v-1} 2^i = 2^v - 1$$

and

$$-x_m = \bar{x}_m + 2, \qquad x_{m,v} = 0 \qquad (8.59)$$
$$-x_m = 1, \qquad x_{m,v} = 1. \qquad (8.60)$$

Hence, negation requires an addition and a complementation together with several auxiliary operations in order to deal with the case $x_{m,v} = 1$.
If we add two integers x_m and h_n, the result

$$c_n = \sum_{i=0}^{v-1} c_{n,i} 2^i + c_{n,v} 2^v + c_{n,v+1} 2^{v+1} \qquad (8.61)$$

cannot be greater than 2^{v+1} and, since $2^v \equiv -1$ and $c_{n,v+1} = 1$ only for $c_{n,i} = 0$, $i = 0, \ldots, v$,

$$c_n \equiv \sum_{i=0}^{v-1} c_{n,i} 2^i - c_{n,v}, \qquad c_{n,v+1} = 0 \qquad (8.62)$$

$$c_n \equiv -2, \qquad c_{n,v+1} = 1. \qquad (8.63)$$

If we multiply an integer x_m by 2^d, we obtain, for $x_{m,v} = 0$, an integer result

$$c_n = \sum_{i=0}^{v-1-d} x_{m,i} 2^{i+d} + 2^v \sum_{i=v-d}^{v-1} x_{m,i} 2^{i+d-v} \qquad (8.64)$$

and, since $2^v \equiv -1$,

$$c_n \equiv \sum_{i=0}^{v-1-d} x_{m,i} 2^{i+d} - \sum_{i=v-d}^{v-1} x_{m,i} 2^{i+d-v}, \qquad x_{m,v} = 0. \qquad (8.65)$$

For $x_{m,v} = 1$, c_n reduces to

$$c_n \equiv -2^d, \qquad x_{m,v} = 1. \qquad (8.66)$$

Therefore, arithmetic modulo a Fermat number is significantly more complex than arithmetic modulo a Mersenne number. However, the practical implementation can be greatly simplified by using various data code translation techniques [8.8–10]. In one of these techniques, the input sequence x_m is first mapped into a new sequence

$$a_m = 2^v - x_m \qquad (8.67)$$

or, by introducing the integer $\bar{x}_m$ obtained by complementing the v least significant bits $x_{m,i}$ of x_m,

$$a_m = \bar{x}_m + 1 + x_{m,v}\, 2^v, \tag{8.68}$$

which indicates that the coded samples a_m are obtained by simply complementing the v least significant bits of x_m and adding 1 to $\bar{x}_m + x_{m,v}2^v$.

With this technique, the input data stream is encoded only once prior to transform computation and all operations are performed on the coded sequences with a single decoding operation on the final result, this decoding operation being also defined by (8.68). We now demonstrate that arithmetic modulo Fermat numbers on the coded sequences a_m is much simpler than on the original sequences.

Consider first negation. If $a_{m,v} = 1$, then $x_m = 0$ and no modification is required. If $a_{m,v} = 0$, coding the complement $\bar{a}_m$ of a_m yields

$$\bar{a}_m = 2^v - 1 - a_m \tag{8.69}$$

or, with (8.67) and $-1 \equiv 2^v$,

$$\bar{a}_m \equiv 2^v + x_m, \tag{8.70}$$

which shows, by comparison with (8.67) that $\bar{a}_m$ is the coded representation of $-x_m$. Thus, negation is performed on the coded samples by a simple complementation except when $a_{m,v} = 1$. If a_m and b_m are the coded values of two integers x_m and h_n, the sum of a_m and b_m is given by

$$c_m = \sum_{i=0}^{v-1} c_{m,i}\, 2^i + c_{m,v}\, 2^v \tag{8.71}$$

with

$$c_m = 2^{v+1} - x_m - h_n. \tag{8.72}$$

The coded value d_m of $x_m + h_n$ is defined by

$$d_m = 2^v - x_m - h_n. \tag{8.73}$$

Thus,

$$d_m = c_m - 2^v \equiv \sum_{i=0}^{v-1} c_{m,i}\, 2^i + 1 - c_{m,v}, \tag{8.74}$$

which indicates that addition in the transposed system is executed with ordinary adders, but with high order carry fed back, after complementation, to the least significant carry input of the adder. If one or both bits $a_{m,v}$ and $b_{m,v}$ are zero, one or both of the operands x_m and h_n are zero. In this case, the operation must be inhibited.

It can be verified easily, from the rules of addition, that multiplication by 2^d corresponds in the transposed system to a simple d-bit rotation around the v-bit word, with complementation of the overflow bits. When $a_{m,v} = 1$, we have $x_m = 0$ and the inversion of the overflow bits is inhibited. The process is illustrated in Fig. 8.2 for a multiplication by 2.

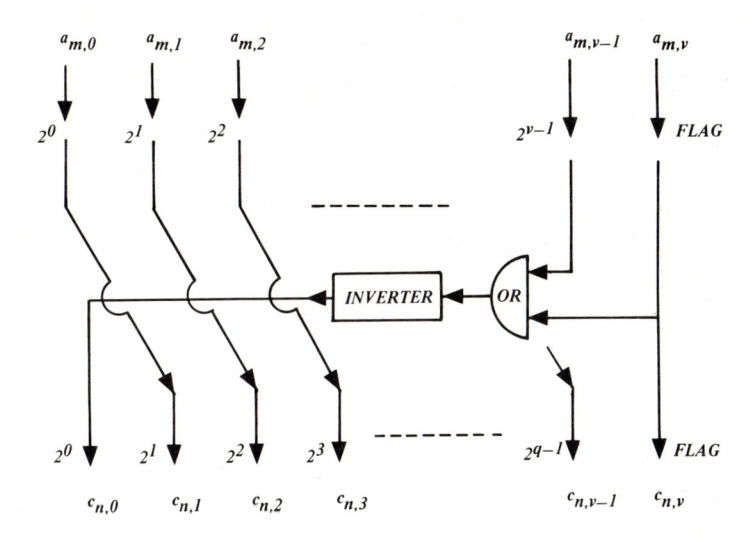

Fig. 8.2. Multiply-by-two circuit in the transposed Fermat number system

Therefore, the foregoing code translation technique reduces the arithmetic operations to one's complement arithmetic, with the exception that the overflow bits are complemented before around-carry and that some additional hardware must be used for treating separately the cases corresponding to zero-valued input data items. With this approach, arithmetic modulo a Fermat number is only slightly more complex than arithmetic modulo a Mersenne number.

8.3.3 Computation of Complex Convolutions by FNTs

We now consider a complex circular convolution $y_l + j\hat{y}_l$ defined modulo $(2^v + 1)$,

$$y_l + j\hat{y}_l \equiv \sum_{n=0}^{N-1} (h_n + j\hat{h}_n)(x_{l-n} + j\hat{x}_{l-n}) \text{ modulo } (2^v + 1)$$

$$j = \sqrt{-1}, \qquad\qquad l = 0, ..., N-1, \qquad (8.75)$$

where h_n, x_m and y_l are the real parts of the input and output sequences and $\hat{h}_n$, $\hat{x}_m$, and $\hat{y}_l$ are the imaginary parts of the input and output sequences. With the conventional approach, this complex convolution is calculated by evaluating four real convolutions:

$$y_l \equiv \sum_{n=0}^{N-1} h_n x_{l-n} - \sum_{n=0}^{N-1} \hat{h}_n \hat{x}_{l-n} \text{ modulo } (2^v + 1) \tag{8.76}$$

$$\hat{y}_l \equiv \sum_{n=0}^{N-1} h_n \hat{x}_{l-n} + \sum_{n=0}^{N-1} \hat{h}_n x_{l-n} \text{ modulo } (2^v + 1). \tag{8.77}$$

We shall now show that the evaluation of $y_l + \hat{y}_l$ can be done with only two real convolutions by taking advantage of the special properties of $j = \sqrt{-1}$ in certain rings of integers [8.11, 12]. This is done by noting that in the ring of integers modulo $(2^v + 1)$, with v even, we have $2^v \equiv -1$, which means that $j = \sqrt{-1}$ is congruent to $2^{v/2}$. $y_l + j\hat{y}_l$ is evaluated by first computing the two real auxiliary convolutions a_l and b_l defined by

$$a_l \equiv \sum_{n=0}^{N-1} (h_n + 2^{v/2} \hat{h}_n)(x_{l-n} + 2^{v/2} \hat{x}_{l-n}) \text{ modulo } (2^v + 1) \tag{8.78}$$

$$b_l \equiv \sum_{n=0}^{N-1} (h_n - 2^{v/2} \hat{h}_n)(x_{l-n} - 2^{v/2} \hat{x}_{l-n}) \text{ modulo } (2^v + 1). \tag{8.79}$$

Since $2^v \equiv -1$, we have

$$y_l \equiv (a_l + b_l)/2 \text{ modulo } (2^v + 1) \tag{8.80}$$

$$\hat{y}_l \equiv -2^{v/2}(a_l - b_l)/2 \text{ modulo } (2^v + 1). \tag{8.81}$$

Therefore this method supports the computation of a circular convolution by FNTs with only two multiplications per complex output sample instead of four in the conventional case.

8.4 Word Length and Transform Length Limitations

When a convolution is computed via NTTs, all the calculations are executed on integer data sequences and the convolution product is obtained modulo q without roundoff errors. This feature provides significant advantage over other methods when high accuracy is needed, but can also impose a requirement for relatively long words to ensure that the result remains within the modulo range. In order to analyze these limitations for arithmetic operations modulo q, we assume that the two input sequences x_n and h_n are integer sequences. The output $\tilde{y}_n$ of the ordinary convolution is given by

$$\tilde{y}_n = h_n * x_n. \tag{8.82}$$

The output y_n of the same convolution computed modulo q will numerically equal to $\tilde{y}_n$ if

$$|\tilde{y}_n| < q/2. \tag{8.83}$$

This condition will be met for a length-N circular convolution if

$$|h_n|_{\max} |x_n|_{\max} < q/2N, \tag{8.84}$$

which means that the word length of the original input sequences must be slightly less than half of the word length corresponding to the modulus q. Tighter bounds on input signal amplitudes can be found using the L_p norms [8.13] defined by

$$\|x\|_r = \left(\frac{1}{N} \sum_{n=0}^{N-1} |x_n|^r\right)^{1/r}, \quad r \geqslant 1. \tag{8.85}$$

$\tilde{y}_n$ is bounded by

$$|\tilde{y}_n| \leqslant N\|x\|_r \|h\|_s \tag{8.86}$$

with

$$1/r + 1/s = 1 \quad r, s \geqslant 1. \tag{8.87}$$

These bounds are better than (8.84), especially when the circular convolutions are used to compute aperiodic convolutions by the overlap-add method, with both input sequences padded out with zeros.

Thus, when convolutions are computed by NTTs, the only source of quantization noise is the input quantization noise resulting from scaling and rounding of the input sequences required to avoid overflow [8.14]. This implies that the output signal-to-noise ratio (SNR) is relatively independent of the convolution length N and increases by 3 dB for each increase of word length by one bit. By contrast, when the convolution is evaluated by FFTs with fixed word length, one must account for the roundoff noise incurred in FFT computations and the SNR increases by about 6 dB for each additional bit of word length and decreases by about 2 dB for every doubling of the convolution length. This shows that for fixed word lengths, the computation by NTTs is less noisy than the computation by FFTs for long convolutions. For words of 12 bits, for instance, NTT filtering gives a better SNR than FFT filtering for N greater than about 32.

This motivates one to use NTTs for computing long convolutions. However, for a given modulus q, the maximum number of distinct powers of ± 2 is equal to $2\lceil \log_2 q \rceil$, with $a = \lceil \log_2 q \rceil$, where a is the smallest integer such that $a \geqslant \log_2 q$. Thus, for NTTs computed without multiplications, there is a rigid relationship between word length and maximum convolution length, and long convolutions imply long word lengths, even if these long word lengths far exceed the desired accuracy.

One solution to this problem consists of simply computing the convolutions $y_{1,n}$ and $y_{2,n}$ of two consecutive blocks simultaneously. Assuming, for instance, that h_n is a fixed input sequence of positive integers and that $x_{1,n}$ and $x_{2,n}$ are two consecutive positive integer sequences, the two length-N convolutions $y_{1,n}$ and $y_{2,n}$ can be computed in a single step by

$$x_n = x_{1,n} + 2^e x_{2,n} \tag{8.88}$$

$$y_n \equiv h_n * x_n \quad \text{modulo } q \tag{8.89}$$

$$y_{1,n} \equiv y_n \quad \text{modulo } 2^e \tag{8.90}$$

$$y_{2,n} = 2^{-e}(y_n - y_{1,n}) \tag{8.91}$$

with

$$e = (\lceil \log_2 q \rceil)/2. \tag{8.92}$$

With this method, the transform length is doubled for a given accuracy and there is no overflow, provided $|y_{1,n}|, |y_{2,n}| < (\sqrt{q})/2$.

Another solution to computing long convolutions with NTTs using moderate word lengths is possible by mapping the one-dimensional convolution of length N into multidimensional convolutions using one of the methods discussed in Chap. 3. For instance, if N is the product of d distinct Mersenne numbers N_1, $N_2, \ldots, N_d$ with $N = N_1 N_2 \ldots N_d$, we can map the length N convolution into a d-dimensional convolution of size $N_1 \times N_2 \times \ldots \times N_d$ by using the Agarwal-Cooley algorithm (Sect. 3.3.1). This is always possible because all Mersenne numbers are mutually prime (theorem 2.15). The nested convolutions are then calculated with Mersenne transforms defined modulo $N_1, N_2 \ldots N_d$ and the convolution product y_l is obtained without overflow provided that $|y_i| < N_i/2$, where N_i is the smallest Mersenne number.

8.5 Pseudo Transforms

We have seen that Mersenne transforms defined modulo $(2^p - 1)$, with p prime, and Fermat number transforms defined modulo $(2^{2^t} + 1)$ can be used to compute circular convolutions. Both transforms are computed without multiplication but suffer serious limitations which relate mainly to the lack of an FFT-type algorithm for Mersenne transforms and to the problems associated with arithmetic modulo $(2^{2^t} + 1)$ for FNTs. It would seem difficult to consider the use of any modulus other than a Mersenne or a Fermat number because of the problems associated with the corresponding arithmetic. In the following, however, we shall show that these difficulties can be circumvented by defining NTTs modulo integers q_i which are factors of pseudo Mersenne numbers q, with $q = 2^p - 1$, p composite or of pseudo Fermat numbers q, with $q = 2^v + 1$, $v \neq 2^t$. In both cases, q is composite and can always be defined as the product of two factors

$$q = q_1 q_2. \tag{8.93}$$

Under these conditions, if an N length NTT, which supports circular convolution, can be defined modulo q_2, this NTT computes the N-length convolution y_l modulo q_2, with

$$y_l \equiv \sum_{n=0}^{N-1} h_n x_{l-n} \text{ modulo } q_2 \tag{8.94}$$

The difficulty of performing the arithmetic operations modulo q_2 can be circumvented by exploiting the fact that q_2 is a factor of q. Thus y_l can be computed modulo q, with just one final reduction modulo q_2,

$$y_l \equiv (\sum_{n=0}^{N-1} h_n x_{l-n} \text{ modulo } q) \text{ modulo } q_2 \tag{8.95}$$

With this method [8.15, 16], if q is a pseudo Mersenne number, all operations but the last reduction are done in one's complement arithmetic. The price to be paid for use of this approach is that all operations modulo $(2^p - 1)$ must be executed on word lengths longer than that of the final result. However, the increase in number of operations is very limited when q_1 is small and the penalty is more than offset by the fact that p needs no longer to be a prime.

8.5.1 Pseudo Mersenne Transforms

We shall first consider the use of pseudo Mersenne transforms defined modulo $q = 2^p - 1$, with p composite. For p even, $q = (2^{p/2} - 1)(2^{p/2} + 1)$, and the transform length cannot be longer than that which is possible for $2^{p/2} - 1$. Thus, we need be concerned only with the cases corresponding to p odd. In order to specify the pseudo Mersenne transforms, we shall use the following theorem introduced by Erdelsky [8.4].

Theorem 8.7: Given a prime number p_1 and two integers u and g such that $u \geqslant 1$, $|g| \geqslant 2$, $g \not\equiv 1$ modulo p_1, the NTT of length $N = p_1^u$ and of root g supports circular convolution. This NTT is defined modulo $q_2 = (g^{p_1^u} - 1)/(g^{p_1^{u-1}} - 1)$.

In order to demonstrate this theorem, we must establish that the three conditions of theorem 8.4 are satisfied.

$$g^N \equiv 1 \text{ modulo } q_2 \tag{8.96}$$

$$N N^{-1} \equiv 1 \text{ modulo } q_2 \tag{8.97}$$

$$((g^{p_1^{u-1}} - 1), q_2) = 1. \tag{8.98}$$

The condition (8.96) follows immediately from the fact that $g^N \equiv 1$ modulo

$(g^{p_1^*} - 1)$, with q_2 factor of $(g^{p_1^*} - 1)$. For the condition (8.97) we note that Fermat's theorem implies that

$$g^{p_1} \equiv g \text{ modulo } p_1 \tag{8.99}$$

Hence, by repeated multiplications, $g^{p_1^*} = g^N \equiv g$ modulo p_1 and $g^{p_1^{*-1}} \equiv g$ modulo p_1. Since $g \not\equiv 1$ modulo p_1, we have $q_2 \equiv 1$ modulo p_1. Therefore, $(p_1, q_2) = 1$ and $(N, q_2) = 1$, which implies that N has an inverse modulo q_2.

In order to establish condition (8.98), note that

$$q_2 = 1 + g^{p_1^{*-1}} + \ldots + g^{(p_1-1)p_1^{*-1}} \tag{8.100}$$

or

$$q_2 = (g^{p_1^{*-1}} - 1)[g^{(p_1-2)p_1^{*-1}} + 2g^{(p_1-3)p_1^{*-1}} + \ldots p_1 - 1] + p_1, \tag{8.101}$$

which implies that $g^{p_1^{*-1}} - 1$ is mutually prime with q_2 if $[(g^{p_1^{*-1}} - 1), p_1] = 1$. This condition is immediately established, because $g^{p_1^{*-1}} \equiv g$ modulo p_1 and $g \not\equiv 1$ modulo p_1. Therefore (8.98) is proved and this completes the demonstration of the theorem.

We can now derive two classes of pseudo Mersenne transforms from the-

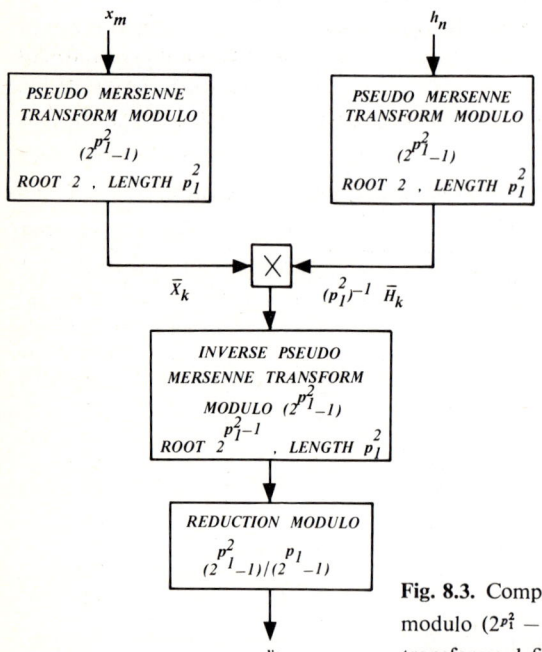

Fig. 8.3. Computation of a circular convolution modulo $(2^{p_1^2} - 1)/(2^{p_1} - 1)$ by pseudo Mersenne transforms defined modulo $(2^{p_1^2} - 1)$. p_1 prime

orem 8.7. One set of pseudo Mersenne transforms is obtained by setting $g = 2$ and $u = 2$. This yields pseudo Mersenne transforms of length $N = p_1^2$, p_1 prime, and defined modulo q_2 where

$$q_2 = (2^{p_1^2} - 1)/(2^{p_1} - 1). \tag{8.102}$$

Using these pseudo transforms, computation is executed modulo q, with $q = 2^{p_1^2} - 1$, on data words of p_1^2 bits and the final result is obtained by a final reduction modulo q_2, giving words of approximately $p_1^2 - p_1$ bits. Thus, the length $-p_1^2$ circular convolution modulo$(2^{p_1^2} - 1)/(2^{p_1} - 1)$ is computed as shown in Fig. 8.3, with the pseudo Mersenne transform

$$\bar{X}_k \equiv \sum_{m=0}^{p_1^2-1} x_m \, 2^{mk} \text{ modulo } (2^{p_1^2} - 1) \tag{8.103}$$

with similar definitions for the transform $\bar{H}_k$ of h_n, and for the inverse transform.

Another class of pseudo Mersenne transforms is derived from theorem 8.7 by setting $u = 1$, $g = 2^{p_2}$ and $p = p_1 p_2$. This gives pseudo Mersenne transforms of length $N = p_1$, p_1 prime, and defined modulo q_2, where

$$q_2 = (2^{p_1 p_2} - 1)/(2^{p_2} - 1). \tag{8.104}$$

With this pseudo transform, computation is executed modulo $(2^{p_1 p_2} - 1)$ on data word lengths of $p_1 p_2$ bits and the final result is obtained modulo $(2^{p_1 p_2} - 1)/(2^{p_2} - 1)$ on words of approximately $p_2(p_1 - 1)$ bits.

Table 8.1 lists the parameters for various pseudo Mersenne transforms defined modulo $(2^p - 1)$, with p odd. The most interesting transforms are those which have a composite number of terms and a useful word length as close as

Table 8.1. Parameters for various pseudo Mersenne transforms defined modulo $(2^p - 1)$ and convolutions defined modulo q_2, with q_2 factor of $2^p - 1$

p	Prime factorization of $2^p - 1$	Modulus q_2	Transform length N	Root g	Effective word length Nb of bits
15	7·31·151	$(2^{15} - 1)/7$	5	2^3	12
21	7^2·127·337	$(2^{21} - 1)/49$	7	2^3	15
25	31·601·1801	$(2^{25} - 1)/31$	25	2	20
27	7·73·262657	$(2^{27} - 1)/511$	27	2	18
35	31·71·127·122921	$(2^{35} - 1)/3937$	35	2	23
35	31·71·127·122921	$(2^{35} - 1)/(127)$	5	2^7	28
35	31·71·127·122921	$(2^{35} - 1)/31$	7	2^5	30
45	7·31·73·151·631·23311	$(2^{45} - 1)/511$	5	2^9	36
49	127·4432676798593	$(2^{49} - 1)/127$	7	2^7	42
49	127·4432676798593	$(2^{49} - 1)/127$	49	2	42

possible to p. In particular, the transform of 49 terms defined modulo $(2^{49} - 1)$ seems to be particularly interesting, since it can be computed with a 2-stage FFT-type algorithm and an effective word length which is only about 15% shorter than the computation word length.

8.5.2 Pseudo Fermat Number Transforms

Pseudo Fermat number transforms are defined modulo q, with $q = 2^v + 1$, $v \neq 2^t$. In order to specify the pseudo Fermat number transforms, we shall use the following theorem introduced by Erdelsky [8.4].

Theorem 8.8: Given an odd prime integer v_1 and two integers u and g such that $u \geqslant 1$, $g \geqslant 2$, $g \not\equiv -1$ modulo v_1, the NTT of length $N = 2v_1^u$ and of root g supports circular convolution. This NTT is defined modulo $q_2 = (g^{v_1^u} + 1)/(g^{v_1^{u-1}} + 1)$.

In order to prove this theorem, we must demonstrate that the three conditions of theorem 8.4 are satisfied.

$$g^N \equiv 1 \text{ modulo } q_2 \tag{8.105}$$

$$N N^{-1} \equiv 1 \text{ modulo } q_2 \tag{8.106}$$

$$[(g^{v_1^u} - 1), q_2] = 1 \quad \text{and} \quad [(g^{2v_1^{u-1}} - 1), q_2] = 1. \tag{8.107}$$

Since $g^{v_1^u} \equiv -1$ modulo $(g^{v_1^u} + 1)$, we have $g^N \equiv 1$ modulo $(g^{v_1^u} + 1)$ and therefore, $g^N \equiv 1$ modulo q_2, because q_2 is a factor of $g^{v_1^u} + 1$. To establish the condition (8.106), we note that Fermat's theorem implies that

$$g^{v_1} \equiv g \text{ modulo } v_1 \tag{8.108}$$

Hence, by repeated multiplications, $g^{v_1^{u-1}} \equiv g^{v_1^u} \equiv g$ modulo v_1. Thus, $q_2 \equiv 1$ modulo v_1, since $g \not\equiv -1$ modulo v_1. Therefore, $(v_1, q_2) = 1$ and we have $(N, q_2) = 1$ provided that q_2 is odd. When g is even, q_2 is obviously odd. When g is odd, we have

$$q_2 = (g^{v_1^{u-1}} + 1) [g^{(v_1 - 2)v_1^{u-1}} - 2g^{(v_1 - 3)v_1^{u-1}} + \ldots - (v_1 - 1)] + v_1, \tag{8.109}$$

which implies that q_2 is odd, since v_1 is odd. Thus we have $(N, q_2) = 1$ and N has an inverse modulo q_2.

In order to establish that the condition $[(g^{v_1^u} - 1), q_2] = 1$ corresponding to (8.107) is met, we note that

$$g^{v_1^u} - 1 = g^{v_1^u} + 1 - 2 = q_2(g^{v_1^{u-1}} + 1) - 2, \tag{8.110}$$

which implies that $[(g^{v_1^u} - 1), q_2] = (q_2, 2)$ and, since q_2 is odd,

$$[(g^{v_1^u} - 1), q_2] = 1, \qquad (8.111)$$

In order to establish that the condition $[(g^{2v_1^{u-1}} - 1), q_2] = 1$ is satisfied, we note that, since $g^{2v_1^{u-1}} - 1 = (g^{v_1^{u-1}} + 1)(g^{v_1^{u-1}} - 1)$, this condition corresponds to $[(g^{v_1^{u-1}} + 1), q_2] = 1$ and $[(g^{v_1^{u-1}} - 1), q_2] = 1$. The condition (8.110) implies that $[(g^{v_1^{u-1}} - 1), q_2] = 1$, since $(g^{v_1^{u-1}} - 1)$ is a factor of $(g^{v_1^u} - 1)$. We note also that (8.109) implies

$$[(g^{v_1^{u-1}} + 1), q_2] = [(g^{v_1^{u-1}} + 1), v_1] \qquad (8.112)$$

and, since $g \not\equiv -1$ modulo v_1 and $g^{v_1^{u-1}} \equiv g$ modulo v_1, we have $[(g^{v_1^{u-1}} + 1), q_2] = 1$, which completes the proof of the theorem.

An immediate application of theorem 8.8 is that, if $g = 2$ and $u = 1$, we can define for $v_1 \neq 3$ a NTT of length $N = 2v_1$ which supports circular convolution. This NTT is defined modulo q_2, with $q_2 = (2^{v_1} + 1)/3$.

Similarly, for $u = 2$ and $g = 2$, we have an NTT of length $N = 2v_1^2$. This NTT has the circular convolution property and is defined modulo q_2, with $q_2 = (2^{v_1^2} + 1)/(2^{v_1} + 1)$.

A systematic application of theorems 8.7 and 8.8 yields a large number of pseudo Fermat number transforms. We summarize the main characteristics of some of these transforms for v even and v odd, respectively, in Tables 8.2 and 8.3. It can be seen that there is a large choice of pseudo Fermat number transforms having a composite number of terms. This allows one to select word lengths that are more taylored to meet the needs of particular applications than when word lengths are limited to powers of two, as with FNTs.

The same pseudo transform technique can also be applied to moduli other than $2^p - 1$ and $2^v + 1$, and the cases corresponding to moduli $2^{2p} - 2^p + 1$

Table 8.2. Parameters for various pseudo Fermat number transforms defined modulo $(2^v + 1)$ and convolutions defined modulo q_2, with q_2 factor of $2^v + 1$. v even

v	Prime factorization of $2^v + 1$	Modulus q_2	Transform length N	Root g	Effective word length Nb of bits
20	17·61681	$(2^{20} + 1)/17$	40	2	16
22	5·397·2113	$(2^{22} + 1)/5$	44	2	19
24	97·257·673	$(2^{24} + 1)/257$	48	2	16
26	5·53·157·1613	$(2^{26} + 1)/5$	52	2	24
28	17·15790321	$(2^{28} + 1)/17$	56	2	24
34	5·137·953·26317	$(2^{34} + 1)/5$	68	2	32
38	5·229·457·525313	$(2^{38} + 1)/5$	76	2	36
40	257·4278255361	$(2^{40} + 1)/257$	80	2	32
44	17·353·2931542417	$(2^{44} + 1)/17$	88	2	40
46	5·277·1013·1657·30269	$(2^{46} + 1)/5$	92	2	44

8. Number Theoretic Transforms

Table 8.3. Parameters for various pseudo Fermat number transforms defined modulo $(2^v + 1)$ and convolutions defined modulo q_2, with q_2 factor of $2^v + 1$. v odd

v	Prime factorization of $2^v + 1$	Modulus q_2	Transform length N	Root g	Effective word length Nb of bits
15	$3^2 \cdot 11 \cdot 331$	$(2^{15} + 1)/9$	10	2^3	12
21	$3^2 \cdot 43 \cdot 5419$	$(2^{21} + 1)/9$	14	2^3	18
25	$3 \cdot 11 \cdot 251 \cdot 4051$	$(2^{25} + 1)/33$	50	2	20
27	$3^4 \cdot 19 \cdot 87211$	$(2^{27} + 1)/1539$	54	2	16
29	$3 \cdot 59 \cdot 3033169$	$(2^{29} + 1)/3$	58	2	27
33	$3^2 \cdot 67 \cdot 683 \cdot 20857$	$(2^{33} + 1)/9$	22	2^3	30
35	$3 \cdot 11 \cdot 43 \cdot 281 \cdot 86171$	$(2^{35} + 1)/33$	14	2^5	30
41	$3 \cdot 83 \cdot 8831418697$	$(2^{41} + 1)/3$	82	2	39
45	$3^3 \cdot 11 \cdot 19 \cdot 331 \cdot 18837001$	$(2^{45} + 1)/513$	10	2^9	36
49	$3 \cdot 43 \cdot 4363953127297$	$(2^{49} + 1)/129$	98	2	41

are discussed in [8.17]. These moduli are factors of $2^{6p} - 1$ and NTTs of dimension $6p$ and root 2 can be defined modulo some factors of $2^{2p} - 2^p + 1$.

When a convolution is calculated by pseudo transforms, the computation is performed modulo q and the final result is obtained modulo q_2, with $q = q_1 q_2$. Since $q_2 < q$, it is possible to detect overflow conditions by simply comparing the result of the calculations modulo q with the convolution product defined modulo q_2 [8.18].

8.6 Complex NTTs

We consider a complex integer $x + j\hat{x}$, where x and $\hat{x}$ are defined in the field $GF(q)$ of the integers defined modulo a prime q. Thus, x and $\hat{x}$ can take any integer value between 0 and $q - 1$. We also assume that $j = \sqrt{-1}$ is not a member of $GF(q)$, which means that -1 is a quadratic nonresidue modulo q. Then, the Gaussian integers $x + j\hat{x}$ are similar to ordinary complex numbers, with real and imaginary parts treated separately and addition and multiplication defined by

$$(x_1 + j\hat{x}_1) + (x_2 + j\hat{x}_2) = x_1 + x_2 + j(\hat{x}_2 + \hat{x}_2) \qquad (8.113)$$

$$(x_1 + j\hat{x}_1)(x_2 + j\hat{x}_2) = x_1 x_2 - \hat{x}_1 \hat{x}_2 + j(\hat{x}_1 x_2 + x_1 \hat{x}_2). \qquad (8.114)$$

Since each integer $\hat{x}$ and x can take only q distinct values, the Gaussian integers $x + j\hat{x}$ can take only q^2 distinct values and are said to pertain to the *extension field* $GF(q^2)$. Since $x + j\hat{x}$ pertain to a finite field, the successive powers of $x + j\hat{x}$ given by $(x + j\hat{x})^n$, $n = 0, 1, 2, \ldots$ yield a sequence which is reproduced with a periodicity N, and we can define roots of order N in $GF(q^2)$. In order to specify the permissible orders of the various roots, we note first that

$$(x + j\hat{x})^q = \sum_{i=0}^{q} C_k^i \, x^i \, \hat{x}^{q-i} \, j^{q-i}, \tag{8.115}$$

where the C_k^i are the binomial coefficients

$$C_k^i = \frac{q!}{i!(q-i)!}. \tag{8.116}$$

Since these coefficients are integers, $i!(q-i)!$ must divide $q!$. However $i!(q-i)!$ cannot divide q because q is a prime. Therefore $i!(q-i)!$ divides $(q-1)!$ for $i \neq 0, q$ and (8.115) reduces to

$$(x + j\hat{x})^q \equiv x^q + j^q \, \hat{x}^q \text{ modulo } q \tag{8.117}$$

and, thus, via Fermat's theorem,

$$(x + j\hat{x})^q \equiv x + j^q \, \hat{x} \text{ modulo } q. \tag{8.118}$$

If we now raise $x + j^q \, \hat{x}$ to the qth power, we obtain

$$(x + j\hat{x})^{q^2} \equiv x + j^{q^2} \, \hat{x} \text{ modulo } q. \tag{8.119}$$

Since $j^{q^2} = j$ for $q^2 \equiv 1$ modulo 4, we have in this case

$$(x + j\hat{x})^{q^2} \equiv x + j\hat{x} \text{ modulo } q. \tag{8.120}$$

This implies that, when $q^2 \equiv 1$ modulo 4, any root of order N in $GF(q^2)$ must satisfy the condition

$$N \mid (q^2 - 1). \tag{8.121}$$

The Diophantine equation $q^2 \equiv 1$ modulo 4 has only the two solutions $q_1 \equiv 1$ modulo 4 and $q_2 \equiv 3$ modulo 4. We know however that j cannot be a member of $GF(q)$. This implies that -1 must be a quadratic nonresidue modulo q and, therefore, that $(-1/q) = -1$, where $(-1/q)$ is the Legendre symbol. A consequence of theorem 2.11 is that

$$(-1/q) = (-1)^{(q-1)/2}. \tag{8.122}$$

This imposes the condition

$$q \equiv 3 \text{ modulo } 4. \tag{8.123}$$

This condition is established easily for Mersenne and pseudo Mersenne transforms defined modulo $(2^p - 1)$, because, in this case

$$q = 2^p - 1 \equiv -1 \equiv 3 \text{ modulo } 4. \tag{8.124}$$

If we now consider a simple Mersenne transform of dimension p with root 2, we can then define complex Mersenne transforms of lengths $4p$ and $8p$ by replacing the real root 2 with the complex roots g_1 and g_2

$$g_1 = 2j \tag{8.125}$$

$$g_2 = 2(1 + j). \tag{8.126}$$

Since p is an odd prime, we have $g_1^{4p} \equiv 1$ modulo q and $g_2^{8p} \equiv 1$ modulo q, with g_1^n and g_2^n taking, respectively, $4p$ and $8p$ distinct values for $n = 0, \ldots, 4p - 1$ and $n = 0, \ldots, 8p - 1$. Thus, we can define complex Mersenne transforms [8.12, 15] of length $4p$ and $8p$ which support the circular convolution by

$$\bar{X}_k \equiv \sum_{m=0}^{4p-1} x_m (2j)^{mk} \text{ modulo } (2^p - 1) \tag{8.127}$$

$$\bar{X}_k \equiv \sum_{m=0}^{8p-1} x_m (1+j)^{mk} \text{ modulo } (2^p - 1). \tag{8.128}$$

The advantage of these complex Mersenne transforms over the corresponding real transforms is that, for the same word length, the transform length is increased to $4p$ and $8p$ instead of $2p$ and that the transforms can be computed by a 3-stage FFT-type algorithm, without multiplications.

A similar approach is applicable to pseudo Mersenne transforms defined modulo $(2^p - 1)$ with p odd [8.15] and to pseudo Fermat number transforms defined modulo $(2^v + 1)$, with v odd [8.16]. Complex Mersenne transforms are particularly interesting because they can be implemented in one's complement arithmetic and their maximum length is both large and highly factorizable, leading to an efficient FFT-type implementation. This is exemplified by the transform defined modulo $(2^{49} - 1)$ which operates on data words of 49 bits with an effective word length of 42 bits. This transform has a maximum transform length equal to 392 and can be computed by utilizing 3-stage radix-2 and 2-stage radix-7 FFT-type algorithms.

Reed and Truong [8.19, 20] have investigated the general case of complex Mersenne transforms. They have shown that complex transforms which support the circular convolution can be defined modulo q, with $q = 2^p - 1$ and p prime for any length N such that $N | (q^2 - 1)$. Thus, we have

$$N | 2^{p+1}(2^{p-1} - 1). \tag{8.129}$$

For transforms of length N, with $N = 2^{p+1}$, the roots are given by

$$g = a + jb \tag{8.130}$$

with

$$a \equiv \pm 2^{2^{p-2}} \text{ modulo } (2^p - 1) \tag{8.131}$$

$$b \equiv \pm (-3)^{2^{p-2}} \text{ modulo } (2^p - 1). \tag{8.132}$$

This specifies relatively large transforms which operate in one's complement arithmetic and which can be computed entirely by a radix-2 FFT-type algorithm. Unfortunately, the roots are not simple and some general multiplications are required in the computation of the transform.

8.7 Comparison with the FFT

There is a large choice of NTT candidates for the computation of convolutions. From the standpoint of arithmetic operations count, the most useful NTTs are those which can be computed without multiplications and which can be calculated with an FFT-type algorithm while allowing the longest possible transform length for a given word length. In practice, this means that the best NTTs are Fermat number transforms and complex pseudo Mersenne transforms [8.21].

The maximum lengths for multiplication-free transforms is 128 for an FNT defined modulo ($2^{32} + 1$), with data words of 32 bits, and is 392 for a complex pseudo Mersenne transform defined modulo ($2^{49} - 1$), with effective word length of 42 bits. Thus, NTTs seem to be best suited for computing the convolution of short and medium length sequences in computers where the cost of multiplications is significantly higher than the cost of additions and when a high accuracy is required. A particularly interesting application concerns the evaluation of data sequences such as phase angles, which are by nature defined modulo an integer. In this case, there is no need to scale down the input sequence data to prevent overflow and both the input and output sequences are defined with the same number of bits.

In all other cases, the input sequences must be scaled down in order to avoid overflow and the number of bits corresponding to the input sequences must be less than half the number of bits of the output sequences. Since all calculations must be performed on words of length equal to that of the input sequence, the price to be paid for obtaining an exact answer with NTTs is the use of word lengths which are approximately twice that of those corresponding to other methods. For real convolutions, however, NTTs use only real arithmetic instead of complex arithmetic for FFTs so that the hardware requirements are about the same. A practical comparison between FFTs and FNTs has been reported in [8.7]. It has been shown that for convolutions of lengths in the range 32–2048 computed with FNTs on words of 32 bits, the computer execution times (IBM

370/155) were about 2 to 4 times shorter than with an efficient FFT program. In this comparison, the convolutions of lengths above 128 are computed by two-dimensional FNTs.

An interesting aspect of number theoretic transforms is their analogy with discrete Fourier transforms. NTTs are defined with roots of unity g of order N modulo an integer q, while DFTs are defined with complex roots of unity W of order N in the field of complex numbers. Hence NTTs can be viewed as DFTs defined in the ring of numbers modulo q. In fact, NTTs can also be considered as particular cases of polynomial transforms in which the N-bit words are viewed as polynomials. This is particularly apparent for polynomial transforms of length 2^{t+1} defined modulo $(Z^{2^t} + 1)$. Such transforms compute a circular convolution of length 2^{t+1} on polynomials of length 2^t. If the 2^{t+1} input polynomials are defined as words of 2^t bits, the polynomial transform reduces to an FNT of length 2^{t+1}, of root 2 and defined modulo $(2^{2^t} + 1)$. Thus, polynomial transforms and NTTs are DFTs defined in finite fields and rings of polynomials or integers. Their main advantage over DFTs is that systematic advantage is taken from the operation in finite fields or rings to define simple roots of unity which allow one to eliminate the multiplications for transform computation and to replace complex arithmetic by real arithmetic.

Appendix A Relationship Between DFT and Convolution Polynomial Transform Algorithms

The main objective of this chapter is to present an inverse polynomial transform algorithm for the computation of multidimensional DFTs and to show that the multidimensional convolution algorithms discussed in Chap. 6 may be viewed as a combination of this inverse algorithm with the direct reduced DFT algorithm introduced in Chap. 7. This leads to develop new multidimensional convolution algorithms based on other combinations of the direct and inverse DFT polynomial transform algorithms and which have both a simple structure and reduced computational complexity. We conclude this chapter by giving direct and inverse polynomial transform algorithms for the computation of multidimensional cosine transforms.

A.1 Computation of Multidimensional DFTs by the Inverse Polynomial Transform Algorithm

We consider a two-dimensional DFT, $\bar{X}_{k_1,k_2}$, of size $N_1 \times N_2$

$$\bar{X}_{k_1,k_2} = \sum_{n_1=0}^{N_1-1} \sum_{n_2=0}^{N_2-1} x_{n_1,n_2} W_1^{2n_1 k_1} W_2^{2n_2 k_2},$$

$$W_1 = e^{-j\pi/N_1}, \quad W_2 = e^{-j\pi/N_2}, \quad j = \sqrt{-1},$$

$$k_1 = 0, \ldots, N_1-1, \quad k_2 = 0, \ldots, N_2-1. \tag{A.1}$$

This two-dimensional DFT may be computed by one of the polynomial transform algorithms of Chap. 7. We shall now show that this DFT may also be evaluated by other polynomial transform algorithms that are similar to the algorithms of Chap. 7, but with all operations performed in reverse order and with polynomial transform replaced by inverse polynomial transforms.

A.1.1 The Inverse Polynomial Transform Algorithm

We shall introduce the inverse polynomial transform algorithm by using modified rings of polynomials [A.1-2], as in Sect. 7.1.5, with $N_1 = 2^{t_1}$ and $N_2 = 2^{t_2}$. We first rewrite (A.1) as

Appendix A

$$\bar{X}_{k_1,k_2} = W_2^{-k_2} \sum_{n_1=0}^{N_1-1} \sum_{n_2=0}^{N_2-1} x_{n_1,n_2} W_1^{2n_1 k_1} W_2^{(2n_2+1)k_2}. \quad (A.2)$$

Since $2n_2 + 1$ is odd and n_1 is defined modulo 2^{t_1}, the permutation $(2n_2+1)n_1$ modulo N_1, with $N_1 = 2^{t_1}$, maps all values of n_1. With this permutation, $\bar{X}_{k_1,k_2}$ becomes

$$\bar{X}_{k_1,k_2} = W_2^{-k_2} \sum_{n_1=0}^{N_1-1} \sum_{n_2=0}^{N_2-1} x_{(2n_2+1)n_1,n_2} W_1^{2(2n_2+1)n_1 k_1} W_2^{(2n_2+1)k_2}. \quad (A.3)$$

We begin the calculation of $\bar{X}_{k_1,k_2}$ by evaluating the N_1 odd-time DFTs of size N_2

$$\bar{X}_{n_1,k_2} = \sum_{n_2=0}^{N_2-1} x_{(2n_2+1)n_1,n_2} W_2^{(2n_2+1)k_2}. \quad (A.4)$$

The $N_1 N_2$ output samples $\bar{X}_{n_1,k_2}$ are organized as N_1 polynomials of N_2 terms with

$$X_{n_1}(z) = \sum_{k_2=0}^{N_2-1} \left[\sum_{n_2=0}^{N_2-1} x_{(2n_2+1)n_1,n_2} W_2^{(2n_2+1)k_2} \right] z^{k_2}. \quad (A.5)$$

We now show that a multiplication of $X_{n_1}(z)$ by $W_2^{(2n_2+1)d}$ is equivalent to a multiplication modulo $(z^{N_2}+1)$ of $X_{n_1}(z)$ by z^{-d}.

Multiplying $X_{n_1}(z)$ by z^{-d} yields

$$z^{-d} X_{n_1}(z) = \sum_{k_2=0}^{N_2-1} \left[\sum_{n_2=0}^{N_2-1} x_{(2n_2+1)n_1,n_2} W_2^{(2n_2+1)k_2} \right] z^{k_2-d} \quad (A.6)$$

$$z^{-d} X_{n_1}(z) = \sum_{k_2=d}^{N_2-1} \left[\sum_{n_2=0}^{N_2-1} x_{(2n_2+1)n_1,n_2} W_2^{(2n_2+1)(k_2+d)} \right] z^{k_2}$$

$$+ \sum_{k_2=0}^{d-1} \left[\sum_{n_2=0}^{N_2-1} x_{(2n_2+1)n_1,n_2} W_2^{(2n_2+1)(k_2+d+N_2)} \right] z^{k_2+N_2}. \quad (A.7)$$

Since $W_2^{N_2} = -1$ and $z^{N_2} \equiv -1$ modulo $(z^{N_2}+1)$, we have

$$z^d X_{n_1}(z) \equiv W^{(2n_2+1)d} X_{n_1}(z) \text{ modulo } (z^{N_2}+1). \quad (A.8)$$

Once $X_{n_1}(z)$ is evaluated, $\bar{X}_{k_1,k_2}$ is computed by multiplications of $X_{n_1}(z)$ by $W_1^{2(2n_2+1)n_1 k_1}$

$$\bar{X}_{k_1}(z) = W_2^{-k_2} \sum_{n_1=0}^{N_1-1} X_{n_1}(z) W_1^{2(2n_2+1)n_1 k_1}. \quad (A.9)$$

A.1 Computation of Multidimensional DFTs

However, (A.8) means that the multiplications by $W_1^{2(2n_2+1)n_1k_1}$ may be replaced by multiplications by $z^{-2n_1k_1N_2/N_1}$ modulo $(z^{N_2}+1)$. Thus, (A.9) may be replaced by (A.10) with

$$\bar{X}_{k_1}(z) \equiv \sum_{n_1=0}^{N_1-1} X_{n_1}(z) z^{-2n_1k_1N_2/N_1} \text{ modulo } (z^{N_2}+1) \tag{A.10}$$

$$\bar{X}_{k_1}(z) = \sum_{k_2=0}^{N_2-1} \bar{X}_{k_1,k_2} W_2^{k_2} z^{k_2}. \tag{A.11}$$

Equation (A.10) represents an inverse polynomial transform of size N_1 which has roots that are simple powers of z if $2N_2 \geqslant N_1$. Therefore the inverse polynomial transform is computed without multiplications if $2N_2 \geqslant N_1$. Thus, the DFT of size $N_1 \times N_2$ is computed as shown in Fig. A.1 with input row-column permutations followed by N_1 odd-time DFTs of size N_2, one inverse polynomial transform of size N_1 and N_1N_2 output multiplications by $W_2^{-k_2}$. It may be seen that the computation procedure is the same as that corresponding to the direct polynomial transform algorithm represented in Fig. 7.4, except that

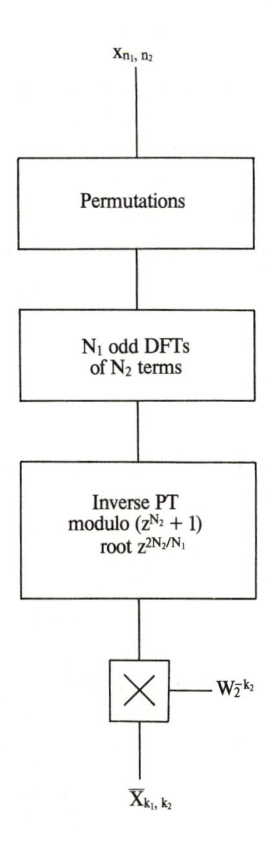

Fig. A.1. Computation of a DFT of size $N_1 \times N_2$ by the inverse polynomial transform algorithm

the order of the operations is reversed and the polynomial transform is replaced by an inverse polynomial transform. Hence the computational complexity of the inverse polynomial transform algorithm is identical to that of the direct polynomial transform algorithm.

By using the general approach developped in Sect. 7.1, it may be seen easily that the computation may be split into two parts corresponding respectively to n_2 odd and n_2 even [A.2]. This partial decomposition method, which is patterned after the technique shown in Fig. 7.5, reduces the number of postmultiplications by W^{-k_2} to only $N_1 N_2/2$ and replaces the inverse polynomial transform defined modulo $(z^{N_2}+1)$ and the N_1 odd-time DFTs of size N_2 by two inverse polynomial transforms defined modulo $(z^{N_2}+1)$ and $2N_1$ odd-time DFTs of size $N_2/2$. The postmultiplications by W^{-k_2} may be eliminated altogether by resorting to a complete decomposition process similar to that shown in Fig. 7.2, but with operations performed in reverse order, polynomial transforms replaced by inverse polynomial transforms and odd-frequency DFTs replaced by odd-time DFTs. In the same way, the inverse polynomial transform may be generalized to any multidimensional DFT, provided some common factors exist among at least two dimensions.

It would seem that the inverse polynomial transform algorithms do not bring significant advantages over the corresponding direct transform algorithms, since they have the same computational complexity. However, we shall show that the inverse polynomial transform algorithms is less susceptible to round-off noise than the direct polynomial transform algorithm and that the two algorithms may be combined to produce new convolution algorithms that combine a good structural simplicity with a low computational complexity. Prior to developping these points, we first discuss some interesting polynomial transform algorithms which are based on rings of polynomials modulo (z^N+j) or modulo (z^N-j).

A.1.2 Complex Polynomial Transform Algorithms

In the preceding section, we have seen that a DFT of size $N_1 \times N_2$, with $N_1 = 2^{t_1}$, $N_2 = 2^{t_2}$, is computed by an inverse polynomial transform of size N_1 defined modulo $(z^{N_2}+1)$. The inverse polynomial transform is computed without multiplications provided $2N_2 \geqslant N_1$. If this condition is not satisfied, then we have $2N_1 \geqslant N_2$, and the obvious solution is to permute the role of the lines and the columns. However, if the conditions $2N_2 \geqslant N_1$ and $2N_1 \geqslant N_2$ are satisfied simultaneously, that is to say if the DFTs are of sizes $2N \times N$, $N \times N$ and $N \times 2N$, the algorithm may be implemented either with an inverse polynomial transform of size N_1 defined modulo $(z^{N_2}+1)$ or with an inverse polynomial transform of size N_2 defined modulo $(z^{N_1}+1)$. In the first case, the DFT is mapped into N_1 odd-time DFTs of size N_2 while the second case corresponds to N_2 odd-time DFTs of size N_1. It is therefore clear that one must choose the algorithm that results in the largest polynomial transform in order to reduce the size of the odd-time DFTs. We shall show now that the condition $2N_2 \geq N_1$ may be relaxed to $4N_2 \geq N_1$ by using complex polynomial transforms that can still be computed without multiplications [A.2] and which map the DFT more efficiently than real polynomial transforms. We shall illustrate this point by considering a DFT of size $N \times 4N$ computed by the inverse polynomial transform algorithm. Assuming $N = 2^t$,

$$\bar{X}_{k_1,k_2} = \sum_{n_1=0}^{4N-1} \sum_{n_2=0}^{N-1} x_{n_1,n_2} W^{n_1 k_1} W^{4n_2 k_2}$$

$$W = e^{-j2\pi/4N}, \quad k_1 = 0, \ldots, 4N-1, \quad k_2 = 0, \ldots, N-1, \quad j = \sqrt{-1}. \quad (A.12)$$

By introducing postmultiplications by W^{-k_2} and using the permutation $(4n_2+1)n_1$, the DFT may be modified with

$$\bar{X}_{k_1,k_2} = W^{-k_2} \sum_{n_1=0}^{4N-1} \sum_{n_2=0}^{N-1} x_{(4n_2+1)n_1,n_2} W^{(4n_2+1)n_1 k_1} W^{(4n_2+1)k_2}. \quad (A.13)$$

We begin the computation of $\bar{X}_{k_1,k_2}$ by evaluating $4N$ odd-time DFTs of size N and by organizing the output samples of these DFTs into $4N$ polynomials of N terms

$$X_{n_1}(z) = \sum_{k_2=0}^{N-1} \left(\sum_{n_2=0}^{N-1} x_{(4n_2+1)n_1,n_2} W^{(4n_2+1)k_2} \right) z^{k_2}. \quad (A.14)$$

Since $W^{-N} = j$, a multiplication of $X_{n_1}(z)$ by z^{-d} modulo $(z^N - j)$ is equivalent to a multiplication of $X_{n_1}(z)$ by $W^{(4n_2+1)d}$ and $\bar{X}_{k_1,k_2}$ may be derived from $X_{n_1}(z)$ by

$$\bar{X}_{k_1}(z) \equiv \sum_{n_1=0}^{4N-1} X_{n_1}(z) z^{-n_1 k_1} \quad \text{modulo } (z^N - j) \quad (A.15)$$

$$\bar{X}_{k_1}(z) = \sum_{k_2=0}^{N-1} \bar{X}_{k_1,k_2} W^{k_2} z^{k_2} \quad (A.16)$$

where (A.15) defines an inverse polynomial transform modulo $(z^N - j)$. The computation process is summarized in Fig. A.2. The evaluation of the inverse polynomial transform (A.15) is done with polynomial additions and multiplications by powers of z modulo $(z^N - j)$. Since these multiplications are simple polynomial rotations where the overflow words are multiplied by j, the computation of the inverse polynomial transform defined modulo $(z^N - j)$ do not require any non-trivial multiplications. It is shown easily [A.2] that if the odd DFTs are computed by a conventional radix-2 FFT algorithm where the trivial multiplications by ± 1, $\pm j$ in the two first stages are not counted, the complex polynomial transform algorithm computes the DFT of size $N \times 4N$ with

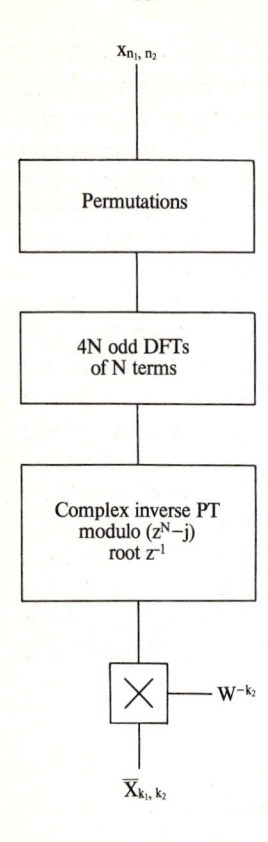

Fig. A.2. Computation of a DFT of size $N \times 4N$ by a complex inverse polynomial transform algorithm

$8N^2(2+\log_2 N)$ real multiplications and $4N^2(6+5\log_2 N)$ real additions. If we had used the algorithm using real polynomial transforms, as in the preceding section, the DFT of size $N \times 4N$ would have been mapped into N odd DFTs of length $4N$ by an inverse polynomial transform of size N defined modulo $(z^{4N}+1)$. In this case, the DFT of size $N \times 4N$ would have been computed with $8N^2(4+\log_2 N)$ real multiplications and $4N^2(8+5\log_2 N)$ real additions. Thus, the use of a complex polynomial transforms of length $4N$ instead of a real polynomial transform of size N saves $16N^2$ real multiplications and $8N^2$ real additions.

Complex polynomial transform may be used in a similar way with the direct polynomial transform algorithm and with a partial or complete decomposition technique.

A.1.3 Round-off Error Analysis

When a two-dimensional DFT of size $N_1 \times N_2$ is computed either by the direct or inverse polynomial transform algorithm with complete decomposition, as in Fig. 7.2, the computation process may be viewed as similar to a row-column procedure in which the DFTs calculated on the lines or columns are replaced by polynomial transforms. Hence the round-off noise analysis may be performed as in Sect. 4.2.4, on the $\log_2 N_1 + \log_2 N_2$ stages of the FFT decomposition. However, the FFT stages that are replaced by polynomial transforms have only trivial multiplications which do not produce any round-off noise. Therefore the only noise associated with these stages relates to the additions. This means that the overall noise produced by the polynomial transform algorithm is less than the noise associated with the conventional row-column method. The fixed-point error analysis has been done in [A.2]. It has been shown that the inverse polynomial transform algorithm is superior to the direct polynomial transform algorithm from the standpoint of round-off noise. This is due to the fact that in the fixed-point FFT computation of a DFT, the first stages contribute proportionally less to the total round-off noise because of the scaling introduced at each stage. It is therefore advantageous, from the standpoint of round-off noise, to move the polynomial transforms at the end of the algorithm, as is done with the inverse polynomial transform method. In the case of a DFT of size $N \times N$, with $N = 2^t$ and large values of N, the analysis shows that the ratio rms(error)/rms(signal) is about 1.7 times less with the inverse polynomial transform method than with the conventional row-column method.

A.2 Computation of Multidimensional Convolutions by a Combination of the Direct and Inverse Polynomial Transform Methods

We consider now the computation of a two-dimensional circular convolution of size $N_1 \times N_2$

$$y_{u_1,u_2} = \sum_{n_1=0}^{N_1-1} \sum_{n_2=0}^{N_2-1} h_{n_1,n_2} x_{u_1-n_1, u_2-n_2}. \tag{A.17}$$

In order to simplify the presentation, we will restrict the following discussion to the cases corresponding to $N_1 = 2^{t_1}$, $N_2 = 2^{t_2}$, although the method which we will describe applies to any multidimensional circular convolutions with some common factors among at least two dimensions.

The circular convolution (A.17) may be evaluated by using the circular convolution property of the DFT. In this case, the convolution is evaluated by computing the two-dimensional DFTs $\bar{X}_{k_1,k_2}$ and $\bar{H}_{k_1,k_2}$ of the input sequences x_{m_1,m_2} and h_{n_1,n_2}, and y_{u_1,u_2} is obtained by calculating the inverse DFT of $\bar{X}_{k_1,k_2} \bar{H}_{k_1,k_2}$. Since the two-dimensional DFTs of Size $N_1 \times N_2$ and the two-dimensional inverse DFT of size $N_1 \times N_2$ have common factors in both dimensions, their computation may be simplified by using a polynomial transform algorithm. In the following, we will show that by using various combinations of different polynomial transform algorithms for the evaluation of the direct DFTs and the inverse DFT, it is possible to derive several new polynomial transform algorithms for the computation of convolutions. We will begin this discussion by indicating how the conventional convolution polynomial transform algorithms relate to DFT polynomial transform algorithms.

A.2.1 Computation of Convolutions by DFT Polynomial Transform Algorithms

We assume here that the two-dimensional circular convolution of size $N_1 \times N_2$ is computed with two DFTs of size $N_1 \times N_2$ and an inverse DFT of size $N_1 \times N_2$ where the DFTs are evaluated by the direct polynomial transform algorithm of Sect. 7.1.5 and the inverse DFT is calculated by the inverse polynomial transform algorithm of Sect. A.1.1. Under these conditions, the permuted output samples $\bar{H}_{(2k_2+1)k_1,k_2}$ of the DFT of h_{n_1,n_2} are obtained by

$$H_{n_1}(z) = \sum_{n_2=0}^{N_2-1} h_{n_1,n_2} W_2^{-n_2} z^{n_2} \tag{A.18}$$

$$\bar{H}_{(2k_2+1)k_1}(z) \equiv \sum_{n_1=0}^{N_1-1} H_{n_1}(z) z^{2n_1 k_1 N_2/N_1} \quad \text{modulo } (z^{N_2}+1) \tag{A.19}$$

$$\bar{H}_{(2k_2+1)k_1,k_2} \equiv \bar{H}_{(2k_2+1)k_1}(z) \quad \text{modulo } (z - W_2^{2k_2+1}), \tag{A.20}$$

with $W_1 = e^{-j\pi/N_1}$, $W_2 = e^{-j\pi/N_2}$, and similar relations for the permuted DFT $\bar{X}_{(2k_2+1)k_1,k_2}$ of x_{m_1,m_2}. We now compute the inverse DFT of $\bar{H}_{k_1,k_2}\bar{X}_{k_1k_2}$ by the inverse polynomial transform method of Sect. A.1.1. This gives

$$Y_{k_1}(z) = \sum_{u_2=0}^{N_2-1} \left[\sum_{k_2=0}^{N_2-1} \frac{1}{N_1 N_2} \bar{H}_{(2k_2+1)k_1,k_2} \bar{X}_{(2k_2+1)k_1,k_2} W_2^{-(2k_2+1)u_2} \right] z^{u_2} \tag{A.21}$$

$$\bar{Y}_{u_1}(z) \equiv \sum_{k_1=0}^{N_1-1} Y_{k_1}(z) z^{-2u_1 k_1 N_2/N_1} \quad \text{modulo } (z^{N_2}+1) \tag{A.22}$$

$$\bar{Y}_{u_1}(z) = \sum_{u_2=0}^{N_2-1} y_{u_1,u_2} W_2^{-u_2} z^{u_2}. \tag{A.23}$$

Since the output row-column permutations in (A.20) are identical to the input row-column permutations in (A.21), these permutations cancel and may be ignored. Thus, the convolution of size $N_1 \times N_2$ is computed with $2N_1 N_2$ multiplications by $W_2^{-n_2}$ and $W_2^{-m_2}$, two direct polynomial transforms defined modulo $(z^{N_2}+1)$, $2N_1$ odd-frequency DFTs of size N_2, $N_1 N_2$ multiplications,

N_1 odd-time inverse DFTs of size N_2, an inverse polynomial transform and $N_1 N_2$ multiplications by W^{u_2}. It may be seen that this convolution algorithm is identical to the method described in Sect. 6.5 and shown schematically in Fig. 6.6 which computes the convolution with premultiplications by $W_2^{-n_2}$, $W_2^{-m_2}$, two polynomial transforms, N_1 polynomial products modulo $(z^{N_2} + 1)$, an inverse polynomial transform and postmultiplications by $W_2^{u_2}$. The only difference between the two methods is that the polynomial products are computed here by odd DFTs.

It can be shown easily that the other convolution techniques using polynomial transforms which are described in Chap. 6 may be viewed as derived from the direct and inverse DFT polynomial transform algorithms. We shall show now that the use of different combinations of the direct and inverse DFT polynomial transform algorithms yields new convolution techniques with interesting properties.

A.2.2 Convolution Algorithms Based on Polynomial Transforms and Permutations

In the preceding section, the two-dimensional convolution is computed by two-dimensional DFTs evaluated by direct polynomial transform algorithms and by a two-dimensional inverse DFT calculated by the inverse polynomial transform algorithm. The choice of the direct polynomial transform algorithm for the direct DFTs and of the inverse polynomial transform algorithm for the inverse DFT is totally arbitrary, and one may choose any of the two polynomial transform algorithm for the evaluation of the two-dimensional DFTs and inverse DFTs. Moreover, each of these algorithms may be based on polynomial transforms defined on the lines or on the columns. This yields a large number of possible convolution algorithms which provide interesting variations over the basic convolution technique using polynomial transforms. We shall show here that the premultiplications and postmultiplications in the preceding algorithm may be replaced by simple permutations when the DFTs are calculated by the inverse polynomial transform algorithm and the inverse DFT is evaluated by the direct polynomial transform algorithm. In this case, the DFT of the input sequence x_{m_1, m_2} is computed as Sect. A.1.1 by

$$X_{m_1}(z) = \sum_{k_2=0}^{N_2-1} \left[\sum_{m_2=0}^{N_2-1} x_{(2m_2+1)m_1, m_2} W_2^{(2m_2+1)k_2} \right] z^{k_2} \qquad (A.24)$$

$$\bar{X}_{k_1}(z) \equiv \sum_{m_1=0}^{N_1-1} X_{m_1}(z) z^{-2m_1 k_1 N_2/N_1} \quad \text{modulo } (z^{N_2}+1) \qquad (A.25)$$

$$\bar{X}_{k_1}(z) = \sum_{k_2=0}^{N_2-1} \bar{X}_{k_1, k_2} W_2^{k_2} z^{k_2},$$

$W_1 = e^{-j\pi/N_1}, W_2 = e^{-j\pi/N_2}, k_1 = 0, \ldots, N_1-1, k_2 = 0, \ldots, N_2-1 \qquad (A.26)$

with similar relations for the DFT $\bar{H}_{k_1,k_2}$ of the second input sequence h_{n_1,n_2}. Using the direct polynomial transform method for the inverse length $(N_1 \times N_2)$ DFT of $\bar{H}_{k_1,k_2} \bar{X}_{k_1,k_2}$ yields

$$Y_{k_1}(z) = \sum_{k_2=0}^{N_2-1} \frac{1}{N_1 N_2} \bar{H}_{k_1,k_2} \bar{X}_{k_1,k_2} W_2^{k_2} z^{k_2} \tag{A.27}$$

$$\bar{Y}_{(2u_2+1)u_1}(z) \equiv \sum_{k_1=0}^{N_1-1} Y_{k_1}(z) z^{2u_1 k_1 N_2/N_1} \quad \text{modulo } (z^{N_2}+1) \tag{A.28}$$

$$y_{(2u_2+1)u_1,u_2} \equiv \bar{Y}_{(2u_2+1)u_1}(z) \quad \text{modulo } (z - W_2^{-2k_2-1}). \tag{A.29}$$

Since the postmultiplications by $W_2^{-k_2}$ in (A.26) and the premultiplications by $W_2^{k_2}$ in (A.27) cancel, the two-dimensional convolution of size $N_1 \times N_2$ may be computed as shown in Fig. A.3. When the sequence h_{n_1,n_2} is fixed, the DFT $\bar{H}_{k_1,k_2}$ may be precomputed and the premultiplications and postmultiplications are eliminated. It may be seen that this algorithm replaces these multipli-

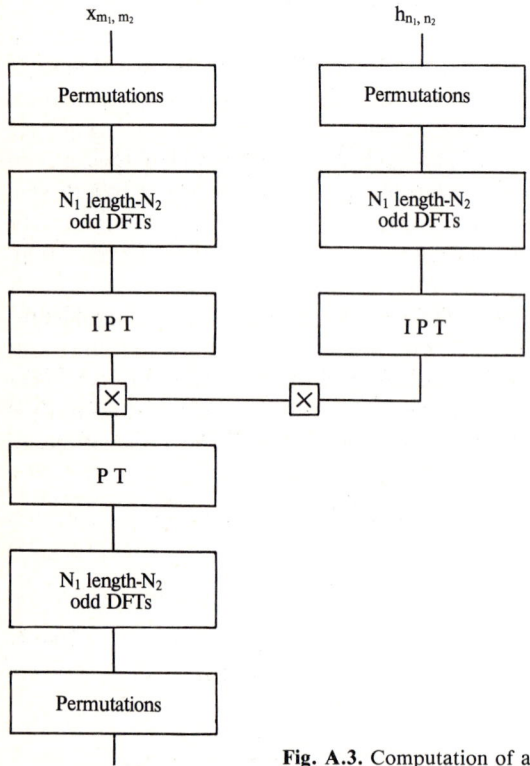

Fig. A.3. Computation of a two-dimensional circular convolution by polynomial transforms and permutations

cations by permutations. Since these permutations may be merged with the bit reversals in the DFTs, their cost is usually less than the cost of the auxiliary multiplications and this procedure is faster than the method discussed in the preceding section.

Similar techniques may be used to reduce the number of auxiliary multiplications in the partitioned algorithms.

A.3 Computation of Multidimensional Discrete Cosine Transforms by Polynomial Transforms

Multidimensional DCT (discrete cosine transforms) are used primarily for image compression. We will show in this section that the computation of multidimensional DCTs may be simplified by the use of direct and inverse polynomial transform algorithms.

A.3.1 Computation of Direct Multidimensional DCTs

The two-dimensional DCT $\bar{X}_{k_1,k_2}$ of a $(N_1 \times N_2)$-point sequence x_{n_1,n_2} may be defined by

$$\bar{X}_{k_1,k_2} = 4 \sum_{n_1=0}^{N_1-1} \sum_{n_2=0}^{N_2-1} x_{n_1,n_2} \cos\left[\frac{\pi k_1(2n_1+1)}{2N_1}\right] \cos\left[\frac{\pi k_2(2n_2+1)}{2N_2}\right]$$

$$k_1 = 0,\ldots,N_1-1, \quad k_2 = 0,\ldots,N_2-1, \tag{A.30}$$

where x_{n_1,n_2} may be either real or complex. In the following, we will assume that $N_1 = 2^{t_1}$ and $N_2 = 2^{t_2}$.

In a first step, we use a slightly modified version of the method introduced in [A.3] to convert the $(N_1 \times N_2)$-point DCT into a $(N_1 \times N_2)$-point DFT. This is done by defining the auxiliary sequence y_{n_1,n_2} which is derived from x_{n_1,n_2} by

$$y_{n_1,n_2} = \begin{bmatrix} x_{2n_1,2n_2}, & 0 \leqslant n_1 < \frac{N_1}{2}, & 0 \leqslant n_2 < \frac{N_2}{2} \\ x_{2N_1-2n_1-1,2n_2}, & \frac{N_1}{2} \leqslant n_1 < N_1, & 0 \leqslant n_2 < \frac{N_2}{2} \\ x_{2n_1,2N_2-2n_2-1}, & 0 \leqslant n_1 < \frac{N_1}{2}, & \frac{N_2}{2} \leqslant n_2 < N_2 \\ x_{2N_1-2n_1-1,2N_2-2n_2-1}, & \frac{N_1}{2} \leqslant n_1 < N_1, & \frac{N_2}{2} \leqslant n_2 < N_2. \end{bmatrix} \tag{A.31}$$

We now compute the odd-time DFT $\bar{y}_{k_1,k_2}$ of y_{n_1,n_2}

$$\bar{y}_{k_1,k_2} = W_1^{k_1} W_2^{k_2} \sum_{n_1=0}^{N_1-1} \sum_{n_2=0}^{N_2-1} y_{n_1,n_2} W_1^{4n_1 k_1} W_2^{4n_2 k_2},$$

$$W_1 = e^{-j2\pi/4N_1}, \quad W_2 = e^{-j2\pi/4N_2}. \tag{A.32}$$

The DCT $\bar{X}_{k_1,k_2}$ is obtained from $\bar{y}_{k_1,k_2}$ by the simple relations

$$\bar{X}_{k_1,k_2} = \bar{y}_{k_1,k_2} + j\bar{y}_{N_1-k_1,k_2} + j\bar{y}_{k_1,N_2-k_2} - \bar{y}_{N_1-k_1,N_2-k_2}, \quad k_1, k_2 \neq 0 \tag{A.33}$$

$$\bar{X}_{0,0} = 4\bar{y}_{0,0}$$

$$\bar{X}_{0,k_2} = 2(\bar{y}_{0,k_2} + j\bar{y}_{0,N_2-k_2}), \quad k_2 \neq 0 \tag{A.34}$$

$$\bar{X}_{k_1,0} = 2(\bar{y}_{k_1,0} + j\bar{y}_{N_1-k_1,0}), \quad k_1 \neq 0. \tag{A.35}$$

These relations may be verified easily by substituting $\bar{y}_{k_1,k_2}$ defined by (A.32) into (A.33–35). The most difficult part of the procedure corresponds to the calculation of $\bar{y}_{k_1,k_2}$ which we will now simplify by using the inverse poly-

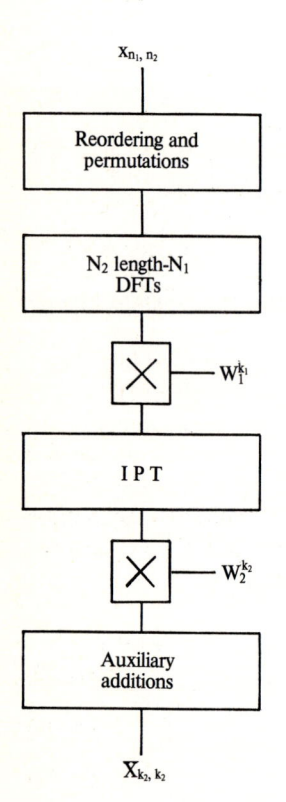

Fig. A.4. Computation of a two-dimensional DCT by an inverse polynomial transform

nomial transform algorithm [A.4]. This is done by first performing the row-column permutation $(4n_1+1)n_2$ modulo N_2 and by computing N_2 odd DFTs along the lines, with

$$Y_{n_2}(z) = \sum_{k_1=0}^{N_1-1} \left[\sum_{n_1=0}^{N_1-1} y_{n_1,(4n_1+1)n_2} W_1^{(4n_1+1)k_1} \right] z^{k_1} . \tag{A.36}$$

$\bar{y}_{k_1,k_2}$ is then obtained by a complex inverse polynomial transform

$$\bar{Y}_{k_2}(z) \equiv \sum_{n_2=0}^{N_2-1} Y_{n_2}(z) z^{-4n_2 k_2 N_1/N_2} \quad \text{modulo } (z^{N_1} - j) \tag{A.37}$$

$$\bar{Y}_{k_2}(z) = \sum_{k_1=0}^{N_1-1} \bar{y}_{k_1,k_2} W_2^{-k_2} z^{k_1} . \tag{A.38}$$

The computation procedure is shown schematically in Fig. A.4. When $4N_1 \geq N_2$, the inverse polynomial transform is evaluated without multiplications and this polynomial transform method saves for large DCTs about 50% of the multiplications and 15% of additions over the conventional row-column method.

A.3.2 Computation of Inverse Multidimensional DCTs

The inverse DCT of the input sequence $\bar{X}_{k_1,k_2}$ may be defined by

$$x_{n_1,n_2} = \frac{1}{N_1 N_2} \sum_{k_1=0}^{N_1-1} \sum_{k_2=0}^{N_2-1} v_{k_1,k_2} \cos\left[\frac{\pi(2n_1+1)k_1}{2N_1}\right] \cos\left[\frac{\pi(2n_2+1)k_2}{2N_2}\right]$$

$$v_{k_1,k_2} = \bar{X}_{k_1,k_2}, \quad k_1, k_2 \neq 0$$

$$v_{0,0} = \bar{X}_{0,0}/4$$

$$v_{k_1,0} = \bar{X}_{k_1,0}/2, \quad v_{0,k_2} = \bar{X}_{0,k_2}/2 . \tag{A.39}$$

The inverse DCT is computed with a procedure which is similar to the method used in the preceding section, by evaluating the inverse odd DFT y_{n_1,n_2} of the auxiliary sequence Y_{k_1,k_2}, with

$$Y_{k_1,k_2} = \bar{X}_{k_1,k_2} - j\bar{X}_{N_1-k_1,k_2} - j\bar{X}_{k_1,N_2-k_2} - \bar{X}_{N_1-k_1,N_2-k_2}, \quad k_1, k_2 \neq 0 \tag{A.40}$$

$$Y_{0,0} = \bar{X}_{0,0} \tag{A.41}$$

$$Y_{0,k_2} = \bar{X}_{0,k_2} - j\bar{X}_{0,N_2-k_2}, \quad k \neq 0 \tag{A.42}$$

$$Y_{k_1,0} = \bar{X}_{k_1,0} - j\bar{X}_{N_1-k_1,0}, \quad k_1 \neq 0 \tag{A.43}$$

$$y_{n_1,n_2} = \frac{1}{N_1 N_2} \sum_{k_1=0}^{N_1-1} \sum_{k_2=0}^{N_2-1} Y_{k_1,k_2} W_1^{-k_1} W_2^{-k_2} W_1^{-4n_1 k_1} W_2^{-4n_2 k_2}. \tag{A.44}$$

It can then be shown easily, by substituting (A.40–43) into (A.44), that the DCT output samples x_{n_1,n_2} are derived from y_{n_1,n_2} by

$$x_{2n_1,2n_2} = y_{n_1,n_2}/4, \quad n_1, n_2 = 0, \ldots, \frac{N}{2} - 1$$

$$x_{2n_1-1,2n_2} = y_{N_1-n_1,n_2}/4$$

$$x_{2n_1,2n_2-1} = y_{n_1,N_2-n_2}/4$$

$$x_{2n_1-1,2n_2-1} = y_{N_1-n_1,N_2-n_2}/4. \tag{A.45}$$

The conputation of the odd-frequency DFT y_{n_1,n_2} may be simplified by using a direct polynomial transform algorithm with

$$Y_{k_1}(z) = \frac{1}{4} \sum_{k_2=0}^{N_2-1} Y_{k_1,k_2} W_1^{-k_1} z^{k_2} \tag{A.46}$$

$$\bar{Y}_{(4n_2+1)n_1}(z) \equiv \sum_{k_1=0}^{N_1-1} Y_{k_1}(z) z^{4n_1 k_1 N_2/N_1} \quad \text{modulo } (z^{N_2} - j) \tag{A.47}$$

$$y_{(4n_2+1)n_1,n_2} \equiv \bar{Y}_{(4n_2+1)n_1}(z) \quad \text{modulo } [z - W_2^{-(4n_2+1)}]. \tag{A.48}$$

This shows that when $4N_2 \geqslant N_1$, the polynomial transform is computed without multiplications and that the only multiplications which relate to the evaluation of the two-dimensional DCT are the $N_1 N_2$ premultiplications by $W_1^{-k_1}$ and the multiplications which correspond to the computation of the N_1 odd DFTs if size N_2 defined by (A.47).

Appendix B   Short Polynomial Product Algorithms

This appendix completes the set of algorithms given in section 3.7 by listing the algorithms with compute the polynomial products modulo (z^4+1) and modulo $(z^5-1)/(z-1)$ with 7 multiplications and by detailing the two-dimensional polynomial product algorithms defined modulo $(z_1^2+z_1+1)(z_2^5-1)/(z_2-1)$, modulo $(z_1^4+1)(z_2^2+z_2+1)$ and modulo $(z_1^2+1)(z_2^5-1)/(z_2-1)$. As in section 3.7, the input sequences are labelled x_m and h_n for one-dimensional polynomial products and x_{m_1,m_2} and h_{n_1,n_2} for two-dimensional polynomial products. Since h_n and h_{n_1,n_2} are assumed to be fixed, expressions involving h_n or h_{n_1,n_2} are presumed to be precomputed and stored. The output sequence is y_l for one-dimensional polynomial products and y_{l_1,l_1} for two-dimensional polynomial products. Input and output additions must be executed in the index numerical order.

B.1 Polynomial Product Modulo (z^4+1)

7 multiplications, 41 additions

$a_0 = x_0 + x_2$
$a_1 = x_1 + x_3$
$a_2 = x_2 + x_2$
$a_3 = a_2 + a_2$
$a_4 = x_3 + x_3$
$a_5 = a_4 + a_4$
$a_6 = x_0 + a_3$
$a_7 = x_1 + a_5$
$a_8 = x_0 + x_0 + x_1$
$a_9 = a_8 + a_8 + x_2$

$a_{10} = a_9 + a_9 + x_3$
$a_{11} = a_7 + a_7$
$a_{12} = x_0$
$a_{13} = x_3$
$a_{14} = a_0 + a_1$
$a_{15} = a_0 - a_1$
$a_{16} = a_6 + a_{11}$
$a_{17} = a_6 - a_{11}$
$a_{18} = a_{10}$

$b_0 = (3h_0 - 10h_1 + 5h_2 + 6h_3)/4$
$b_1 = (10h_0 - 5h_1 - 6h_2 + 3h_3)/2$
$b_2 = (h_0 - 9h_1 - 4h_2 + 2h_3)/6$
$b_3 = (3h_0 + 7h_1 - 12h_2 + 6h_3)/18$
$b_4 = (-3h_0 + 4h_1 + 3h_2)/72$
$b_5 = (-5h_0 + 5h_2 - 4h_3)/120$
$b_6 = (5h_1 - 3h_3)/90$

$$m_k = a_{k+12} b_k \qquad k = 0, \ldots, 6$$

$u_0 = m_6 + m_6$
$u_1 = u_0 + u_0$
$u_2 = m_2 + m_3 + u_1$
$u_3 = m_4 + m_5$
$u_4 = (u_3 + u_3) + 2u_3$

$u_5 = -m_2 + m_3$
$u_6 = -m_4 + m_5$
$u_7 = u_6 + u_6$
$u_8 = (u_7 + u_7) + 2u_7$

$y_0 = m_0 + u_2 + u_3 + u_1$
$y_1 = -m_1 + u_5 + u_8 - m_6$

$y_2 = -u_2 - u_4 + u_0$
$y_3 = u_5 + u_7 - u_1$

B.2 Polynomial Product Modulo $(z^5 - 1)/(z - 1)$

7 multiplications, 46 additions

$a_0 = x_0 + x_2$
$a_1 = x_1 + x_3$
$a_2 = x_2 + x_2$
$a_3 = a_2 + a_2$
$a_4 = x_3 + x_3$
$a_5 = a_4 + a_4$
$a_6 = x_0 + a_3$
$a_7 = x_1 + a_5$
$a_8 = x_0 + x_0 - x_1$
$a_9 = a_8 + a_8 + x_2$

$a_{10} = a_9 + a_9 - x_3$
$a_{11} = a_7 + a_7$
$a_{12} = x_0$
$a_{13} = x_3$
$a_{14} = a_0 + a_1$
$a_{15} = a_0 - a_1$
$a_{16} = a_6 + a_{11}$
$a_{17} = a_6 - a_{11}$
$a_{18} = a_{10}$

$b_0 = (6h_0 + h_1 - 10h_2 - 5h_3)/4$
$b_1 = (h_0 - 10h_1 - 5h_2 + 8h_3)/2$
$b_2 = (-28h_0 - 33h_1 + 17h_2 + 42h_3)/18$
$b_3 = (2h_0 - h_1 - 9h_2 + 4h_3)/6$
$b_4 = (2h_0 + 5h_1 - 5h_2)/120$
$b_5 = (-2h_0 + 3h_1 + 4h_2 - 3h_3)/72$
$b_6 = (-h_0 + 5h_2)/90$

$$m_k = a_{k+12} b_k \qquad k = 0, \ldots, 6$$

$u_0 = m_4 + m_4$
$u_1 = u_0 + m_4$
$u_2 = u_1 + u_1$
$u_3 = m_5 + m_5$
$u_4 = u_3 + m_5$
$u_5 = u_4 + u_4$
$u_6 = u_2 - u_5$

$u_7 = m_1 + u_6$
$u_8 = m_6 + m_6$
$u_9 = u_8 + m_6$
$u_{10} = u_9 + u_9$
$u_{11} = m_3 + m_3$
$u_{12} = u_{11} + u_{10} + u_5$
$u_{13} = m_0 - m_4$

$y_0 = u_{13} - u_4 + u_{10} + u_{12}$
$y_1 = -u_7 - u_{13} - m_2 + m_3 - u_2 - u_2 + m_5 - m_6$
$y_2 = u_7 + u_9$
$y_3 = u_{12} + u_0$

B.3 Polynomial Product Modulo $(z_1^2 + z_1 + 1)(z_2^5 - 1)/(z_2 - 1)$

21 multiplications, 83 additions

$a_0 = x_{0,0} + x_{3,0}$
$a_1 = x_{0,1} + x_{3,1}$
$a_2 = x_{0,0} - x_{3,0}$
$a_3 = x_{0,1} - x_{3,1}$
$a_4 = x_{1,0} + x_{2,0}$
$a_5 = x_{1,1} + x_{2,1}$
$a_6 = x_{1,0} - x_{2,0}$
$a_7 = x_{1,1} - x_{2,1}$
$a_8 = a_1 + a_5$
$a_9 = a_0 + a_4$
$a_{10} = a_9 - a_8$
$a_{11} = a_3 - a_7$
$a_{12} = a_2 - a_6$
$a_{13} = a_{12} - a_{11}$
$a_{14} = x_{0,1}$

$a_{15} = x_{0,0}$
$a_{16} = a_{15} - a_{14}$
$a_{17} = x_{3,1}$
$a_{18} = x_{3,0}$
$a_{19} = a_{18} - a_{17}$
$a_{20} = a_6 + a_1 - x_{1,1}$
$a_{21} = a_0 - a_7 - x_{2,0}$
$a_{22} = a_{21} - a_{20}$
$a_{23} = -a_4 + a_3 + x_{1,1}$
$a_{24} = a_2 + a_5 - x_{2,0}$
$a_{25} = a_{24} - a_{23}$
$a_{26} = -a_6 + a_1 - x_{2,1}$
$a_{27} = a_0 + a_7 - x_{1,0}$
$a_{28} = a_{27} - a_{26}$

$b_0 = (4h_{0,0} - h_{1,0} - 4h_{1,1} - 6h_{2,0} + h_{2,1} - h_{3,0} + 6h_{3,1} - 4h_{0,1})/6$
$b_1 = (4h_{0,1} + 5h_{1,0} - h_{1,1} + 5h_{2,0} - 6h_{2,1} - 5h_{3,0} - h_{3,1})/6$
$b_2 = b_0 + b_1$
$b_3 = (h_{1,0} - 2h_{1,1} - 2h_{2,0} + h_{2,1} - h_{3,0} + 2h_{3,1})/6$
$b_4 = (h_{1,0} + h_{1,1} + h_{2,0} - 2h_{2,1} - h_{3,0} - h_{3,1})/6$
$b_5 = b_3 + b_4$
$b_6 = -h_{0,0} + h_{0,1} + h_{2,0} + h_{1,1} + h_{3,0} - 2h_{3,1}$
$b_7 = -h_{0,1} - h_{1,0} - h_{2,0} + h_{2,1} + h_{3,0} + h_{3,1}$
$b_8 = b_6 + b_7$
$b_9 = -h_{0,0} + h_{0,1} + h_{1,1} + 2h_{2,0} - h_{2,1} - h_{3,1}$
$b_{10} = -h_{0,1} - h_{1,0} - h_{2,0} + 2h_{2,1} + h_{3,0}$
$b_{11} = b_9 + b_{10}$
$b_{12} = (2h_{0,0} - 2h_{0,1} - h_{1,0} - h_{1,1} + h_{2,1} - h_{2,0} - h_{3,0} + h_{3,1})/6$
$b_{13} = (2h_{0,1} + 2h_{1,0} - h_{1,1} - h_{2,1} - h_{3,1})/6$
$b_{14} = b_{12} + b_{13}$
$b_{15} = (-h_{1,0} - h_{1,1} - h_{2,0} - h_{2,1} + h_{3,0} + h_{3,1})/6$

$b_{16} = (2h_{1,0} - h_{1,1} + 2h_{2,0} - h_{2,1} - 2h_{3,0} + h_{3,1})/6$
$b_{17} = b_{15} + b_{16}$
$b_{18} = (h_{0,0} - h_{0,1} - h_{2,0} + h_{3,1})/3$
$b_{19} = (h_{0,1} + h_{2,0} - h_{2,1} - h_{3,0})/3$
$b_{20} = b_{18} + b_{19}$

$$m_k = a_{k+8} b_k \qquad k = 0, \ldots, 20$$

$u_0 = m_{12} + m_{14}$
$u_1 = m_{12} + m_{13}$
$u_2 = m_{15} + m_{17}$
$u_3 = m_{15} + m_{16}$
$u_4 = m_{18} + m_{20}$
$u_5 = m_{18} + m_{19}$
$v_0 = u_2 + u_2$
$u_6 = u_3 + u_3$
$u_7 = u_6 - v_0$
$u_8 = u_0 + u_4$
$u_9 = u_1 + u_5$
$u_{10} = m_0 + u_7$
$u_{11} = u_{10} + m_2$
$u_{12} = u_{10} + m_1$
$u_{13} = m_3 + m_5$
$u_{14} = m_3 + m_4$

$u_{15} = m_6 + m_8 + u_8$
$u_{16} = m_6 + m_7 + u_9$
$u_{17} = m_9 + m_{11}$
$u_{18} = m_9 + m_{10}$
$u_{19} = u_{11} + u_{15} + u_{17}$
$u_{20} = u_{12} + u_{16} + u_{18}$
$u_{21} = u_{19} + u_{13}$
$u_{22} = u_{20} + u_{14}$
$u_{23} = u_{19} - u_{13}$
$u_{24} = u_{20} - u_{14}$
$u_{25} = u_0 - u_1$
$u_{26} = u_2 + u_3$
$u_{27} = u_4 - u_5$
$u_{28} = u_{26} - u_{27}$

$y_{0,0} = u_{23} + u_{17} + u_8$
$y_{0,1} = u_{24} + u_{18} + u_9 - u_6$
$y_{1,0} = u_{21} - u_{25} + u_3 - u_5$
$y_{1,1} = u_{22} - u_{28} - u_0$

$y_{2,0} = u_{23} + u_{28} - u_1$
$y_{2,1} = u_{24} + u_{25} - u_2 - u_4$
$y_{3,0} = u_{21} + u_{15} + v_0$
$y_{3,1} = u_{22} + u_{16}$

B.4 Polynomial Product Modulo $(z_1^4 + 1)(z_2^2 + z_2 + 1)$

21 multiplications, 76 additions

$a_0 = x_{0,0} + x_{2,0}$
$a_1 = x_{1,0} + x_{3,0}$
$a_2 = x_{0,1} + x_{2,1}$
$a_3 = x_{1,1} + x_{3,1}$
$a_4 = x_{2,1} - x_{2,0}$
$a_5 = x_{1,0} - x_{1,1}$
$a_6 = a_4 + x_{0,0}$
$a_7 = a_5 + x_{3,1}$

$a_8 = x_{3,0} - x_{1,1}$
$a_9 = x_{2,0} - x_{0,1}$
$a_{10} = a_2 + a_3$
$a_{11} = a_0 + a_1$
$a_{12} = a_{11} - a_{10}$
$a_{13} = a_2 - a_3$
$a_{14} = a_0 - a_1$
$a_{15} = a_{14} - a_{13}$

B.4 Polynomial Product Modulo $(z_1^4+1)(z_2^2+z_2+1)$

$a_{16} = a_7 - a_9$
$a_{17} = a_6 + a_8$
$a_{18} = a_{17} - a_{16}$
$a_{19} = -a_7 - a_9$
$a_{20} = a_6 - a_8$
$a_{21} = a_{20} - a_{19}$
$a_{22} = x_{0,1}$
$a_{23} = x_{0,0}$

$a_{24} = a_{23} - a_{22}$
$a_{25} = x_{3,1}$
$a_{26} = x_{3,0}$
$a_{27} = a_{26} - a_{25}$
$a_{28} = x_{3,1} - a_4 - x_{1,0} + x_{0,1}$
$a_{29} = x_{0,0} - a_5 - x_{2,1} + x_{3,0}$
$a_{30} = a_{29} - a_{28}$

$b_{10} = (2h_{0,0} - h_{0,1} - 2h_{1,1} + 2h_{1,0} + h_{2,0} - 2h_{2,1} - h_{3,0} - h_{3,1})/6$
$b_{11} = (-h_{0,0} + 2h_{0,1} + 2h_{1,1} + h_{2,0} + h_{2,1} + 2h_{3,0} - h_{3,1})/6$
$b_{12} = b_{10} + b_{11}$
$b_{13} = (2h_{0,0} - h_{0,1} + h_{2,0} - 2h_{2,1} - h_{3,0} - h_{3,1})/6$
$b_{14} = (-h_{0,0} + 2h_{0,1} + h_{2,0} + h_{2,1} + 2h_{3,0} - h_{3,1})/6$
$b_{15} = b_{13} + b_{14}$
$b_{16} = (h_{0,0} + h_{0,1} - 2h_{1,1} + 2h_{1,0} - h_{2,0} - h_{2,1})/6$
$b_{17} = (-2h_{0,0} + h_{0,1} + 2h_{1,1} + 2h_{2,0} - h_{2,1})/6$
$b_{18} = b_{16} + b_{17}$
$b_{19} = (h_{0,0} + h_{0,1} - h_{2,0} - h_{2,1} - 2h_{3,0} - 2h_{3,1})/6$
$b_{20} = (-2h_{0,0} + h_{0,1} + 2h_{2,0} - h_{2,1} + 4h_{3,0} - 2h_{3,1})/6$
$b_{21} = b_{19} + b_{20}$
$b_{22} = -2h_{0,0} + h_{0,1} + h_{1,1} - h_{1,0} + h_{2,1} + h_{3,0} + h_{3,1}$
$b_{23} = h_{0,0} - 2h_{0,1} - h_{1,1} - h_{2,0} - 2h_{3,0} + h_{3,1}$
$b_{24} = b_{22} + b_{23}$
$b_{25} = -h_{0,0} + h_{1,1} - h_{1,0} - h_{2,0} + 2h_{2,1} + h_{3,0} + h_{3,1}$
$b_{26} = h_{0,0} - h_{0,1} - h_{1,1} - h_{2,0} - h_{2,1} - 2h_{3,0} + h_{3,1}$
$b_{27} = b_{25} + b_{26}$
$b_{28} = (h_{1,0} - h_{1,1} - h_{3,0} - h_{3,1})/3$
$b_{29} = (h_{1,1} + 2h_{3,0} - h_{3,1})/3$
$b_{30} = b_{28} + b_{29}$

$$m_k = a_{k+10} b_{k+10} \quad k = 0, \ldots, 20$$

$u_0 = m_0 + m_2$
$u_1 = m_0 + m_1$
$u_2 = m_3 + m_5$
$u_3 = m_3 + m_4$
$u_4 = m_6 + m_8$
$u_5 = m_6 + m_7$
$u_6 = m_9 + m_{11}$
$u_7 = m_9 + m_{10}$
$u_8 = m_{12} + m_{14}$
$u_9 = m_{12} + m_{13}$
$u_{10} = m_{15} + m_{17}$
$u_{11} = m_{15} + m_{16}$

$u_{12} = m_{18} + m_{20}$
$u_{13} = m_{18} + m_{19}$
$u_{14} = u_0 + u_2$
$u_{15} = u_1 + u_3$
$u_{16} = u_0 - u_2$
$u_{17} = u_1 - u_3$
$u_{18} = u_4 + u_6$
$u_{19} = u_5 + u_7$
$u_{20} = u_4 - u_6$
$u_{21} = u_5 - u_7$
$u_{22} = u_{12} - u_{13}$

$y_{0,0} = -u_{14} - u_{18} - u_8 - u_{12}$
$y_{0,1} = -u_{15} - u_{19} - u_9 - u_{13}$
$y_{1,0} = u_{16} + u_{10} + u_{20} + u_{12}$
$y_{1,1} = u_{17} + u_{21} + u_{11} + u_{13}$

$y_{2,0} = u_{14} + u_{19} - u_{18} - u_{13}$
$y_{2,1} = u_{15} - u_{18} + u_{22}$
$y_{3,0} = u_{16} - u_{21} - u_{22}$
$y_{3,1} = u_{17} - u_{21} + u_{20} - u_{12}$

B.5 Polynomial Product Modulo $(z_1^2 + 1)(z_2^5 - 1)/(z_2 - 1)$

21 multiplications, 76 additions

$a_0 = x_{0,0} + x_{2,0}$
$a_1 = x_{0,0} - x_{2,0}$
$a_2 = x_{1,0} + x_{3,0}$
$a_3 = x_{1,0} - x_{3,0}$
$a_4 = x_{0,1} + x_{2,1}$
$a_5 = x_{0,1} - x_{2,1}$
$a_6 = x_{1,1} + x_{3,1}$
$a_7 = x_{1,1} - x_{3,1}$
$a_8 = a_4 + a_6$
$a_9 = a_0 + a_2$
$a_{10} = a_8 + a_9$
$a_{11} = a_4 - a_6$
$a_{12} = a_0 - a_2$
$a_{13} = a_{11} + a_{12}$
$a_{14} = x_{0,1}$

$a_{15} = x_{0,0}$
$a_{16} = x_{0,0} + x_{0,1}$
$a_{17} = x_{3,1}$
$a_{18} = x_{3,0}$
$a_{19} = x_{3,0} + x_{3,1}$
$a_{20} = a_5 + a_3$
$a_{21} = a_1 - a_7$
$a_{22} = a_{20} + a_{21}$
$a_{23} = a_5 - a_3$
$a_{24} = a_1 + a_7$
$a_{25} = a_{23} + a_{24}$
$a_{26} = a_{12} - a_{24} + x_{0,1} + x_{3,1} - x_{3,0}$
$a_{27} = -a_{11} + a_{23} + x_{0,0} + a_{19}$
$a_{28} = a_{26} + a_{27}$

$b_0 = (h_{0,0} + h_{0,1} + h_{1,0} + h_{1,1} - 2h_{2,1} - 3h_{3,0} - h_{3,1})/4$
$b_1 = (-h_{0,0} + h_{0,1} - h_{1,0} + h_{1,1} + 2h_{2,0} + h_{3,0} - 3h_{3,1})/4$
$b_2 = (b_0 - b_1)/2$
$b_3 = (-h_{0,0} - h_{0,1} + h_{1,0} + h_{1,1} - 2h_{2,1} - h_{3,0} + h_{3,1})/4$
$b_4 = (h_{0,0} - h_{0,1} - h_{1,0} + h_{1,1} + 2h_{2,0} - h_{3,0} - h_{3,1})/4$
$b_5 = (b_3 - b_4)/2$
$b_6 = (2h_{2,1} + 2h_{3,0})/2$
$b_7 = (-2h_{2,0} + 2h_{3,1})/2$
$b_8 = (b_6 - b_7)/2$
$b_9 = -h_{1,0} - h_{1,1} - h_{2,0} + h_{2,1} + 2h_{3,0}$
$b_{10} = h_{1,0} - h_{1,1} - h_{2,0} - h_{2,1} + 2h_{3,1}$
$b_{11} = (b_9 - b_{10})/2$
$b_{12} = (5h_{0,0} - 5h_{0,1} - 5h_{1,0} - 5h_{1,1} - 4h_{2,0} + 2h_{2,1} - h_{3,0} + 3h_{3,1})/20$
$b_{13} = (5h_{0,0} + 5h_{0,1} + 5h_{1,0} - 5h_{1,1} - 2h_{2,0} - 4h_{2,1} - 3h_{3,0} - h_{3,1})/20$
$b_{14} = (b_{12} - b_{13})/2$
$b_{15} = (-h_{0,0} + h_{0,1} - h_{1,0} - h_{1,1} - 2h_{2,1} + h_{3,0} + h_{3,1})/4$
$b_{16} = (-h_{0,0} - h_{0,1} + h_{1,0} - h_{1,1} + 2h_{2,0} - h_{3,0} + h_{3,1})/4$

B.5 Polynomial Product Modulo $(z_1^2+1)(z_2^5-1)/(z_2-1)$

$b_{17} = (b_{15} - b_{16})/2$
$b_{18} = (2h_{2,0} - h_{2,1} - 2h_{3,0} + h_{3,1})/5$
$b_{19} = (h_{2,0} + 2h_{2,1} - h_{3,0} - 2h_{3,1})/5$
$b_{20} = (b_{18} - b_{19})/2$

$$m_k = a_{k+8} b_k \qquad k = 0, \ldots, 20$$

$u_0 = m_2 - m_0$
$u_1 = m_2 + m_1$
$u_3 = m_5 - m_3$
$u_4 = m_5 + m_4$
$u_5 = m_8 - m_6$
$u_6 = m_8 + m_7$
$u_7 = m_{11} - m_9$
$u_8 = m_{11} + m_{10}$
$u_9 = m_{14} - m_{12}$
$u_{10} = m_{14} + m_{13}$
$u_{11} = m_{17} - m_{15}$
$u_{12} = m_{17} + m_{16}$
$u_{13} = m_{20} - m_{18}$

$u_{14} = m_{20} + m_{19}$
$u_{15} = u_7 + u_{13} - u_{14}$
$u_{16} = u_8 + u_{13} + u_{14}$
$u_{17} = u_0 + u_5 + u_{15}$
$u_{18} = u_1 + u_6 + u_{16}$
$u_{19} = u_{17} + u_3$
$u_{20} = u_{18} + u_4$
$u_{21} = u_{17} - u_3 + u_{14}$
$u_{22} = u_{18} - u_4 - u_{13}$
$u_{23} = u_9 + u_{11}$
$u_{24} = u_{10} + u_{12}$
$u_{25} = u_9 - u_{11}$
$u_{26} = u_{10} - u_{12}$

$y_{0,0} = u_{21} + u_{15} + u_{26}$
$y_{0,1} = u_{22} + u_{16} - u_{25}$
$y_{1,0} = u_{19} - u_{23} - u_{13}$
$y_{1,1} = u_{20} - u_{24} - u_{14}$

$y_{2,0} = u_{21} - u_{26}$
$y_{2,1} = u_{22} + u_{25}$
$y_{3,0} = u_{19} + u_5 + u_{23}$
$y_{3,1} = u_{20} + u_6 + u_{24}$

Problems

Chapter 2

2.1 Find the GCD of 65 and 104 by Euclid's algorithm. Solve the Diophantine equation
$$65x + 104x = 143$$

2.2 Find the residue of 3^{62} modulo 19.

2.3 Verify that $2^{25} - 1$ is divisible by 601.

2.4 Prove that 3 divides every pseudo-Fermat number $2^v + 1$, with v odd.

2.5 Compute $\emptyset(2205)$.

2.6 Find the maximum order of an integer modulo 11. Give an example of such an integer.

2.7 Determine whether 63 is a quadratic residue or non-residue of 209.

2.8 Find the number n defined modulo 992, knowing that $n \equiv 29$ modulo 32 and $n \equiv 5$ modulo 31.

2.9 Compute the residue of the polynomial $z^{32} - 5z^{13} + z^4 + 3z^2 + 1$ modulo $(z^2 + z + 1)$.

2.10 Find the polynomial $A(z)$ defined modulo $(z^4 - 1)$, knowing that $A(z) \equiv 2z - 2$ modulo $(z^2 + 1)$, $A(z) \equiv 16$ modulo $(z - 1)$, $A(z) \equiv -8$ modulo $(z + 1)$. Apply first the Chinese remainder theorem directly. Then, show that the computation may be simplified by proceeding in two stages.

Chapter 3

3.1 Show how a cyclic convolution of length 60 may be computed by split nesting. Evaluate the computational complexity by using the short algorithms of Sect. 3.7.

3.2 Give the flow diagram of a circular convolution of length 6 computed by the Agarwal-Cooley algorithm with the short convolution algorithms of Sect.

3.7. Verify the result with the sequences $\{x_m\} = \{1, 2, 0, 3, 1, 4\}$ and $\{h_n\} = \{0, 2, 1, 0, 3, 4\}$.

3.3 Show how to compute a two-dimensional polynomial product modulo $(x^3 - 1/x - 1, y^5 - 1/y - 1)$ with 21 multiplications by using the algorithms of Sect. 3.7. Evaluate the number of additions required by the algorithm.

3.4 Given the polynomial product algorithm modulo $(z^3 - 1)/(z - 1)$ of Sect. 3.7.2, derive a polynomial product algorithm modulo $(z^2 - z + 1)$.

3.5 When the aperiodic convolution of two sequences of lengths N_1 and N_2 is computed by circular convolutions of N terms, the input sequences must be augmented to length N by concatenating 0's, and N must be such that $N \geq N_1 + N_2 - 1$. Show that if $N = N_1 + N_2 - 2$ or $N = N_1 + N_2 - 3$, the same method may be used with some modifications. Apply this technique to the evaluation of the aperiodic convolution of two 3-point sequences.

3.6 Show that the aperiodic convolution of two sequences of lengths L_1 and L_2 may be computed by two circular convolutions of lengths N_1 and N_2, with $N_1 + N_2 = L_1 + L_2$, and $(N_1, N_2) = 1$.

3.7 Find two real polynomials of degree 2 which are the two factors of $z^4 + 1$. Using this property, devise an algorithm which computes a polynomial product modulo $(z^4 + 1)$ with 12 multiplications. Evaluate the number of multiplications when this algorithm is nested with a circular convolution of size N by the Agarwal-Cooley technique. Compare the computational complexity with an approach using the polynomial product algorithm modulo $(z^4 + 1)$ of Sect. 3.7.2.

3.8 Devise two algorithms which compute the polynomial products modulo $(z^2 + 2z + 2)$ and $(z^2 - 2z + 2)$ respectively, with 3 multiplications each.

3.9 Using the results of problem 3.8, develop an algorithm which computes the aperiodic convolution of two sequences of 4 terms with 9 multiplications.

3.10 Evaluate the computational complexity of a 60-point circular convolution computed by the Agarwal-Cooley technique and distributed arithmetic for the 4-point circular convolutions.

Chapter 4

4.1 Show how to compute the length-N DFT of a real sequence with a DFT of size $N/2$ and auxiliary operations. $N = 2^t$.

4.2 For a real input sequence x_n, devise a method for computing with a real $(N/2)$-point DFT the N-point DFT $\bar{X}_k$, where only the odd harmonics are present. ($\bar{X}_k = 0$ for k even). $N = 2^t$.

Problems

4.3 Show how to compute a DFT of size N, $N = 2^{2^t}$ with an FFT algorithm using successive radices $N_1 = \sqrt{N}$, $N_2 = \sqrt{N_1}$, ... Evaluate the number of operations.

4.4 Show how a DFT of size $N \times N$, with $N = 2^t$, may be computed by a radix-2 FFT algorithm which operates simultaneously on the rows and on the columns. Evaluate the computational complexity of this algorithm and compare it with that of the conventional row-column method.

4.5 Show how a DFT of size N^2 may be converted into a two-dimensional DFT of size $N \times N$, plus some auxiliary computations.
Hint: begin with a FFT decomposition.

4.6 The input sequence of a DFT used for the overlap-add computation of an aperiodic convolution has zeros in its last terms. Discuss the properties of the DFT of a sequence of N terms, $N = 2^t$, where the $N/2$ last terms are zero. Show how the computation may be simplified.

4.7 Show that some simplifications are possible when one computes simultaneously the circular convolution and the circular correlation of the two N-point sequences x_m and h_n by N-point DFTs.

4.8 In the computation of a N-point DFT, with $N = 2^t$, by the Rader-Brenner technique, the complex twiddle factors are replaced by the real twiddle factors $1/\cos(2\pi k/N)$. When N is large, the multiplications by $1/\cos(2\pi k/N)$ yield relatively large round-off errors. Show that, with sligth modifications, the circular shifting theorem may be used to replace the complex twiddle factors by the real twiddle factors $\cos(2\pi k/N)$ instead of $1/\cos(2\pi k/N)$. Discuss the implications on the round-off errors.

4.9 The computation of a 3-point DFT requires non-trivial multiplications when the computations are performed in the field of complex numbers. Show that the 3-point DFT is computed with only trivial multiplications when the operations are performed in $R(\mu)$, with μ being a cube root of unity, that is to say when the numbers are represented by $a + b\mu$. Evaluate the computational complexity of a DFT of size 3^t calculated with this technique.

4.10 Evaluate the computational complexity of a radix-16 FFT algorithm. Compare with the computational complexity of radix-2 and radix-4 algorithms.

Chapter 5

5.1 Show that a DFT of size 272 may be computed with 17 DFTs of 16 points, one DFT of size 16×6, plus some auxiliary operations.

5.2 Define an algorithm which computes a DFT of size $N_1 N_2 N_3$ with $N_1 N_2$ DFTs of N_3 points, $N_2 N_3$ DFTs of N_1 points, $N_1 N_3$ circular convolutions of N_2 points and $2 N_1 N_2 N_3$ complex multiplications.

5.3 Define the $(N-1)$-point transforms and inverse transforms which compute a polynomial product modulo $(z^N-1)/(z-1)$, with N prime. *Hint:* a N-point circular convolution is computed by N-point DFTs and inverse DFTs.

5.4 Using the chirp-z transform algorithm, show that a DFT of size N^2 may be converted into $2N$ convolutions of N points plus some auxiliary operations.

5.5 The Winograd Fourier transform algorithm is normally implemented with short DFT algorithms where all multiplications are in the center of the algorithm. Show that this is not necessary and that the Winograd algorithm may be combined with the FFT algorithm. Evaluate the computational complexity of a DFT of size 400 computed with this hybrid technique. Compare with the results obtained by the conventional Winograd technique.

5.6 When a N-point DFT is computed by the chirp-z transform algorithm, the DFT is converted into a N-point complex circular convolution, plus some auxiliary multiplications. Show that most of the complex multiplications in the N-point convolutions may be implemented with two real multiplications when $N = 2^t$.

5.7 Derive a Winograd Fourier transform algorithm using the chirp-z transform algorithm for the small DFTs. Compare with the conventional approach which uses Rader's algorithm.

5.8 The Winograd Fourier transform algorithm normally uses short DFT algorithms which are implemented with multiplications by pure real or pure imaginary numbers. Convert the algorithms of Sect. 5.5 for DFTs of sizes 3, 5 and 16 into new algorithms which require respectively 3, 5 and 16 complex multiplications. Give the number of operations for a DFT of size 240 evaluated by the Winograd algorithm using these short DFTs. Compare with the conventional method.

5.9 Derive a 11-point DFT algorithm from Rader's algorithm.

5.10 Show how to construct a 13-point DFT algorithm with Rader's algorithm. Evaluate the computational complexity.

Chapter 6

6.1 Write a computer program (FORTRAN, PL/1) which computes by an FFT-type algorithm a polynomial transform of size N defined modulo (z^N+1), with $N = 2^t$.

6.2 Write an APL program which computes by an FFT-type algorithm a polynomial transform of size N or $2N$ defined modulo (z^N+1), with $N = 2^t$.

6.3 Show how the complex convolution algorithm of Sect. 3.3.3 may be explained in terms of polynomial transform.

6.4 Show how to compute a polynomial product modulo $(z_1^4 - 1)$, $(z_2^4 + 4)$ with two polynomial transforms of size 4. *Hint:* $z_2^4 + 4$ factors into two real polynomials of degree 2.

6.5 Find the maximum size of the polynomial convolutions defined modulo $(z^8 - z^7 + z^5 - z^4 + z^3 - z + 1)$ that can be computed by polynomial transforms evaluated without multiplications. Define these polynomial transforms. Suggest a simple method for computing the polynomial transform.

6.6 A two-dimensional circular convolution of size 15×15 may be either computed by nesting two convolutions of sizes 3×3 and 5×5 evaluated by polynomial transforms or by using polynomial transforms of size 15. Compare the two methods.

6.7 Demonstrate that the polynomial transform of root z and length $N = 6p$ has the convolution property when defined modulo $z^{2p} - z^p + 1$, with $p = 2^t$.

6.8 Show how to compute simultaneously two 6-point circular convolutions by polynomial transforms defined modulo $(z^2 - z + 1)$. Compare the computational complexity with direct computation by the Agarwal-Cooley algorithm.

6.9 Show how to compute a polynomial transform of length p, with p prime, by Rader's algorithm. Compare the computational complexity with that obtained in Table 6.2 for $p = 5$.

Chapter 7

7.1 Write a computer program (FORTRAN or PL/1) which computes a DFT of size $N \times N$, with $N = 2^t$, by the polynomial transform method of Fig. 7.5.

7.2 Give the algorithm derived from Fig. 7.4 which computes by a real polynomial transform a DFT of size $N_1 \times N_2$, with $N_1 = 2^{t_1}$, $N_2 = 2^{t_2}$. Show that some values of N_1 and N_2 yield two possible methods with different computational complexities.

7.3 Give an algorithm which computes the modified DFT $\bar{X}_{k_1, k_2}$ of size $N_1 \times N_2$ with one polynomial transform of size N_1. Determine the cases where the polynomial transform must be calculated with non-trivial multiplications, as a function of the relative size of N_1 and N_2. $N_1 = 2^{t_1}$, $N_2 = 2^{t_2}$,

$$\bar{X}_{k_1, k_2} = \sum_{n_1=0}^{N_1-1} \sum_{n_2=0}^{N_2-1} x_{n_1, n_2} W^{n_1 k_1} W^{n_2} W^{N_1 n_2 k_2 / N_2}$$

$$W = e^{-j2\pi/N_1}$$

Problems

7.4 Compute the hybrid two-dimensional transform

$$X_{k_1,k_2} = \sum_{n_1=0}^{N-1} \sum_{n_2=0}^{N-1} x_{n_1,n_2} W^{n_1 k_1} \cos(2\pi n_2 k_2/N),$$

$W = e^{-j2\pi/N}$, by polynomial transforms.

7.5 Evaluate the computational complexity of an algorithm which computes a DFT of size 85 by the method of Sect. 7.2. Compare with the conventional Winograd technique for a DFT of size 120.

7.6 Show how a 272-point DFT be computed by the polynomial transform algorithm of Fig. 7.2. *Hint:* Use Good's algorithm. Evaluate the computational complexity.

7.7 Describe an algorithm which computes a DFT of size 15×15 by method of Sect. 7.1 with a polynomial transform of size 15, plus auxiliary calculations.

7.8 The algorithm described in Sect. 7.1.2 (Fig. 7.2) may be viewed as a radix-2 algorithm in which each stage converts a DFT of size $2^i \times 2^i$ into a DFT of size $2^{i-1} \times 2^{i-1}$ by polynomial transforms. Develop a radix-4 algorithm in which each stage converts by polynomial transforms a DFT of size $2^i \times 2^i$ into a DFT of size $2^{i-2} \times 2^{i-2}$.

Chapter 8

8.1 Show how to compute a 17-point DFT by a Fermat number transform. Evaluate the computational complexity.

8.2 Give an algorithm which combines polynomial transforms and number theoretic transforms to compute a two-dimensional convolution of size 32×32. Evaluate the computational complexity.

8.3 Assuming computations are performed in ternary arithmetic (that is to say, multiplications by powers of 3 are "easy"), define useful generalized pseudo-Fermat transforms computed modulo $3^p + 1$.

8.4 Assuming computations are performed in ternary arithmetic (that is to say, multiplications by powers of 3 are "easy"), define useful generalized pseudo-Mersenne transforms computed modulo $(3^p - 1)$.

8.5 Define a way of performing additions and multiplications modulo p^2 when the integers x are given in a two-digit representation a, b, with $x = pa + b$, and $a, b = 0, \ldots, p-1$. Show that the multiplication modulo p^2 may be implemented by table lookups, using two tables of $2p$ elements each.

8.6 Demonstrate that the NTT of root 2 and length $6p$ has the circular convolution property when defined modulo $(2^{2p} - 2^p + 1)$, with $p = 2^t$.

References

Chapter 2
2.1 T. Nagell: *Introduction to Number Theory*, 2nd ed. (Chelsea, New York 1964)
2.2 G. H. Hardy, E. M. Wright: *An Introduction to the Theory of Numbers*, 4th ed. (Oxford University Press, Ely House, London 1960)
2.3 N. H. McCoy: *The Theory of Numbers* (MacMillan, New York 1965)
2.4 J. H. McClellan, C. M. Rader: *Number Theory in Digital Signal Processing* (Prentice-Hall, Englewood Cliffs, N. J. 1979)
2.5 M. Abramowitz, I. Stegun: *Handbook of Mathematical Functions*, 7th ed. (Dover, New York 1970) pp. 864–869
2.6 W. Sierpinski: *Elementary Theory of Numbers* (Polska Akademia Nauk Monographie Matematyczne, Warszawa 1964)
2.7 I. M. Vinogradov: *Elements of Number Theory*, (Dover, New York 1954)
2.8 D. J. Winter: *The Structure of Fields*, Graduate Texts in Mathematics, Vol. 16 (Springer, Berlin, New York, Heidelberg 1974)
2.9 R. C. Agarwal, J. W. Cooley: New algorithms for digital convolution. IEEE Trans. ASSP-**25**, 392–410 (1977)
2.10 J. H. Griesmer, R. D. Jenks: "SCRATCHPAD I. An Interactive Facility for Symbolic Mathematics", in Proc. Second Symposium on Symbolic and Algebraic Manipulation, ACM, New York, 42–58 (1971)
2.11 S. Winograd: On computing the discrete Fourier transform. Math. Comput. **32**, 175–199 (1978)
2.12 S. Winograd: Some bilinear forms whose multiplicative complexity depends on the field of constants. Math. Syst. Th., **10**, 169–180 (1977)

Chapter 3
3.1 T. G. Stockham: "Highspeed Convolution and Correlation", in 1966 Spring Joint Computer Conf., AFIPS Proc. **28**, 229–233
3.2 B. Gold, C. M. Rader, A. V. Oppenheim, T. G. Stockham: *Digital Processing of Signals*, (McGraw-Hill, New York 1969) pp. 203–213
3.3 R. C. Agarwal, J. W. Cooley: "New Algorithms for Digital Convolution", in 1977 Intern. Conf., Acoust., Speech, Signal Processing Proc., p. 360
3.4 I. J. Good: The relationship between two fast fourier Transforms. IEEE Trans. C-**20**, 310–317 (1971)
3.5 R. C. Agarwal, J. W. Cooley: New algorithms for digital convolution. IEEE ASSP-**25**, 392–410 (1977)
3.6 H. J. Nussbaumer: "New Algorithms for Convolution and DFT Based on Polynomial Transforms", in IEEE 1978 Intern. Conf. Acoust., Speech, Signal Processing Proc., pp. 638–641
3.7 H. J. Nussbaumer, P. Quandalle: Computation of convolutions and discrete Fourier transforms by polynomial transforms. IBM J. Res. Dev., **22**, 134–144 (1978)
3.8 R. C. Agarwal, C. S. Burrus: Fast one-dimensional digital convolution by multidimensional techniques. IEEE Trans. ASSP-**22**, 1–10 (1974)
3.9 H. J. Nussbaumer: Fast polynomial transform algorithms for digital convolution. IEEE Trans. ASSP-**28**, 205–215, (1980)

3.10 A. Croisier, D. J. Esteban, M. E. Levilion, V. Riso: Digital Filter for PCM Encoded Signals, US Patent 3777130, Dec. 4, 1973
3.11 C. S. Burrus: Digital filter structures described by distributed arithmetic. IEEE Trans. CAS-**24**, 674–680 (1977)
3.12 D. E. Knuth: *The Art of Computer Programming*, Vol. 2, *Semi-Numerical Algorithms* (Addison-Wesley, New York 1969)

Chapter 4
4.1 B. Gold, C. M. Rader: *Digital Processing of Signals* (McGraw-Hill, New York 1969)
4.2 E. O. Brigham: *The Fast Fourier Transform* (Prentice-Hall, Englewood Cliffs, N. J. 1974)
4.3 L. R. Rabiner, B. Gold: *Theory and Application of Digital Signal Processing* (Prentice-Hall, Englewood Cliffs, N. J. 1975)
4.4 A. V. Oppenheim, R. W. Schafer: *Digital Signal Processing* (Prentice-Hall, Englewood Cliffs, N. J. 1975)
4.5 A. E. Siegman: How to compute two complex even Fourier transforms with one transform step. Proc. IEEE **63**, 544 (1975)
4.6 J. W. Cooley, J. W. Tukey: An algorithm for machine computation of complex Fourier series. Math. Comput. **19**, 297–301 (1965)
4.7 G. D. Bergland: A fast Fourier transform algorithm using base 8 iterations. Math. Comput. **22**, 275–279 (1968)
4.8 R. C. Singleton: An algorithm for computing the mixed radix fast Fourier transform. IEEE Trans. AU-**17**, 93–103 (1969)
4.9 R. P. Polivka, S. Pakin: *APL: the Language and Its Usage* (Prentice-Hall, Englewood Cliffs, N. J. 1975)
4.10 P. D. Welch: A fixed-point fast Fourier transform error analysis. IEEE Trans. AU-**17**, 151–157 (1969)
4.11 T. K. Kaneko, B. Liu: Accumulation of round-off errors in fast Fourier transforms. J. Assoc. Comput. Mach. **17**, 637–654 (1970)
4.12 C. J. Weinstein: Roundoff noise in floating point fast Fourier transform computation. IEEE Trans. AU-**17**, 209–215 (1969)
4.13 C. M. Rader, N. M. Brenner: A new principle for fast Fourier transformation. IEEE Trans. ASSP-**24**, 264–265 (1976)
4.14 S. Winograd: On computing the discrete Fourier transform. Math. comput. **32**, 175–199 (1978)
4.15 K. M. Cho, G. C. Temes: "Real-factor FFT algorithms", in IEEE 1978 Intern. Conf. Acoust., Speech, Signal Processing, pp. 634–637
4.16 H. J. Nussbaumer, P. Quandalle: Fast computation of discrete Fourier transforms using polynomial transforms. IEEE Trans. ASSP-**27**, 169–181 (1979)
4.17 G. Bonnerot, M. Bellanger: Odd-time odd-frequency discrete Fourier transform for symmetric real-valued series. Proc. IEEE **64**, 392–393 (1976)
4.18 G. Bruun: z-transform DFT filters and FFTs. IEEE Trans. ASSP-**26**, 56–63 (1978)
4.19 G. K. McAuliffe: "Fourier Digital Filter or Equalizer and Method of Operation Therefore", US Patent No. 3 679 882, July 25, 1972

Chapter 5
5.1 L. I. Bluestein: "A Linear Filtering Approach to the Computation of the Discrete Fourier Transform", in 1968 Northeast Electronics Research and Engineering Meeting Rec., pp. 218–219
5.2 L. I. Bluestein: A linear filtering approach to the computation of the discrete Fourier transform. IEEE Trans. AU-**18**, 451–455 (1970)
5.3 C. M. Rader: Discrete Fourier transforms when the number of data samples is prime. Proc. IEEE **56**, 1107–1108 (1968)
5.4 S. Winograd: On computing the discrete Fourier transform. Proc. Nat. Acad. Sci. USA **73**, 1005–1006 (1976)

5.5   L. R. Rabiner, R. W. Schafer, C. M. Rader: The Chirp z-transform algorithm and its application. Bell Syst. Tech. J. **48**, 1249–1292 (1969)
5.6   G. R. Nudd, O. W. Otto: Real-time Fourier analysis of spread spectrum signals using surface-wave-implemented Chrip-z transformation. IEEE Trans. **MTT-24**, 54–56 (1975)
5.7   M. J. Narasimha, K. Shenoi, A. M. Peterson: "Quadratic Residues: Application to Chirp Filters and Discrete Fourier Transforms", in IEEE 1976 Acoust., Speech, Signal Processing Proc., pp. 376–378
5.8   M. J. Narasimha: "Techniques in Digital Signal Processing", Tech. Rpt. 3208-3, Stanford Electronics Laboratory, Stanford University (1975)
5.9   J. H. McClellan, C. M. Rader: *Number Theory in Digital Signal Processing* (Prentice-Hall, Englewood Cliffs, N. J. 1979)
5.10 H. J. Nussbaumer, P. Quandalle: Fast computation of discrete Fourier transforms using polynomial transforms. IEEE Trans. **ASSP-27**, 169–181 (1979)
5.11 I. J. Good: The interaction algorithm and practical Fourier analysis. J. Roy. Stat. Soc. **B-20**, 361–372 (1958); **22**, 372–375 (1960)
5.12 I. J. Good: The relationship between two fast Fourier transforms. IEEE Trans. **C-20**, 310–317 (1971)
5.13 D. P. Kolba, T. W. Parks: A prime factor FFT algorithm using high-speed convolution. IEEE Trans. **ASSP-25**, 90–103 (1977)
5.14 C. S. Burrus: "Index Mappings for Multidimensional Formulation of the DFT and Convolution", in 1977 IEEE Intern. Symp. on Circuits and Systems Proc., pp. 662–664
5.15 S. Winograd: "A New Method for Computing DFT", in 1977 IEEE Intern. Conf. Acoust., Speech and Signal Processing Proc., pp. 366–368
5.16 S. Winograd: On computing the discrete Fourier transform. Math. Comput. **32**, 175–199 (1978)
5.17 H. F. Silverman: An introduction to programming the Winograd Fourier transform algorithm (WFTA). IEEE Trans. **ASSP-25**, 152–165 (1977)
5.18 R. W. Patterson, J. H. McClellan: Fixed-point error analysis of Winograd Fourier transform algorithms. IEEE Trans. **ASSP-26**, 447–455 (1978)
5.19 L. R. Morris: A comparative study of time efficient FFT and WFTA programs for general purpose computers. IEEE Trans. **ASSP-26**, 141–150 (1978)

Chapter 6
6.1   H. J. Nussbaumer: Digital filtering using polynomial transforms. Electron. Lett. **13**, 386–387 (1977)
6.2   H. J. Nussbaumer, P. Quandalle: Computation of convolutions and discrete Fourier transforms by polynomial transforms. IBM J. Res. Dev. **22**, 134–144 (1978)
6.3   P. Quandalle: "Filtrage numérique rapide par transformées de Fourier et transformées polynômiales—Etude de l'implantation sur microprocesseurs" Thèse de Doctorat de Spécialité, University of Nice, France (18 mai 1979)
6.4   R. C. Agarwal, J. W. Cooley: New algorithms for digital convolution. IEEE Trans. **ASSP-25**, 392–410 (1977)
6.5   B. Arambepola, P. J. W. Rayner: Efficient transforms for multidimensional convolutions. Electron. Lett. **15**, 189–190 (1979)

Chapter 7
7.1   H. J. Nussbaumer, P. Quandalle: Fast computation of discrete Fourier transforms using polynomial transforms. IEEE Trans. **ASSP-27**, 169–181 (1979)
7.2   H. J. Nussbaumer, P. Quandalle: "New Polynomial Transform Algorithms for Fast DFT Computation", in IEEE 1979 Intern. Acoustics, Speech and Signal Processing Conf. Proc., pp. 510–513
7.3   C. M. Rader: Discrete Fourier transforms when the number of data samples is prime. Proc. IEEE **56**, 1107–1108 (1968)

7.4 G. Bonnerot, M. Bellanger: Odd-time odd-frequency discrete Fourier transform for symmetric real-valued series. Proc. IEEE **64**, 392–393 (1976)
7.5 C. M. Rader, N. M. Brenner: A new principle for fast Fourier transformation. IEEE Trans. ASSP-**24**, 264–266 (1976)
7.6 I. J. Good: The relationship between two fast Fourier transforms. IEEE Trans. C-**20**, 310–317 (1971)
7.7 S. Winograd: On computing the discrete Fourier transform. Math. Comput. **32**, 175–199 (1978)
7.8 H. J. Nussbaumer: DFT computation by fast polynomial transform algorithms. Electron. Lett. **15**, 701–702 (1979)
7.9 H. J. Nussbaumer, P. Quandalle: Computation of convolutions and discrete Fourier transforms by polynomial transforms. IBM J. Res. Dev. **22**, 134–144 (1978)
7.10 R. C. Agarwal, J. W. Cooley: New algorithms for digital convolution. IEEE Trans. ASSP-**25**, 392–410 (1977)

Chapter 8

8.1 I. J. Good: The relationship between two fast Fourier transforms. IEEE Trans. C-**20**, 310–317 (1971)
8.2 J. M. Pollard: The fast Fourier transform in a finite field. Math. Comput. **25**, 365–374 (1971)
8.3 P. J. Nicholson: Algebraic theory of finite Fourier transforms. J. Comput. Syst. Sci. **5**, 524–547 (1971)
8.4 P. J. Erdelsky: "Exact convolutions by number-theoretic transforms"; Rept. No. AD-A013 395, San Diego, Calif. Naval Undersea Center (1975)
8.5 C. M. Rader: Discrete convolutions via Mersenne transforms. IEEE Trans. C-**21**, 1269–1273 (1972)
8.6 R. C. Agarwal, C. S. Burrus: Fast convolution using Fermat number transforms with applications to digital filtering. IEEE Trans. ASSP-**22**, 87–97 (1974)
8.7 R. C. Agarwal, C. S. Burrus: Number theoretic transforms to implement fast digital convolution. Proc. IEEE **63**, 550–560 (1975)
8.8 L. M. Leibowitz: A simplified binary arithmetic for the Fermat number transform. IEEE Trans. ASSP-**24**, 356–359 (1976)
8.9 J. H. McClellan: Hardware realization of a Fermat number transform. IEEE Trans. ASSP-**24**, 216–225 (1976)
8.10 H. J. Nussbaumer: Linear filtering technique for computing Mersenne and Fermat number transforms. IBM J. Res. Dev. **21**, 334–339 (1977)
8.11 H. J. Nussbaumer: Complex convolutions via Fermat number transforms. IBM J. Res. Dev. **20**, 282–284 (1976)
8.12 E. Vegh, L. M. Leibowitz: Fast complex convolutions in finite rings. IEEE Trans. ASSP-**24**, 343–344 (1976)
8.13 L. B. Jackson: On the interaction of round-off noise and dynamic range in digital filters. Bell Syst. Tech. J. **49**, 159–184 (1970)
8.14 P. R. Chevillat, F. H. Closs: "Signal processing with number theoretic transforms and limited word lengths", in IEEE 1978 Intern. Acoustics, Speech and Signal Processing Conf. Proc., pp. 619–623
8.15 H. J. Nussbaumer: Digital filtering using complex Mersenne transforms. IBM J. Res. Dev. **20**, 498–504 (1976)
8.16 H. J. Nussbaumer: Digital filtering using pseudo Fermat number transforms. IEEE Trans. ASSP-**26**, 79–83 (1977)
8.17 E. Dubois, A. N. Venetsanopoulos: "Number theoretic transforms with modulus $2^{2q} - 2^q + 1$", in IEEE 1978 Intern. Acoustics, Speech and Signal Processing Conf. Proc., pp. 624–627

8.18 H. J. Nussbaumer: Overflow detection in the computation of convolutions by some number theoretic transforms. IEEE Trans. ASSP-**26**, 108–109 (1978)
8.19 I. S. Reed, T. K. Truong: The use of finite fields to compute convolutions. IEEE Trans. IT-**21**, 208–213 (1975)
8.20 I. S. Reed, T. K. Truong: Complex integer convolutions over a direct sum of Galois fields. IEEE Trans. IT-**21**, 657–661 (1975)
8.21 H. J. Nussbaumer: Relative evaluation of various number theoretic transforms for digital filtering applications. IEEE Trans. ASSP-**26**, 88–93 (1978)

Appendix A

A.1 H. J. Nussbaumer: *"Inverse Polynomial Transform Algorithms for* DFTs *and Convolutions"*, in IEEE 1981 Intern. Acoustics, Speech and Signal Processing Conf. Proc., pp. 315–318
A.2 H. J. Nussbaumer: New polynomial transform algorithms for multidimensional DFTs and convolutions. IEEE Trans. ASSP-**29**, 74–83 (1981)
A.3 J. Makhoul: A fast cosine transform in one and two dimensions. IEEE Trans. ASSP-**28**, 27–33 (1980)
A.4 H. J. Nussbaumer: "Fast polynomial transform computation of the 2-D DCT", in 1981 Intern. Conf. on Digital Signal Processing, pp. 276–283

Subject Index

Agarwal-Cooley algorithm 43, 202, 230
Agarwal-Burrus algorithm 56
Algorithms
  convolution 66, 78
  DFT 123, 144
  polynomial product 73
  reduced DFT 207
APL FFT program 95

Bezout's relation 6
Bit reversal 94
Butterfly 94
Bruun algorithm 104

Chinese remainder theorem
  integers 9, 33, 35, 43, 116, 125, 215
  polynomials 26, 48, 152, 157, 163, 178
Chirp Z-transform 112
Congruence 7, 14, 26, 56
Convolution 22, 27, 34
  circular 23, 30, 32, 43, 81, 107, 117, 151, 212
  skew-circular 174
  complex 52, 177, 227
Cook-Toom algorithm 27
Correlation 82, 117, 185
Cyclic convolution see circular convolution
Cyclotomic polynomials 30, 36, 37, 152, 157, 182

Decimation
  in frequency 89, 187
  in time 87
Diophantine equations 6, 8, 237
Discrete Fourier transform (DFT) 80, 112, 181
Distributed arithmetic 64
Division
  integers 4
  polynomials 25

Equivalence class 7
Euclid's algorithm 5, 8, 9, 27

Euler
  theorem 9, 13, 214, 215
  totient function 11, 30, 157, 213

Fast Fourier transform (FFT) 85
  computer program 95
Fermat
  number 19, 222, 223
  number transform (FNT) 222
  prime 21, 223
  theorem 13, 215, 217, 232, 234, 237
Field 25, 157
Finite impulse response filter (FIR) 33, 55, 113
Fourier transform see discrete Fourier transform

Galois field (GF) 25, 236
Gauss 15, 18, 236
Greatest common divisor (GCD) 5
Group 23

Identity 24
"In place" computation 94, 143
Interpolation see Lagrange interpolation
Irreducible polynomial 25, 157
Isomorphism 24

Lagrange interpolation 28, 37
Legendre symbol 18, 22, 237

Mersenne
  number 19, 216, 230
  prime 20
  transform 216
Modulus 7
Multidimensional
  convolution 45, 108, 151
  DFT 102, 141, 193
  polynomial transform 178
Mutually prime 5, 20, 21, 26, 43, 170, 201

Nesting 44, 57, 60, 134, 170, 194
Number theoretic transform (NTT) 211

Subject Index

One's complement   20, 227
Odd DFT   102, 187, 207
Order   14, 24
Overlap-add   29, 33, 55
Overlap-save   34

Parseval's theorem   83
Permutation   10, 43, 82, 125, 170, 183
Polynomial   22
   product   30, 34, 47
   transform   151
Prime   5
Prime factor FFT   125, 194
Primitive root   11, 21, 117, 202
Pseudo Fermat transform   234
Pseudo Mersenne transform   231

Quadratic
   non-residue   17, 22, 236
   residue   17, 115
Quantization error (FFT)   96, 142
Quotient   5

Rader algorithm   116, 129, 134, 202, 204, 205, 207

Rader-Brenner FFT   99, 190, 207, 208
Recursive   60, 114
Reduced DFT   102, 121, 182, 207
Relatively prime *see* mutually prime
Remainder   5
Residue   7
   class   7
   reduction   120, 152
   polynomial   25, 152, 163
Ring   24, 173, 196
Roundoff   96, 228
Roots   11, 14, 117, 156, 157, 161, 212, 237
Row-column method   102

Skew-circular convolution   174
Split nesting   47, 139
Split prime factor algorithm   129

Totient function *see* Euler totient function
Twiddle factors   86

Winograd   29, 280
   Fourier transform algorithm (WFTA)   133, 201, 202, 204, 205, 207